Diagnostics and Therapeutic Advances in GI Malignancies

This series will highlight the recent innovations in the diagnostics and therapeutic strategies for different Gastrointestinal (GI) cancers.

Gastrointestinal cancers are a group of cancers that affect the digestive system and include gastric cancer, colorectal cancer, liver cancer, esophageal cancer, and pancreatic cancer. GI cancers are the leading health problem in the world and their burden is increasing in many countries. This heavy burden is due to the lack of effective early detection methods and to the emergence of chemoradioresistance. Attempts at improving the outcome of GI cancers by incorporating cytotoxic agents such as chemo drugs have been so far disappointing. These results indicate that the main challenge remains in the primary resistance of GI cancer cells to chemotherapy in the majority of patients. Therefore, improvement in the outcomes of these malignancies is dependent on the introduction of new agents that can modulate the intrinsic and acquired mechanisms of resistance.

The increased understanding of the biology, metabolism, genetic, epigenetic, and molecular pathways dysregulated in GI cancers has revealed the complexity of the mechanisms implicated in tumor development. These include alterations in the expression of key oncogenic or tumor suppressive miRNAs, modifications in methylation patterns, the upregulation of key oncogenic kinases, etc.

The individual books in this series will focus on the genetic basis of each gastrointestinal cancers, molecular pathophysiology, and different biomarkers to estimate cancer risk, detection of cancer at microscopic dimensions, and suitable and effectiveness of the therapies. In addition, the volumes will discuss the role of various signaling molecules/pathways and transcriptional factors in the regulation of the tumor microenvironment and effect on the tumor growth.

Lastly, it will elaborate the use of molecularly targeted drugs that have been proven to be effective for the treatment of GI cancers, with a focus on the emerging strategies.

This edition will provide researchers and physicians with novel ideas and perspectives for future research that translates the bench to the bedside.

More information about this series at http://www.springer.com/series/16343

Ganji Seeta Rama Raju • L.V.K.S. Bhaskar
Editors

Theranostics Approaches to Gastric and Colon Cancer

Springer

Editors
Ganji Seeta Rama Raju
Department of Energy and Materials Engineering
Dongguk University-Seoul
Jung-gu, Seoul, Republic of Korea

L.V.K.S. Bhaskar
Guru Ghasidas University
Bilaspur, Chhattisgarh, India

ISSN 2662-2688 ISSN 2662-2696 (electronic)
Diagnostics and Therapeutic Advances in GI Malignancies
ISBN 978-981-15-2019-8 ISBN 978-981-15-2017-4 (eBook)
https://doi.org/10.1007/978-981-15-2017-4

This Springer imprint is published by the registered company Springer Nature Singapore Pte Ltd.
The registered company address is: 152 Beach Road, #21-01/04 Gateway East, Singapore 189721, Singapore

This book is dedicated to our families,
our teachers, and friends.

Preface

Gastrointestinal malignancies are one of the most common tumors encountered worldwide. The tumor occurs in the small intestine and becomes symptomatic only when it is stabilized and attains a certain size. Among the gastrointestinal malignancies, colorectal cancer and gastric cancer are the most frequently diagnosed tumors. Colorectal cancer remains the deadliest neoplastic disease and diagnosed annually as the third most common disease in both men and women in the USA, whereas gastric cancer is the second deadly type of neoplasia worldwide that is the leading cause of cancer-related mortality. Despite the availability of various modern advanced therapies, the prognosis rate for cancer is still unsatisfactory and mortality rates remain high. Thus, there is an urgent need for the detection of biomarkers and improvement in the effectiveness of the therapy at the molecular level in order to appropriately target and diagnose the disease at its early stages. Chemotherapy remains the only therapy for cancer; however, the resistance developed by tumor cells against current chemo drugs is a major hurdle. Additionally, the metastatic recurrence after resection is also a major cause of death in patients. Therefore, understanding the underlying mechanism and factors involved in the development of metastasis is crucial and would help in determining specific targets for therapy. Neoadjuvant therapies effectively stifle the evolution of drug resistance and recurrence. Moreover, the use of phytochemicals along with chemodrugs and immunotherapies could support the development of new strategies for cancer therapy. This can be further strengthened by the support of nanotechnology to potentiate target therapy. Thus, many more novel strategies should be developed, which specifically focus on reprogramming immune cells and their pathways to sensitize tumor cells against the drugs used.

The preface illustrates a clear understanding about the molecular mechanism of chemoresistance and helps in designing new drugs, therapeutic strategies, and stronger drug delivery mechanisms in cancer. This book's focus is on novel theragnostic approaches. Theragnostics is a field of science that combines diagnosis and specific targeted therapies. This field also involves nanotechnology to unite the diagnostic and therapeutic application and further uses specific molecular pathways to acquire diagnostic images and targeted therapy for drug delivery. In this book, we have also evaluated and examined the crucial proteins and their signalling pathways involved in carcinogenesis. This book contains 13 chapters that elaborately discuss the biology and signaling pathways of various proteins involved in both colorectal and gastric cancer progression. Additionally, we also examine the therapeutic role of phytochemicals and evaluate the efficacy of phytochemicals, such as resveratrol, curcumin, and quercetin, for their clinical significance as strong anticancer drugs that can sensitize tumor cells. Immunotherapy and nanotechnology are explored in various ways for targeted therapy, drug delivery, and efficacy of drug at the target site. To further confirm the correlation between colorectal cancer susceptibility and alcohol metabolizing and detoxification, gene polymorphisms from previously published association studies were examined in meta-analysis.

It is our pleasure to present a comprehensive summary of the novel fields to the scientific community for a better and holistic understanding that can aid in the future advances in the field of theragnostics for gastrointestinal malignancies. We hope that this book will inspire new research ideas and innovations for the ultimate benefit of patients and their families.

Seoul, Republic of Korea
Bilaspur, Chhattisgarh, India

Ganji Seeta Rama Raju
L.V.K.S. Bhaskar

Contents

About the Series Editor

Dr. Ganji Purnachandra Nagaraju is a faculty member in the Department of Hematology and Medical Oncology at Emory University School of Medicine. Dr. Nagaraju obtained his MSc and his PhD, both in biotechnology, from Sri Venkateswara University in Tirupati, Andhra Pradesh, India. Dr. Nagaraju received his DSc from Berhampur University in Berhampur, Odisha, India. Dr. Nagaraju's research focuses on translational projects related to gastrointestinal malignancies. He has published over 80 research papers in highly reputed international journals and has presented more than 50 abstracts at various national and international conferences. Dr. Nagaraju is author and editor of several published books including (1) *Role of Tyrosine Kinases in Gastrointestinal Malignancies*, (2) *Role of Transcription Factors in Gastrointestinal Malignancies*, (3) *Breaking Tolerance to Pancreatic Cancer Unresponsiveness to Chemotherapy*, (4) *Theranostic Approach for Pancreatic Cancer*, and (5) *Exploring Pancreatic Metabolism and Malignancy*. He serves as editorial board member of several internationally recognized academic journals. Dr. Nagaraju is an associate member of the Discovery and Developmental Therapeutics research program at Winship Cancer Institute. Dr. Nagaraju has received several international awards including FAACC. He also holds memberships with the Association of Scientists of Indian Origin in America (ASIOA), the Society for Integrative and Comparative Biology (SICB), the Science Advisory Board, the RNA Society, the American Association for Clinical Chemistry (AACC), and the American Association of Cancer Research (AACR).

About the Editors

Ganji Seeta Rama Raju is an Assistant Professor at the Department of Energy and Materials Engineering, Dongguk University, Seoul. He received both his M. Phil. and Ph.D. in Physics from Sri Venkateswara University, Tirupati, Andhra Pradesh, India. His research chiefly focuses on nanomaterials, imaging, and drug delivery for various malignancies. To date, he has published over 95 research papers in highly reputed international journals.

L.V.K.S. Bhaskar is a Professor of Zoology and Dean of the School of Life Sciences at Guru Ghasidas Central University, Chhattisgarh. Prior to this appointment, Dr. Bhaskar was the Senior Scientist at Sickle Cell Institute Chhattisgarh. Dr. Bhaskar was awarded the Prof. Pampapathi Rao memorial Gold Medal in 2003. Dr. Bhaskar is an active member of several genetics and biotechnology-related professional and academic bodies. He has conducted research to explore molecular basis of alcoholism, genetics of complex human diseases, population genetics, and cancer biology. Dr. Bhaskar has published more than 135 research papers in reputed national and international journals. Dr. Bhaskar has been a regular reviewer for several national and international journals and also serves on the editorial board of various journals.

Chapter 1
Therapeutic Role of Phytochemicals in Colorectal Cancer

Begum Dariya, Balney Rajitha, Afroz Alam, and Ganji Purnachandra Nagaraju

Abstract Colorectal cancer (CRC) is the universal disease of gastrointestine that has become a common health problem globally and is the principal reason for high mortality and morbidity rate worldwide. The novel conventional therapies including radio and chemotherapy have limitations like developing toxicity in healthy cells and bring about adverse side effects. Poor diet habits and modern life style are the major obstacles determined as the risk for CRC and developing a healthy diet habit and lifestyle may reduce the risk. However studies from epidemiology suggested that fruits and vegetables possess high content of phytochemicals and including them in daily diet would effectively reduce the risk of cancers related to digestive system including colorectal cancer. These natural products possess diverse pharmacological bioactivities and are thus widely explored now as novel agents. Their nontoxic nature and cytotoxicity features make the researchers perform immense research to develop novel drugs against various cancers including pancreatic, breast and colorectal cancer. It was evidenced from in vitro and in vivo, that the phytochemicals are strong anti-inflammatory, anti-oxidants, and anti-cancer agents showing their activity by controlling various signalling pathways that promote proliferation, survival, apoptosis, invasion and metastasis of CRC cells by down regulating and up regulating various proteins involved in these pathways.

Keywords Colorectal cancer · Phytochemicals · Polyphenols · Flavonoids · Isoflavones and cytotoxicity

B. Dariya · B. Rajitha · A. Alam
Department of Bioscience and Biotechnology, Banasthali University, Vanasthali, Rajasthan, India

G. P. Nagaraju (✉)
Department of Hematology and Medical Oncology, Winship Cancer Institute, Emory University, Atlanta, GA, USA
e-mail: pganji@emory.edu

G. S. R. Raju, L.V.K.S. Bhaskar (eds.), *Theranostics Approaches to Gastric and Colon Cancer*, Diagnostics and Therapeutic Advances in GI Malignancies,
https://doi.org/10.1007/978-981-15-2017-4_1

Abbreviations

5-FU	Fluorouracil
ACF	Aberrant crypt foci
Akt	Alpha serine/threonine—protein kinase
AMPK	Adenosine monophosphate activated protein kinase
AOM	Azoxymethane
APC	Adenomatous polyposis coli
Bax	Bcl-2 associated X protein
Bcl-2	B-cell lymphoma
BPIS	Bound polyphenols of inner shell
cAMP	Cyclic adenosine monophosphate
CDK	Cyclin dependent kinase
cIAP	Cellular inhibitors of apoptotic protein
CIMP	CpG island methylator phenotype
CIN	Chromosomal instability
COLO320DM	Colon adenocarcinoma cell lines
COX-2	Cyclooxygenase
CRC	Colorectal cancer
CREB1	cAMP response element binding protein 1
CSC	Cancer stem cells
DiYu	Sanguisorba officinalis
DMH	1,2-dimethyl hydrazine
DNMT	DNA methyltransferase
EA	Ellagic acid
EG	Ellagitannins
EGFR	Epidermal growth factor
EMT	Epithelial mesenchymal transition
ERK	Extracellular signal regulated kinase
ER-β	Estrogen receptor
FDA	Food and drug administration
FOLFOX	Oxaliplatin (Made of folinic acid/chemodrug)
FOXO3	Forkhead box protein O3
Gli	Glioma associated oncogene
Grp78	Glucose regulator protein
GST	Glutathione S transferases
HDAC	Histone deacetylase
IBD	Inflammatory bowel disease
IGF	Insulin growth factor1
IL-6	Interleukin
iNOS	Inducible nitric oxide synthase
LOX	Lipoxygenase
MAPK	Mitogen activated protein kinase
MC37	Mono-carbonyl curcumin analogue

MEK1	Mitogen activated protein kinase
MMP	Matrix metalloproteinase
mPEG	PCL methoxy poly ethylene glycol poly caprolactone
MRP1	Multidrug resistance protein 1
mTOR	Mechanistic target of rapamycin
MUC2	Mucin-2 precursor
NF-κB	Nuclear factor kappa B
NKd2	Naked cuticle 2
NQO1	NADPH quinine oxidoreductase 1
Nrf2	Nuclear factor eryhthroid-2 p45-related factor 2
OCTN2	Organic cation or carnitine transported 2
PARP	Poly ADP ribose polymerase
p-CA	P-Coumaric acid
PCNA	Proliferating cell nuclear antigen
PDK4	Pyruvate dehydrogenase kinase4
PI3K	Phosphoinositide 3-kinase
PPAR γ	Peroxisome proliferator activated receptor γ
Ptch	Patch homolog
ROS	Reactive oxygen species
SFRP2	Secreted frizzled-related protein 2
SGK1	Serine threonine protein kinase
Smo	Smoothened
STAT3	Signal transducers and activators of transcription
tCA	Trans form of cinnamic acid
TGF-β1	Transforming growth factor β
TNF-α	Tumor necrosis factor
TOPK	T-cell originated protein kinase
uPA	Urokinase type plasminogen activator
UPR	Unfolded protein response
WIF1	Wnt inhibitory factor1
XIAP	X-linked inhibitor of apoptosis

1.1 Introduction

Colorectal cancer (CRC) is a most frequent and multi factorial disease ranking third world wide with a survival rate of about 4.49% in men and 4.15% in women. Though the rate of occurrence of CRC has declined, it remains a hazardous disease with elevated risks of mortality and morbidity worldwide. The American cancer society estimated that 97,220 new colon cases and 43,030 new rectal cancer cases are diagnosed in 2018, making CRC the most common gastrointestinal cancer. CRC is a preventable disease influenced by exposure to the environment, certain inflammatory conditions in the digestive systems, life style, dietary factors and etiological transcending genetic factors. Patients are at higher risk to CRC with medical history having diseases including ulcerative colitis, inflammatory bowel disease (IBD) and Crohn's disease. The pathogenesis resulted from the genomic instability including microsatellite, chromosomal instability (CIN) and CpG island methylator phenotype (CIMP) are also prone to CRC. Additionally, CRC is characterized by chromosomal alterations affecting proteins engaged in various signalling pathways, namely WNT, MAPK/PI3K, TGF-β and DNA repair mechanisms, which enhance proliferation and cell survival leading to carcinogenesis (Willett et al. 2013). Furthermore, some studies have suggested that 75% of sporadic CRC can be caused due to less intake of natural veggies and fruits. In Eastern Asia and Europe, fruits and vegetables constitute a staple of the daily routine. However cultural and dietary changes leading to the adoption of American-like diets consisting primarily of red meat and high fat foods has increased the incidence of CRC in these countries. In fact, diets rich in solid foods and meats result in alterations in the composition of the gut microbiome that enrich harmful bacteria (Kim et al. 2013). Comparative studies conducted by O'Keefe et al. between African Americans and rural Americans revealed that butyrate levels, which acts as an antitumor compound, are low in African Americans (Fung et al. 2012; Hofmanova et al. 2013; O'Keefe et al. 2015). Furthermore, they reported over expression of microbial genes playing a major role in the production of secondary bile acids such as lithocholic acid that induce mutations and lead to oxidative stress (Li et al. 2015). These various findings thus depict how the diet can represent both a problem and a solution to GI ailments. Henceforth, phytochemicals in the form of dietary supplements might reduce the risk of cancer. Furthermore, dietary phytochemicals and the intestinal microbiome combined could mediate CRC risk.

Despite important advances in cancer therapeutics including surgery, adjuvant chemo and radio therapy, the survival rate of the patients has not greatly improved yet and tumor relapse due to chemoresistance remains a major obstacle to the eradication of CRC. Interestingly, the studies from epidemiology reported that the risk could be lowered chronic diseases such as CRC by consuming naturally occurring non-nutritive bioactive products called dietary phytochemicals found in grain, fruits and vegetables, which are beneficial to the intestine, especially for the colon. Phytochemicals might also limit the toxicity and resistance towards chemotherapy. Phytochemicals are termed as "nutritional genomics" in plant biochemistry, human nutrition and genomics, since they target the genetic processes and metabolic pathways involved in carcinogenesis (DellaPenna 1999; Ferguson 2009; Kaput and Rodriguez 2004; Milner et al. 2001; Simopoulos and Milner 2010). Phytochemicals are the compounds of non-nutritive nature easily available in plants endowed with chemo-preventive, anticancer and disease preventive properties. These phytochemicals are available in various traditionally available medicinal plants including turmeric, soybean, grapes, tomato, tea and ginger. These chemo-preventive phytochemicals can counter carcinogenesis at different stages including tumor cell proliferation, angiogenesis, metastasis, apoptosis, invasion, mutation and immunity (Bagli et al. 2004; Chen et al. 2008, 2009; Harris et al. 2006; He et al. 2011; Huang et al. 2005; Maher et al. 2011; Moon et al. 2008; Palozza et al. 2010; Wang et al. 2009;

Yao et al. 2008) by blocking the activity of mutagens, suppressing the proliferation of cells (Surh 2003) and protecting against peroxidation of lipids (Ho et al. 2002). Phytochemicals like epigallocatechin gallate extracted from green tea, genistein derived from soybean, resveratrol from grapes, curcumin from turmeric and many more are used widely as anti-cancerous drugs and are chemo-preventive. It can be described as intake of foreign agents that slow down the progression of tumor cells and alter carcinogenesis in the premalignant stage (Jemal et al. 2008). They are available either in natural forms or as synthetic analogues. Each and every phytochemical vary in their mode of action; however they possess common mechanisms comprising DNA methylation, cyclooxygenase inhibition to increase detoxification, induction of apoptosis in tumor cells, PI3 kinase pathway modification, inhibition in invasion and angiogenesis of tumor. Biotransformation, or the fate of the phytochemical in the body after ingestion, occurs as it gets digested and passes through the gastrointestinal epithelium to the circulatory vessels (Rao 2012). However, the bioavailability and mechanism of action of these phytochemicals is poorly understood and remains largely unknown.

1.2 Classification

The phytochemicals are classified based on their origin, biosynthesis and biological properties into polyphenols, terpenoids, organosulfur compounds and phytosterols.

1.2.1 Polyphenols

Polyphenols are the secondary metabolite in plants and constitute the largest and most diverse family of phytochemicals. About 8000 different forms of polyphenol obtained from edible parts of the plants (Fraga et al. 2010) including oilseeds (flax-seeds, olive), legumes, beverages, cereals, fruits and vegetables (Macheix et al. 1991; Velioglu et al. 1998). Structurally, they contain an aromatic ring (Benzene ring) with hydroxyl groups attached to it (Fraga et al. 2010). Due to the occurrence of huge amounts of phenol groups in polyphenols, they are also known as polyhydroxyphenols and exist as glycosides in plants. They show ample variety of properties like antioxidation, anticarcinogenic, antiproliferative and chemoprotective characteristics that induce cell cycle arrest and apoptosis producing detoxify enzymes (Pandey and Rizvi 2009). Pandey et al. (2009) reported that consumption of diet rich in phyto-polyphenols for long-term can protect the body against diseases including cardiovascular disease, diabetes, neuro-degenerative disease, osteoporosis and cancer, regulate the immune system of the host and modify the signalling pathway. As anti-carcinogenic agents they can counter carcinogenic initiation and suppresses cancer cell progression (Muto et al. 2001). The effectiveness of polyphenols against CRC has been tested in vitro, in vivo and also in clinical trials (Gee et al. 2002; Sharma et al.

2001; Wenzel et al. 2000). For instance, phase I metabolizing enzymes including cytochrome P450 and P4502E1 can bio-transform tumor cells rendering them more prone to bind to DNA and promote cellular proliferation. However, polyphenols can inhibit these metabolizing enzymes and prevent them from inducing proliferation. Furthermore, they can increase the effectiveness of phase II metabolic enzymes including glutathione reductase, catalase, glutathione peroxidise, glutathione S-reductase and quinine (Khan et al. 1992; Luceri et al. 2002; Muto et al. 2001). Based on their therapeutic action, origin and chemical structure, polyphynols can thus be classified into many compounds. The most significant polyphenols grouped as: stillbenes, flavonoids, phenolic acids and curcuminoids.

1.2.2 Phenolic Acids

Phenolic acids contribute 30% polyphenols included in the diet taken. These are most commonly extracted from plants derived from hydroxycinnamic or hydroxybenzoic acid. Structurally, they possess phenolic rings and a functional carboxylic acid. Phenolic acids are again sub grouped into compounds based on their origin and include Gallic acid (Green tea), Ellagic acid (Berries and Pomegranate), Caffeic acid (Coffee), Anacardic acid (Cashew) and Rosmarinic acid (Rosemary). These phenolic acids act as an antioxidant and anti-proliferative agents and exert cytotoxic activity against cancer cells.

1.2.3 Derivatives of Benzoic Acid

Meng et al. (2016) determined that ellagic acid and gallic acid are the key components found in Sanguisorba officinalis L. radix (DiYu), traditionally available herb in China. They found that DiYu has anti proliferative activities and that the components gallic and ellagic acids act as an adjunct to 5-FU and promote synergistic cytotoxicity in drug resistant CRC as well chemop-sensitization through ROS mediated, mitochondria-caspase-dependent apoptotic pathways. Ellagic acid is the dimer of gallic acid (Tomás-Barberán and Clifford 2000), is effective in inducing antiproliferative activity, arrest in the cell cycle, and apoptosis via IGF-II and NF-κB inhibition. It alters anti and pro-apoptotic proteins and induces p21/p53 expression (Bell and Hawthorne 2008), whereas gallic acid is involved in up regulating Bax and downregulating Bcl-2 (Faried et al. 2007). Hence, gallic acid demonstrated the highest anti-tumorigenesis activity in cell line and animal models of CRC compared to other benzoic derivatives (Ho et al. 2009; Zheng et al. 2002). Nevertheless, studies investigating the effect of vanillic acid against CRC are still going on (Ho et al. 2009; Zheng et al. 2002).

1.2.4 Derivatives of Cinnamic Acid

Cinnamic acid is an unsaturated organic compound obtained from the deamination of phenylalanine. Functionally, it is an antioxidant, antimicrobial and antidiabetic compound with cytotoxic potential against cancer cells (Budavari 1989; Lafay et al. 2006; Liu et al. 1995; Neogi et al. 2003; Pontiki et al. 2014; Zhu et al. 2016). Wang et al. (2012) compared cinnamic acid with FOLFOX, a combo chemo drug of 5-FU and oxaliplatin, against colon CSCs. They reported that cinnamic acid decreases the population of CSCs dramatically and, together with FOLFOX or alone, is highly effective against CRC. The combination can even sensitize the CSCs by inducing apoptosis against chemo drugs (Huang et al. 2013). Additionally, Zhu et al. (2016) have investigated the trans form of cinnamic acid (tCA) and demonstrated that tCA could be therapeutically effective in inhibiting histone deacetylase (HDAC) in xenografts of nude mice with colon carcinoma. HDAC is an enzyme associated with tumour progression, initiation and metastasis. P-Coumaric acid (p-CA), a cinnamic acid derivative, is naturally available in daily consumed fruits and veggies. Sharma et al. (2018) observed that p-CA can upregulate the unfolded protein response (UPR) and downregulate the glucose regulator protein (Grp78) through the activation of the PERK-elf2α-ATF-4-CHOP and reported its effectiveness in the CRC models in vitro and in vivo. In general, Grp78 is a molecular chaperone whose expression is upregulated in CRC cells due to the activity of 1,2-dimethyl hydrazine (DMH) through the activation of UPR, and promotes tumor progression. UPR can also activate the anti-apoptotic activity of NF-κB, thus promoting angiogenesis. Furthermore, p-CA can also downregulate the expression of p-IκBα and decrease the expression of inflammation-causing proteins including COX-2, IL-6 and TNF-α.

The epidermal growth factor receptor (EGFR) plays a key role in promoting CRC. Roy et al. (2016) have shown that plant-based drugs including p-CA and feruilic acid can downregulate the expression of EGFR by binding to and inhibiting its active site, as revealed in their docking studies. Furthermore, Lu et al. (2018) determined that phenolic acids (Bound polyphenols of inner shell (BPIS)) extracted from foxtail millet can reverse the process of multidrug resistance and sensitise CRC to chemotherapeutic drugs including 5-FU, vincristine and oxaliplatin. They have identified 12 phenolic acids of BPIS, of which p-CA and feruilic acid are the active components. These compounds can increase rhodamine-123 accumulation and decrease the expression of p-glycoprotein (p-gp) and multidrug resistance protein 1 (MRP1). Caffeic acid, another cinnamic acid derivative, also possess antitumor activity. Murad et al. (2015) have suggested that caffeic acid along with 5-caffeoylquinic acid could reduce the risk of causing chronic disease including CRC, wherein these acids induce cell cycle arrest, thus inhibiting differentiation, progression of CRC cells and increasing apoptosis. Kang et al. (2011) have suggested that daily consumption of coffee, which contains caffeic acid and chlorogenic acid, can inhibits CRC cells and metastasis to liver by suppressing the activity of T-cell originated protein kinase (TOPK) and mitogen activated protein kinase (MEK1) in an ATP noncompetitive manner by blocking ERK phosphorylation. However, since the bioavailability of chlorogenic acid is low and it undergoes metabolisation to caffeic acid in the colon, caffeic acid is more effective. Peng et al. (2018) proposed a novel idea and used phenolic acid in combinational with supercritical extracts of Angelica sinensis for RC treatment. They found that the combination suppressed tumor growth and proliferation, inflammatory responses and damages in DNA. Ginkgolic acid is another phenolic acid isolated from Ginkgo. It suppresses the growth and invasion of different cancers like liver, pancreatic and breast cancers (Ma et al. 2015; Shilpa et al. 2015; Zhou et al. 2010). Qiao et al. (2017) used ginkgolic acid to suppress the expression of invasion associated proteins (matrix metalloproteinase (MMP-2, MMP-9), C-X-C chemokine receptor type 4 and urinary type plasminogen activator)) and induce the activation of adenosine monophosphate activated protein kinase (AMPK), thus inhibiting the proliferation, migration and invasion of colon cancer cells.

1.2.5 *Flavonoids*

Flavonoids are naturally occurring polyphenols available as dietary forms identified in almost 5000 varieties (Sebastian et al. 2017) of plants including fruits, tea, wine and vegetables. These flavanoids have gained wide attention due to their chemopreventive activity as an anticancer agent against gastrointestinal cancers (Chang et al. 2018; Grosso et al. 2017; Ivey et al. 2017; Petrick et al. 2015). Structurally, they have a pair of benzene rings interconnected together with three carbon atoms to form an oxygenated heterocycle (Ramos 2008). They are divided into six subclasses based on their chemical structure, namely anthocyanins, flavones, flavan-3-ols, isoflavones, flavanones and flavanols.

1.2.6 *Anthocyanins*

Anthocyanins are antioxidant nutraceuticals available in grapes, coloured berries, eggplant and radish. They have diverse functions, including anti-inflammatory and anti-tumor properties, once ingested. They are sub-grouped into: cyanidine, malvidin, pelargonidin, peonidin, delphinidin and petunidin. Chen et al. (2018) used a freeze-dried raspberry powder, incorporated in the diet of mice, to determine if black raspberry anthocyanins could alter the gut commensal microbiota and if it exerts a chemopreventive action against CRC. They reported that the raspberry anthocyanin demethylated the promoter gene, secreted frizzled-related protein 2 (SFRP2), leading to transcript and protein expression of SFRP2. SFRP2 is a gene that encodes SFRP2 proteins, and its methylation is considered as a potential marker of CRC. Additionally, raspberry anthocyanin also downregulated the expression of p-STAT-3 (transcription factor) and DNA methyltransferase (DNMT3A and DNMT3B, which are involved in DNA methylation. de Moraes et al. (2017) showed that anthocyanin/anthocyanidins exert their antitumor activity through multiple mechanisms, including NF-κB pathway inhibition, apoptosis mediated by mitochondria, downregulation of MMPs expression, metastasis inhibition, downregulation of the Akt/mTOR pathway and activation of sirtuin and AMPK pathways. Mazewski et al. (2018) compared the anticancer activity and mode of action of the plant's rich in anthocyanin. They determined that these anthocyanin extracts suppress anti-apoptotic proteins expression such as survivin, cIAP-2 and XIAP, induce cell cycle arrest at G1 phase and apoptosis. Anthocyanins cyanidin have a greater affinity to kinases, whereby cyanidin and delphidin inhibit EGFR.

1.2.7 *Flavones and Flavanones*

Flavones and flavanones are flavonoids highly available in many citrus fruits including oranges. They are also naturally available in vegetables including peppers and lettuce. They comprise different types including apegenin, luteolin, vitexin, nobiletin, wogonin etc., out of which apegenin and luteolin are the most widely investigated against CRC. Apigenin is a naturally edible dietary flavone characterized as an anti-platelet, anti-inflammatory and anti-tumor. It promotes arrest at the cancer cell cycle and induce apoptosis. It can potentiate chemotherapeutic agents as evidenced in vitro and in vivo against CRC. For example, Shao et al. (2013) suggested that apigenin can sensitise CRC cells to Navitoclax, an anti-cancer drug, in vitro and in vivo, inhibiting AKT and ERK pathways. Apigenin also possesses anti-migratory and anti-invasion properties. Chunhua et al. (2013) provided evidence showing that apigenin inhibits tumor growth as well as metastasis to liver and lung by upregulating the expression of transgelin, downregulating MMP-9 expression and suppressing Akt phosphorylation. Transgelin is a tumor suppressor and induces the transformation of sensitive actin-binding protein. Lefort and Blay (2011) also reported that apigenin combined with 5-FU, oxaliplatin and irinotecan can upregulate the expression of CD26 in advanced CRC patients. CD26, which is a cell surface protein involved in activating enzymes such as dipeptidyl peptidase and ecto-adenosine deaminase, plays major role in suppressing the pathways involved in metastasis of tumor. However, apigenin action is fourfold higher in combination with irinotecan s compared to the other chemotherapies. Luteolin is another flavone that has gained importance in cancer therapy, as revealed by epidemiological studies. This flavone exists universally and is safe to use. It acquires biological anti-bacterial, anti-diabetic, anti-proliferative and anti-oxidative activities. Luteolin can modify various signalling pathways involved in CRC. It induces apoptosis by activating mitochondria-mediated caspase pathways, thus regulating the apoptotic protein Bcl-2 (downregulating), Bax (upregulating) and increasing the levels of caspase-9 and caspase-3. In addition, it also maintains oxidative stress by activating ROS. Kang et al. (2017) reported that luteolin stimulates mitogen activated protein kinase signalling pathways (MAPK) to induce apoptosis. Moreover, Luteolin inhibits the epithelial mesenchymal transition (EMT) in CRC, inhibiting the mobility of cancer cells and reducing the mesenchymal marker expression. Liu et al. (2017) revealed that it blocks EMT in CRC by inhibiting the cyclic AMP (cAMP) response element binding protein 1 (CREB1), which affects the cAMP and glucagon signalling pathways. They employed proteomics to determine the effectiveness of luteolin and revealed that this flavone controls CREB1 at its mRNA and protein level and also focus on their target gene in its downstream. Luteolin also sensitizes CRC cells against certain chemotherapeutic drugs including cisplatin and oxaliplatin. In fact, Chian et al. (2014) determined that luteolin sensitises CRC against oxaliplatin through nuclear factor eryhthroid-2 p45-related factor 2 (Nrf2) pathways in a dose dependent manner by activation of Nrf2 and NADPH quinine oxidoreductase 1 (NQO1). Zuo et al. (2018) observed epigenetic modifications in Nrf2, which induce the antioxidative stress pathway downstream, following treatment with luteolin. Qu et al. (2014) also investigated the mechanism through which luteolin sensitizes tumour cells to chemo-drugs, showing that this flavone can potentiates the activity of oxaliplatin and resensitise CRC cells to it through the activation of peroxisome proliferator activated receptor γ (PPAR γ)/organic cation or carnitine transported 2 (OCTN2) pathway. OCTN2 is a solute carrier known as oxaliplatin uptake protein and regulated by the binding of PPARγ to the PPAR response element.

Finally, naringenin is a nontoxic glycoside flavanone available in citrus fruits and used in Chinese traditional medicine because of its pharmacological and anti-cancerous properties. Abaza et al. (2015) showed that naringenin can inhibit the cell growth of tumor including CRC and breast cancer at their S and G2/M phase by downregulating certain cell cycle regulator genes expression including cdk7, cdk6 and cdk4. It can also downregulate Bcl2 and upregulate the expression of Bax, Bak, p18, p19, p21 and caspase proteins. Additionally, naringenin can inhibit cell survival proteins including Akt, PI3k and NFκB in CRC cells. It can also augment the sensitivity of CRC cells to DNA acting drugs. Song et al. (2015) also investigated the anti-proliferative activity of naringenin, suggesting that naringenin downregulates the expression of cyclin D1 through p38 activation in CRC.

1.2.8 Flavanols

Flavanols have also been found to reduce CRC risk. Particularly, catechins have gained wide attention and interest among researchers for the prevention of cancer. The catechins include epicatechin, epicatechin gallate and procyanidin, profusely found in tea and cocoa (Babich et al. 2005; Lamuela-Raventós et al. 2005). They function as antioxidants with apoptotic properties and can arrest the cell cycle in CRC. Ramos et al. (2011) reported that flavanols display different mechanisms of action against CRC cell lines SW480 and caco-2. They determined that epicatechin gallate induces apoptosis through caspase-3 activity and downregulates Bcl-2 proteins in SW480 cell lines and enhances antioxidant activity in caco-2 cell lines, while procyanidine enhances cell survival and proliferation in SW480 cell lines. Owczarek et al. (2017) later found that *Chaenomeles japonica*, a Japanese quince fruit flavanol preparation, rich in procyanidin can inhibit the expression of COX-2, MMP-9 and NF-κB protein in SW480 cells, indicatin their cytotoxic and anti-metastatic nature. Moreover, Papi et al. (2013) reported the cytotoxic effects of a combinational therapy including vitexin, raphasatin and epigallocatechin-3gallate against CRC, suggesting that this combination may act on the mitochondrial pathway, cell cycle arrest at G0/G1 phase and modulate the expression of ROS, Bax, Bcl-2 and caspase-9.

1.3 Isoflavones

Isoflavones are phytoestrogens and secondary metabolites found in soy products. These phytochemicals have various pharmacological effects against many diseases including cardiovascular, osteoporosis and malignancies in colon, prostate and breast (Tse and Eslick 2016) and are susceptible to radiotherapy. Isoflavones employ their antitumor activity by regulating the cell cycle, apoptosis (Prietsch et al. 2014) and by inhibiting certain signalling pathways including MAPK (Chen et al. 2014) and androgen receptor pathways and controlling prostate specific antigen expression (Banerjee et al. 2012). Isoflavones display their cytotoxic effects only in their hydrolyzed forms called aglycones (Andlauer et al. 2000), hence, researchers focus on inducing the enzymatic hydrolysis of isoflavones using microorganisms (Maitan-Alfenas et al. 2014; Yeo and Liong 2010; Yeom et al. 2012). Isoflavones include daidzein, glyctein and genistein, which is the most commonly studied isoflavone.

1.3.1 Genistein

Genistein is the key constituent of soy products and is a tyrosine kinase inhibitor. It regulates various signalling pathways including EGFR, MAPK and Akt (Chen et al. 2014, 2015; Yan et al. 2010), thus affecting proliferation, apoptosis, angiogenesis and metastasis in CRC cells. It also plays major role in attenuating certain proteins that promotes multi drug resistance. Qin et al. (2015) suggested that genistein exerts its anticancer properties in vitro and in vivo by downregulating diverse oncogenic proteins such as miR-95, SGK1 and inhibiting the Akt phosphorylation, which regulates various signalling pathways, in CRC cells. SGK1 is a serine threonine protein kinase also known as serum and glucocorticoid regulated kinase 1, and associated with CRC development and progression (Danielsen et al. 2015; Lang et al. 2010), while miR-95, which is up regulated in tumor cells, is responsible for proliferation and invasion in CRC. They also reported that genistein inhibits the Akt phosphorylation, inducing the pathway involved in mitochondria and resulting in apoptosis of CRC (HCT-116) cells by inhibiting the Bax protein (Qin et al. 2015). Genistein was even found to prevent cell cycle progression by downregulating chk2, cyclin B1, proliferating cell nuclear antigen (PCNA) and upregulating antitumor proteins such as p21 (Fan et al. 2010; Han et al. 2013).

EMT plays a vital role in invasion and metastasis, and correlates with increased vimentin expression and reduced expression of E-cadherin in tumor cells. Zhou et al. (2017) suggested that genistein could regulate metastasis and induce apoptosis by reversing EMT through E-cadherin/Notch1/NF-κB pathway in CRC. Qi et al. (2011) also suggested that genistein can promote the expression of FOXO3 to inhibit expression of EGF, thus suppressing the EGF-induced proliferation of CRC cells through the Akt/PI3K pathway (Fig. 1.1). Furthermore, Genistein can reverse the epigenetic silenced genes responsible for DNA methylation of CpG islands, hence blocking the expression of tumor genes. Zhu et al. (2018) reported that genistein induces demethylation activating wnt inhibitory factor1 (WIF1), in a dose dependent manner, which inhibits the WNT proteins, β-catenin and cyclin-D1. Additionally, they also observed reduction in the expression of MMP9, MMP2 and an increased expression of E-cadherin and metalloproteinase inhibitor 1 after genistein treatment in CRC cells. Moreover, genistein can enhance the efficacy of 5-FU, when used in combination with metformin and lunasin against CSCs in CRC (Montales et al. 2015). Gruca et al. (2014) investigated a genistein derivative in the clinical setting

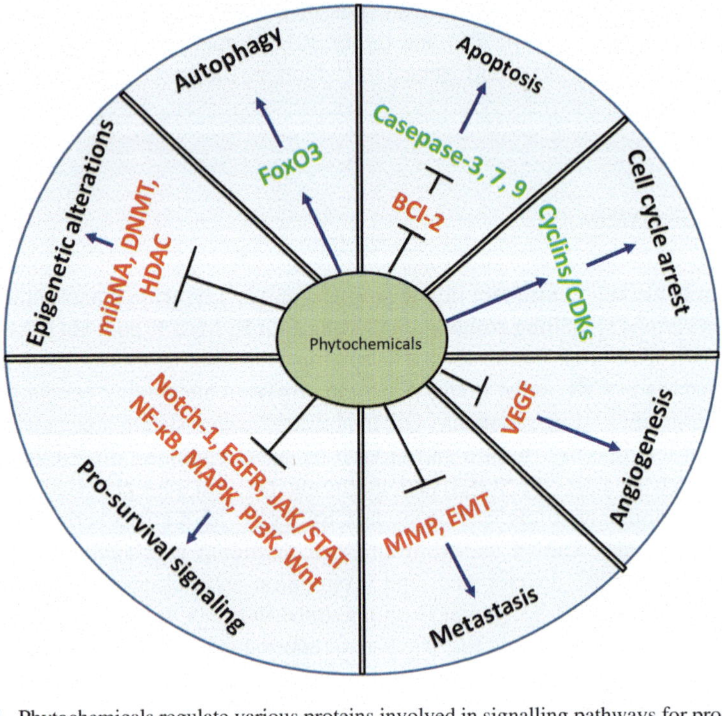

Fig. 1.1 Phytochemicals regulate various proteins involved in signalling pathways for progression of colon cancer

and found that synthetic genistein can sensitise CRC cells to ionizing radiation in radio-therapy, further downregulating EGFR expression.

1.4 Stilbenes

Stilbenes are plant stress metabolites produced in almost all the plants in response to injury, fungal infections or following exposure to UV radiations. They are found in several berries including blueberries, lingonberries and grapes (Rimando et al. 2004). These naturally existing phytochemicals are well known as plant defence compounds and possess diverse cellular and biological functions crucial for human health. Pharmacologically, they increase the sensitivity to insulin, prevent cancer and can thus improve the life span (Rimando and Suh 2008). Among stilbenes, pterostilbene and resveratrol are phenolic compounds with known anti-inflammatory, anticancer, chemopreventive and anti-oxidant properties.

1.5 Pterostilbene

Pterostilbene's are naturally occurring and structurally related to the stilbenes found in blueberries. Biologically they are effective against proliferation, invasion and metastasis of cancer cells (Pan et al. 2007, 2009; Rimando et al. 2002). They have anti-inflammatory properties and are effective in reversing age in rats (Joseph et al. 2008; Pan et al. 2008; Paul et al. 2009). As an anti-inflammatory agent pterostilbene down regulates the expression of inflammatory markers including cyclooxygenase-2 (COX-2) and inducible nitric oxide synthase (iNOS), reducing the risk of CRC by reducing mRNA levels of pro-inflammatory cytokines TNF α, IL 6 and IL-1β (Paul et al. 2009). Paul et al. (2009) investigated some upstream signalling pathways including MAPK, NF-κB and JAK-STAT pathways and determined that pterostilbene inhibits the formation of COX-2 and iNOS mediated by p38 MAP kinase. Moreover, Paul et al. (2010) suggested that pterostilbene can regulate the β-catenin/Wnt signalling pathway, thus inhibiting tumorigenesis in the colon. It can also control the binding potentiality of NF-κB to the nucleus by down regulating the phosphorylation of p65, thus affecting many genes responsible for inflammatory responses (Ganchi et al. 1993; Mosialos 1993). Cyclin D1 and c-MYC are overexpressed in CRC, that promotes proliferation of tumor cells, and are controlled by β-catenin and wnt pathways. Additionally, the estrogen receptor (ER-β) present in the colorectal epithelium regulates cell proliferation and activates apoptosis (Barone et al. 2010). However, in CRC patients, the expression of ER-β is down regulated, thus impairing its protective role in CRC (Caiazza et al. 2015). Pterostilbene can potentiate its chemopreventive activity by enhancing the expression of detoxifying enzymes, facilitating the excretion of xenobiotics (carcinogens), inducing GST and enhancing the activity of NADPH quinone oxidoreductase (NQO1) (Harun and Ghazali 2012). Suh et al. (2007) suggested that pterostilbene can suppressed the aberrant crypt foci (ACF) of CRC in azoxymethane (AOM)-induced proliferation in the colonic cells of experimental animal. They also observed that pterostilbene can inhibit the expression of iNOS and enhance the expression of mucin (MUC2). ACFs correspond to lesions of the early preneoplasm detected in induced colon carcinogenesis in experimental rats (Wargovich et al. 1995). Chiou et al. (2011) used AOM induced colon tumorigenesis to test whether resveratrol and pterostilbene could prevent AOM-induced tumorigenesis in the colon and determined that pterostilbene is more effective than resveratrol in reducing ACF and lymphoid nodules. It can also more effectively reduce the activation of NF-κB, thus inhibiting the phosphorylation of protein kinase-β2. However, when combined with resveratrol, it can also enhance the expression of antioxidant enzymes including glutathione reductase and heme oxygenase-1 through the activation of the Nrf2 signalling pathway. Additionally, pterostilbene is also used in combinational therapy with quercetin to enhance the efficiency of chemo (FOLFOX) and radiotherapy (X-rays) in vivo, thus inhibiting Bcl-2 expression, promoting the expression of super oxide dismutase and sensitizing tumor cells to vascular endothelium-induced cytotoxicity (Priego et al. 2008).

1.5.1 Resveratrol

Resveratrol is the most studied and preclinically investigated polyphenol against CRC, and is abundantly found in grapes and mostly in wine. This stilbene is also known as a phyto-antibiotic, phytoalexin and for the concept of the 'French Paradox' (Bishayee 2009). It harbours various biological properties and can act against various diseases including cancer, pulmonary, anti-aging, diabetes, coronary and arthritis. Resveratrol controls various signalling pathways such as NF-κB, PI3K/Akt and ER involved in CRC cell cycle progression, proliferation, apoptosis, angiogenesis and metastasis. Reddivari et al. (2016) determined that resveratrol in grape extracts induces apoptosis mediated by the mitochondria, which enhances the expression of cleaved PARP, p53 and maintains a ratio between Bax/Bcl-2. Because of its photosensitivity and metabolic instability, Kim et al. (2017) studied the synthetic analogue of resveratrol, HS-1793, and tested its apoptotic nature in CRC cell lines. They determined that the analogue inhibited cell growth, induced apoptosis and G2/M cell cycle arrest through the inhibition of the PI3K/Akt pathway. The combination also inhibited the wnt/β catenin signalling pathway, cyclinD1 and c-myc. Resveratrol promotes cell cycle arrest at the G1/S phase and in low dose at the G2/S phase, inducing apoptosis by activating caspase dependent cyclin-CDK pathway in CRC cells (Liu et al. 2014; Wolter et al. 2001) (Fig. 1.2). As an anti-inflammatory phytochemical, resveratrol inhibits COX-2 expression and counteracts the pro-inflammatory mechanism of NF-κB (Tili and Michaille 2011). Resveratrol modulates about 46 miRNAs, as revealed in a study by Tilli et al. (2010). It also downregulates the expression of TGFβ1 in miR-663 (Langenskiold et al. 2008). Karimi Dermani et al. (2017) showed that resveratrol can upregulate the expression

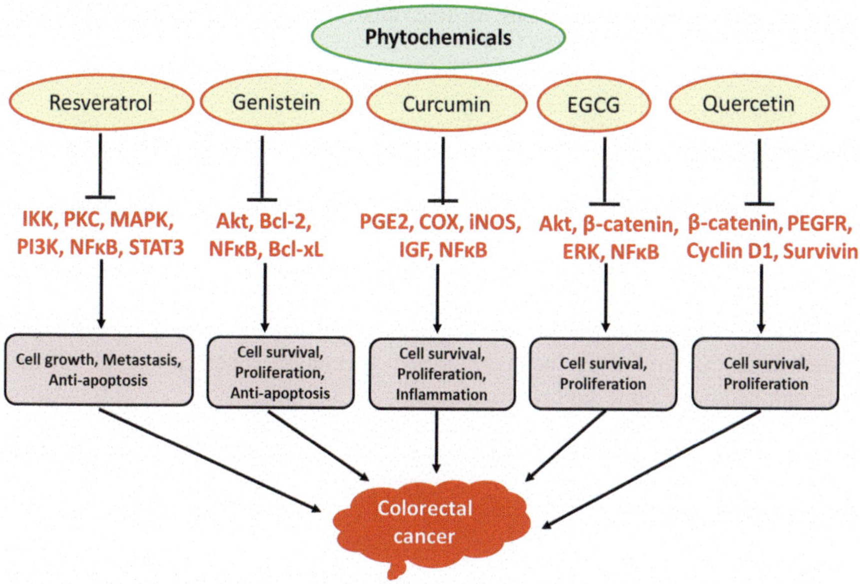

Fig. 1.2 The molecular targets of various phytochemicals promoting apoptosis and inhibiting cell proliferation, survival, and metastasis in colorectal cancer

of miR-200c, which plays a major role as a tumor suppressor by inhibiting EMT and thus enhances apoptosis and prevents tumor invasion. Du et al. (2016) determined the inhibitory effect of resveratrol against the hedgehog signalling pathway and its downstream proteins including patch homolog (ptch), smoothened (Smo) and glioma associated oncogenes (Gli), responsible for increasing the viability, migration and suppressing apoptosis of CRC cell in a time and dose dependent manner. Resveratrol can also be used in combinational therapy in order to potentiate the chemotherapy activity and overcome the limitation of multi drug resistance developed by tumor cells. For instance, Khaleel et al. (2016) assessed the effectiveness of resveratol and didox in modulating doxorubicin. These two phytochemicals synergistically potentiated doxorubicin by blocking the g-glycoprotein pump efflux activity and sensitized CRC cells to doxorubicin. Additionally, they observed increase in the expression of p53 and Bax genes and a decrease in the expression of Bcl-XL genes, thus facilitating apoptosis of CRC cells. Resveratrol can also act effectively when combined with 5-FU, inducing apoptosis in CRC cell lines through the activation of the MAPK signalling pathway (Kumazaki et al. 2013). Moreover, Buhrmann et al. (2015) showed that resveratrol chemosenstitizes CRC cells to 5-FU and potentiates its anti tumor nature by downregulating the expression of NF-κB and inhibiting EMT in tumor cells. Further evidence by Chung et al. (2018) suggests that the combination of resveratrol and 5-FU can inhibit the phosphorylation of STAT3 by binding to the human telomerase transcriptase at its promoter region, thus potentiating pro-apoptotic activity and resensitizing the cells to 5-FU. However further clinical studies is required to determine the effectiveness and molecular mechanism of resveratrol in CRC.

1.6 Curcuminoid

Curcuminoid is a yellow coloured crystalline spice powder obtained from the dried rhizome of Curcuma longa, and is called turmeric. It is traditionally used for healing wounds, rheumatism, sinusitis, anorexia and disorders like hepatic and biliary conditions. Major curcuminoids also include demethoxycurcumin, bisdemethoxycurcumin, cyclocurcumin and curcumin (Shehzad et al. 2010), curcumin being the most studied compound.

1.6.1 Curcumin

Curcumin is an active component used traditionally against various diseases including arthritis, cholangiocarcinoma, medulloblastoma, colorectal and pancreatic cancer. About 83 articles have been published reporting in vitro and in vivo preclinical studies that describe the effectiveness of curcumin against CRC. It is a hydrophobic polyphenol with various biological properties including antioxidant, anti-inflammatory and antitumor activity. It targets various proteins involved in carcinogenesis, affecting various signalling pathways including PI3K/Akt, NF-κB/COX-2, WNT, NOTCH and STAT. It can actively inhibit tumor initiation, differentiation, progression, invasion and angiogenesis and induce apoptosis in CRC cells. Tumor progression might sometimes be due to mutations in genes involved in the Wnt signalling pathway including β-catenin and APC (Bahrami et al. 2017). Dou et al. (2017) determined that curcumin can inhibit the wnt signalling pathway by downregulating the expression of β-catenin and upregulating NKd2, which is a negative regulator of the wnt cascade. They also observed that miR-21 and miR-130a, which play major contribute to proliferation, differentiation, migration, cell cycle arrest and anti-apoptosis in CRC are also affected by curcumin, indicating that miR-130a may serve as a novel target in cancer therapy (Fig. 1.1). The yellow polyphenol was also found to be effective against stages I and II colon cancer cells. Hence, Dasiram et al. (2017) determined the effect of curcumin against stage III metastatic Duke's type colon cancer by targeting p53 mutated colon adenocarcinoma cells of COLO320DM cell lines in a dose and time dependant manner. They observed cell cycle arrest at G1 phase and a decline in the population of cancer cells at the S phase, suggesting that curcumin induced apoptosis in colon adenocarcinoma cells.

Liang et al. (2017) synthesized a novel analogue of curcumin named monocarbonyl curcumin (MC37), which can act as a multi target agent against various cancers including CRC. This compound can inhibit the assembly of microtubules in cancer cells, induce cell cycle arrest at the G2/M phase thus altering CDK1expression, inhibited NF-κB activity and induce apoptosis by cleaving the mitochondrial membrane, activate caspase 9/3 protein and increase the ratio of Bax/Bcl-2. Curcumin can also prevent NF-κB from binding to the DNA. Tong et al. (2016) determined that curcumin activates 5′ AMPK and inhibits the activity of NF-κB by suppressing p65 expression, thus countering its binding to DNA. They also detected the inhibition of urokinase-type plasminogen activator (uPA) and MMP9, which are contribute to the invasive properties of CRC cells.

Curcumin was also found to inhibit proteins involved in inflammatory responses such as COX-2, iNOS and LOX through by suppressing NF-κB (Gupta et al. 2012). Additionally, curcumin can effectively reverse EMT, which results in the emergence of CSCs and invasion, by altering the expression of key miRNAs, thus preventing cancer cells from developing resistance against chemotherapy (Simental-Mendia et al. 2017). Furthermore, it can effectively reverse MDR and is used extensively in combinational therapy to potentiate the anticancer nature of the chemotherapeutics against resistant tumor cells. When combined with 5-FU, curcumin can induce apoptosis, suppress the ratio of Bcl-2/Bax by decreasing Nrf2 expression and subsequently reversing MDR (Zhang et al. 2018). Yang et al. (2018) also suggested that curcumin can sensitize colon cancer cells to irradiation, as determined by the apoptosis of cancer cells. Novel technologies like nanotechnology can play a key role in targeting tumor cells and sensitizing them to chemo drugs. Accordingly, Javadi et al. (2018) combined nanotechnology with combinational therapy to enhance sensitization by loading the nanoparticle, methoxy poly ethylene glycol poly caprolactone (mPEG-PCL) with curcumin and erlotinib, separately. Erlotinib is an EGFR inhibitor, but its efficiency can be hindered by the development of drug resistance. To address this issue, they targeted αvβ3 integrin and pyruvate dehydrogenase kinase 4 (PDK4), which can contribute to tumor resistance against erlotinib. Their results showed a decrease and increase in the expression of αvβ3 integrin and PDK4 respectively. Insulin growth factor 1 (IGF 1) and survivin are important factors because they inhibit mitochondrial mediated pathways and contribute to the prolonged proliferation of colon cancer cells by blocking apoptosis. In contrast, curcumin can activate the apoptotic signals by inducing p53 expression and downregulating the expression of IGF1, survivin and TNF-α (He et al. 2011). Other proteins that contribute to the emergence of drug resistance include IGFR, Akt, cyclin D1 and COX-2. Curcumin has been shown to effectively downregulate the expression of these molecules and to sensitize CRC cells to 5-FU and oxaliplatin (Patel et al. 2010). Further research is necessary to determining the bioavailability and efficiency of curcumin in the clinical setting. There are many other phyto chemicals including terpenoids, organosulfur compounds, allyl sulphur compounds and phytosterols which possess anti tumor activities effecting signalling pathways involved in proliferation and apoptosis of various cancers including CRC.

The discovery of phytochemicals such as apigenin, genistein, curcumin, resveratrol and ellagic acid, characterized by their anticancer activity both in vitro and in vivo, has led to their investigation in clinical trials, including early phase I/phase I/Phase II studies summarized in Table 1.1. These trials are currently investigating

Table 1.1 Phytochemicals and their status in clinical trials

Phytochemical	Source	Identifier (NCT number)	Phase	Summary	Drug	Outcome measures
Ellagic acid (EA) and Ellagitannins (EG)	Pomogranate	NCT01916239	Phase I & II	The metabolized urolithins exerts anti inflammatory and anti cancer activities	Dietary supplement	Changes in the levels of IGF-1, carcino embryonic antigen (CEA) and miRNA
Apigenin & epigallocatechin gallate (flavonoids)	Green tea & chamomile	NCT00609310	Phase II	Preventing the recurrence of neoplasia in resected CRC patients	Dietary supplement	Recurrence rate of CRC & metastasis CRC tested. (time frame 3 years)
Genistein (Isoflavone)	Soy	NCT01985763	Phase I & II	May inhibit Wnt signaling pathway, enhance inhibition of growth when combined with 5-FU and FOLFOX	Genistein	Response rate for CEA and imaging is calculated with routine follow up
Resveratrol (res) (Stilbene)	Grapes	NCT00620803	Phase I	Safety, tolerability, pharmacokinetic profile in hepatic metastatic CRC patient and pharmacodynamics are tested	Placebo & SRT501	Safe and will be assessed continually
Resveratrol (Stilbene)	Grapes	NCT00433576	Phase I	Res may inhibit the growth of CRC cells by blocking the enzymes involved in cell growth	Resveratrol	Pharmacodynamic of res, concentration of biomarkers, COX-2 and MiG are tested
Resveratrol (Stilbene)	Grapes	NCT00256334	Phase I	Modulates Wnt signaling pathway	Resveratrol	Tested in vivo for alteration of Wnt signaling
Curcuminoids	Turmeric	NCT00118989	Not applicable	Decreases proliferation of cells, colorectal mucosa COX-2 expression and increases apoptosis in adenomatous colonic polyps	Curcuminoids (C3 complex)/ placebo	Expression of COX-2, apoptosis of cancer cells and proliferation of cells are determined

Curcumin	Turmeric	NCT00027495	Phase I	Dose amount of curcumin can be tolerated in healthy individuals to prevent colon cancer	Curcumin	Not provided
Curcumin	Turmeric	NCT02724202	Early phase I	Clinical response tested for combination therapy of curcumin with 5-FU	Curcumin & 5-FU	Tested for the safety in metastatic colon cancer patient. DNA methylation status and miRNA profile are tested

Source: www.clinicaltrials.gov

the safety and efficacy of the compounds for CRC treatment. However, further improvements in the biological or chemical nature of these naturally available polyphenols are required to enhance their bioavailability and activity against cancer.

1.7 Conclusion

The use of phytochemicals as therapeutic drugs might be a promising approach for the partial or complete diminution of CRC with minimum or no adverse effects when used in combination with irradiation or chemotherapy. However, more studies are required to determine the binding efficiency of these phytochemicals to the targets. Phytochemical therapy might also be an attractive tool for the therapy and prevention and treatment of cancers like pancreatic, lung and colorectal cancers. The bioavailability of these compounds remains their major limitation, although the use of analogues and derivatives might provide a solution to this shortcoming. Advances in nanotechnology, in combination with target therapy, is another way to bypass this obstacle. Future clinical and preclinical studies are crucial for exploring the effectiveness and molecular mechanisms of these phytochemicals, and evaluating their clinical effectiveness for CRC treatment.

References

Abaza MS, Orabi KY, Al-Quattan E, Al-Attiyah RJ (2015) Growth inhibitory and chemosensitization effects of naringenin, a natural flavanone purified from Thymus vulgaris, on human breast and colorectal cancer. Cancer Cell Int 15:46

Andlauer W, Kolb J, Stehle P, Furst P (2000) Absorption and metabolism of genistein in isolated rat small intestine. J Nutr 130:843–846

Babich H, Krupka ME, Nissim HA, Zuckerbraun HL (2005) Differential in vitro cytotoxicity of (−)-epicatechin gallate (ECG) to cancer and normal cells from the human oral cavity. Toxicol In Vitro 19:231–242

Bagli E, Stefaniotou M, Morbidelli L, Ziche M, Psillas K, Murphy C, Fotsis T (2004) Luteolin inhibits vascular endothelial growth factor-induced angiogenesis; inhibition of endothelial cell survival and proliferation by targeting phosphatidylinositol 3′-kinase activity. Cancer Res 64:7936–7946

Bahrami A, Amerizadeh F, ShahidSales S, Khazaei M, Ghayour-Mobarhan M, Sadeghnia HR, Maftouh M, Hassanian SM, Avan A (2017) Therapeutic potential of targeting wnt/beta-catenin pathway in treatment of colorectal cancer: rational and progress. J Cell Biochem 118:1979–1983

Banerjee S, Ali S, Azmi A, Kong D, Ahmad A, Bao B, Sarkar F (2012) Abstract 2698. In: Proceedings of the 103rd Annual Meeting of the American Association for Cancer Research; 2012 Mar 31–Apr 4; Chicago, IL. Philadelphia (PA): AACR; Cancer Res 2012; 72(8 Suppl):Abstract nr 2698. doi: https://doi.org/10.1158/1538-7445.AM2012-2698

Barone M, Scavo MP, Papagni S, Piscitelli D, Guido R, Di Lena M, Comelli MC, Di Leo A (2010) ERbeta expression in normal, adenomatous and carcinomatous tissues of patients with familial adenomatous polyposis. Scand J Gastroenterol 45:1320–1328

Bell C, Hawthorne S (2008) Ellagic acid, pomegranate and prostate cancer—a mini review. J Pharm Pharmacol 60:139–144

Bishayee A (2009) Cancer prevention and treatment with resveratrol: from rodent studies to clinical trials. Cancer Prev Res (Phila) 2:409–418

Budavari S (1989) An encyclopedia of chemicals, drugs, and biologicals. The Merck Index, 246

Buhrmann C, Shayan P, Kraehe P, Popper B, Goel A, Shakibaei M (2015) Resveratrol induces chemosensitization to 5-fluorouracil through up-regulation of intercellular junctions, epithelial-to-mesenchymal transition and apoptosis in colorectal cancer. Biochem Pharmacol 98:51–68

Caiazza F, Ryan EJ, Doherty G, Winter DC, Sheahan K (2015) Estrogen receptors and their implications in colorectal carcinogenesis. Front Oncol 5:19

Chang H, Lei L, Zhou Y, Ye F, Zhao G (2018) Dietary flavonoids and the risk of colorectal cancer: an updated meta-analysis of epidemiological studies. Nutrients 10:950

Chen Y, Huang C, Zhou T, Chen G (2008) Genistein induction of human sulfotransferases in HepG2 and Caco-2 cells. Basic Clin Pharmacol Toxicol 103:553–559

Chen J, Hou R, Zhang X, Ye Y, Wang Y, Tian J (2014) Calycosin suppresses breast cancer cell growth via ERbeta-dependent regulation of IGF-1R, p38 MAPK and PI3K/Akt pathways. PLoS One 9:e91245

Chen J, Lin C, Yong W, Ye Y, Huang Z (2015) Calycosin and Genistein induce apoptosis by inactivation of HOTAIR/p-Akt signaling pathway in human breast cancer MCF-7 cells. Cell Physiol Biochem 35:722–728

Chen L, Jiang B, Zhong C, Guo J, Zhang L, Mu T, Zhang Q, Bi X (2018) Chemoprevention of colorectal cancer by black raspberry anthocyanins involved the modulation of gut microbiota and SFRP2 demethylation. Carcinogenesis 39:471–481

Cheng C-W, Shieh P-C, Lin Y-C, Chen Y-J, Lin Y-H, Kuo D-H, Liu J-Y, Kao J-Y, Kao M-C, Way T-D (2009) Indoleamine 2, 3-dioxygenase, an immunomodulatory protein, is suppressed by (−)-epigallocatechin-3-gallate via blocking of γ-interferon-induced JAK-PKC-δ-STAT1 signaling in human oral cancer cells. J Agric Food Chem 58:887–894

Chian S, Li YY, Wang XJ, Tang XW (2014) Luteolin sensitizes two oxaliplatin-resistant colorectal cancer cell lines to chemotherapeutic drugs via inhibition of the Nrf2 pathway. Asian Pac J Cancer Prev: APJCP 15:2911–2916

Chiou YS, Tsai ML, Nagabhushanam K, Wang YJ, Wu CH, Ho CT, Pan MH (2011) Pterostilbene is more potent than resveratrol in preventing azoxymethane (AOM)-induced colon tumorigenesis via activation of the NF-E2-related factor 2 (Nrf2)-mediated antioxidant signaling pathway. J Agric Food Chem 59:2725–2733

Chung SS, Dutta P, Austin D, Wang P, Awad A, Vadgama JV (2018) Combination of resveratrol and 5-flurouracil enhanced anti-telomerase activity and apoptosis by inhibiting STAT3 and Akt signaling pathways in human colorectal cancer cells. Oncotarget 9:32943–32957

Chunhua L, Donglan L, Xiuqiong F, Lihua Z, Qin F, Yawei L, Liang Z, Ge W, Linlin J, Ping Z, Kun L, Xuegang S (2013) Apigenin up-regulates transgelin and inhibits invasion and migration of colorectal cancer through decreased phosphorylation of AKT. J Nutr Biochem 24:1766–1775

Danielsen SA, Eide PW, Nesbakken A, Guren T, Leithe E, Lothe RA (2015) Portrait of the PI3K/AKT pathway in colorectal cancer. Biochim Biophys Acta 1855:104–121

Dasiram JD, Ganesan R, Kannan J, Kotteeswaran V, Sivalingam N (2017) Curcumin inhibits growth potential by G1 cell cycle arrest and induces apoptosis in p53-mutated COLO 320DM human colon adenocarcinoma cells. Biomed Pharmacother [Biomedecine & Pharmacotherapie] 86:373–380

de Sousa Moraes LF, Sun X, MdCG P, Zhu M-J (2017) Anthocyanins/anthocyanidins and colorectal cancer: what is behind the scenes? Crit Rev Food Sci Nutr 59:1–13

DellaPenna D (1999) Nutritional genomics: manipulating plant micronutrients to improve human health. Science 285:375–379

Dou H, Shen R, Tao J, Huang L, Shi H, Chen H, Wang Y, Wang T (2017) Curcumin suppresses the colon cancer proliferation by inhibiting wnt/beta-catenin pathways via miR-130a. Front Pharmacol 8:877

Du Z, Zhou F, Jia Z, Zheng B, Han S, Cheng J, Zhu G, Huang P (2016) The hedgehog/Gli-1 signaling pathways is involved in the inhibitory effect of resveratrol on human colorectal cancer HCT116 cells. Iran J Basic Med Sci 19:1171–1176

Fan Y-z, Li G-h, Wang Y-h, Ren Q-y, Shi H-j (2010) Effects of genistein on colon cancer cells in vitro and in vivo and its mechanism of action. Zhonghua Zhong Liu Za Zhi 32:4–9

Faried A, Kurnia D, Faried L, Usman N, Miyazaki T, Kato H, Kuwano H (2007) Anticancer effects of gallic acid isolated from Indonesian herbal medicine, Phaleria macrocarpa (Scheff.) Boerl, on human cancer cell lines. Int J Oncol 30:605–613

Ferguson LR (2009) Nutrigenomics approaches to functional foods. J Am Diet Assoc 109:452–458

Fraga CG, Galleano M, Verstraeten SV, Oteiza PI (2010) Basic biochemical mechanisms behind the health benefits of polyphenols. Mol Asp Med 31:435–445

Fung KY, Cosgrove L, Lockett T, Head R, Topping DL (2012) A review of the potential mechanisms for the lowering of colorectal oncogenesis by butyrate. Br J Nutr 108:820–831

Ganchi PA, Sun SC, Greene WC, Ballard DW (1993) A novel NF-kappa B complex containing p65 homodimers: implications for transcriptional control at the level of subunit dimerization. Mol Cell Biol 13:7826–7835

Gee JM, Hara H, Johnson IT (2002) Suppression of intestinal crypt cell proliferation and aberrant crypt foci by dietary quercetin in rats. Nutr Cancer 43:193–201

Grosso G, Godos J, Lamuela-Raventos R, Ray S, Micek A, Pajak A, Sciacca S, D'Orazio N, Del Rio D, Galvano F (2017) A comprehensive meta-analysis on dietary flavonoid and lignan intake and cancer risk: level of evidence and limitations. Mol Nutr Food Res 61:1600930

Gruca A, Krawczyk Z, Szeja W, Grynkiewicz G, Rusin A (2014) Synthetic genistein glycosides inhibiting EGFR phosphorylation enhance the effect of radiation in HCT 116 colon cancer cells. Molecules (Basel Switzerland) 19:18558–18573

Gupta SC, Patchva S, Koh W, Aggarwal BB (2012) Discovery of curcumin, a component of golden spice, and its miraculous biological activities. Clin Exp Pharmacol Physiol 39:283–299

Han J, Kurita Y, Isoda H (2013) Genistein-induced G2/M cell cycle arrest of human intestinal colon cancer Caco-2 cells is associated with Cyclin B1 and Chk2 down-regulation. Cytotechnology 65:973–978

Harris GK, Qian Y, Leonard SS, Sbarra DC, Shi X (2006) Luteolin and chrysin differentially inhibit cyclooxygenase-2 expression and scavenge reactive oxygen species but similarly inhibit prostaglandin-E2 formation in RAW 264.7 cells. J Nutr 136:1517–1521

Harun Z, Ghazali AR (2012) Potential chemoprevention activity of pterostilbene by enhancing the detoxifying enzymes in the HT-29 cell line. Asian Pac J Cancer Prev: APJCP 13:6403–6407

He Z-Y, Shi C-B, Wen H, Li F-L, Wang B-L, Wang J (2011) Upregulation of p53 expression in patients with colorectal cancer by administration of curcumin. Cancer Investig 29:208–213

Ho JW, Leung Y, Chan C (2002) Herbal medicine in the treatment of cancer. Curr Med Chem: Anti-Cancer Agents 2:209–214

Ho K, Yazan LS, Ismail N, Ismail M (2009) Apoptosis and cell cycle arrest of human colorectal cancer cell line HT-29 induced by vanillin. Cancer Epidemiol 33:155–160

Hofmanova J, Hyrslova Vaculova A, Kozubik A (2013) Regulation of the metabolism of polyunsaturated fatty acids and butyrate in colon cancer cells. Curr Pharm Biotechnol 14:274–288

Huang S-C, Ho C-T, Lin-Shiau S-Y, Lin J-K (2005) Carnosol inhibits the invasion of B16/F10 mouse melanoma cells by suppressing metalloproteinase-9 through down-regulating nuclear factor-kappaB and c-Jun. Biochem Pharmacol 69:221–232

Huang Y, Zeng F, Xu L, Zhou J, Liu X, Le H (2013) Anticancer effects of cinnamic acid in lung adenocarcinoma cell line h1299-derived stem-like cells. Oncol Res 20:499–507

Ivey KL, Jensen MK, Hodgson JM, Eliassen AH, Cassidy A, Rimm EB (2017) Association of flavonoid-rich foods and flavonoids with risk of all-cause mortality. Br J Nutr 117:1470–1477

Javadi S, Rostamizadeh K, Hejazi J, Parsa M, Fathi M (2018) Curcumin mediated down-regulation of alphaV beta3 integrin and up-regulation of pyruvate dehydrogenase kinase 4 (PDK4) in Erlotinib resistant SW480 colon cancer cells. Phytother Res: PTR 32:355–364

Jemal A, Siegel R, Ward E, Hao Y, Xu J, Murray T, Thun MJ (2008) Cancer statistics, 2008. CA Cancer J Clin 58:71–96

Joseph JA, Fisher DR, Cheng V, Rimando AM, Shukitt-Hale B (2008) Cellular and behavioral effects of stilbene resveratrol analogues: implications for reducing the deleterious effects of aging. J Agric Food Chem 56:10544–10551

Kang NJ, Lee KW, Kim BH, Bode AM, Lee H-J, Heo Y-S, Boardman L, Limburg P, Lee HJ, Dong Z (2011) Coffee phenolic phytochemicals suppress colon cancer metastasis by targeting MEK and TOPK. Carcinogenesis 32:921–928

Kang KA, Piao MJ, Ryu YS, Hyun YJ, Park JE, Shilnikova K, Zhen AX, Kang HK, Koh YS, Jeong YJ, Hyun JW (2017) Luteolin induces apoptotic cell death via antioxidant activity in human colon cancer cells. Int J Oncol 51:1169–1178

Kaput J, Rodriguez RL (2004) Nutritional genomics: the next frontier in the postgenomic era. Physiol Genomics 16:166–177

Karimi Dermani F, Saidijam M, Amini R, Mahdavinezhad A, Heydari K, Najafi R (2017) Resveratrol inhibits proliferation, invasion, and epithelial-mesenchymal transition by increasing miR-200c expression in HCT-116 colorectal cancer cells. J Cell Biochem 118:1547–1555

Khaleel SA, Al-Abd AM, Ali AA, Abdel-Naim AB (2016) Didox and resveratrol sensitize colorectal cancer cells to doxorubicin via activating apoptosis and ameliorating P-glycoprotein activity. Sci Rep 6:36855

Khan SG, Katiyar SK, Agarwal R, Mukhtar H (1992) Enhancement of antioxidant and phase II enzymes by oral feeding of green tea polyphenols in drinking water to SKH-1 hairless mice: possible role in cancer chemoprevention. Cancer Res 52:4050–4052

Kim E, Coelho D, Blachier F (2013) Review of the association between meat consumption and risk of colorectal cancer. Nutr Res 33:983–994

Kim DH, Kim MJ, Sung B, Suh H, Jung JH, Chung HY, Kim ND (2017) Resveratrol analogue, HS-1793, induces apoptotic cell death and cell cycle arrest through downregulation of AKT in human colon cancer cells. Oncol Rep 37:281–288

Kumazaki M, Noguchi S, Yasui Y, Iwasaki J, Shinohara H, Yamada N, Akao Y (2013) Anti-cancer effects of naturally occurring compounds through modulation of signal transduction and miRNA expression in human colon cancer cells. J Nutr Biochem 24:1849–1858

Lafay S, Gil-Izquierdo A, Manach C, Morand C, Besson C, Scalbert A (2006) Chlorogenic acid is absorbed in its intact form in the stomach of rats. J Nutr 136:1192–1197

Lamuela-Raventós RM, Romero-Pérez AI, Andrés-Lacueva C, Tornero A (2005) Review: health effects of cocoa flavonoids. Food Sci Technol Int 11:159–176

Lang F, Perrotti N, Stournaras C (2010) Colorectal carcinoma cells—regulation of survival and growth by SGK1. Int J Biochem Cell Biol 42:1571–1575

Langenskiold M, Holmdahl L, Falk P, Angenete E, Ivarsson ML (2008) Increased TGF-beta 1 protein expression in patients with advanced colorectal cancer. J Surg Oncol 97:409–415

Lefort EC, Blay J (2011) The dietary flavonoid apigenin enhances the activities of the anti-metastatic protein CD26 on human colon carcinoma cells. Clin Exp Metastasis 28:337–349

Li W, Chang J, Wang S, Liu X, Peng J, Huang D, Sun M, Chen Z, Zhang W, Guo W (2015) miRNA-99b-5p suppresses liver metastasis of colorectal cancer by down-regulating mTOR. Oncotarget 6:24448

Liang B, Liu Z, Cao Y, Zhu C, Zuo Y, Huang L, Wen G, Shang N, Chen Y, Yue X, Du J, Li B, Zhou B, Bu X (2017) MC37, a new mono-carbonyl curcumin analog, induces G2/M cell cycle arrest and mitochondria-mediated apoptosis in human colorectal cancer cells. Eur J Pharmacol 796:139–148

Liu L, Hudgins WR, Shack S, Yin MQ, Samid D (1995) Cinnamic acid: a natural product with potential use in cancer intervention. Int J Cancer 62:345–350

Liu B, Zhou Z, Zhou W, Liu J, Zhang Q, Xia J, Liu J, Chen N, Li M, Zhu R (2014) Resveratrol inhibits proliferation in human colorectal carcinoma cells by inducing G1/Sphase cell cycle arrest and apoptosis through caspase/cyclinCDK pathways. Mol Med Rep 10:1697–1702

Liu M-p, Liao M, Dai C, Chen J-f, Yang C-j, Liu M, Chen Z-g, Yao M-c (2016) Sanguisorba officinalis L synergistically enhanced 5-fluorouracil cytotoxicity in colorectal cancer cells by promoting a reactive oxygen species-mediated, mitochondria-caspase-dependent apoptotic pathway. Sci Rep 6:34245

Liu Y, Lang T, Jin B, Chen F, Zhang Y, Beuerman RW, Zhou L, Zhang Z (2017) Luteolin inhibits colorectal cancer cell epithelial-to-mesenchymal transition by suppressing CREB1 expression revealed by comparative proteomics study. J Proteome 161:1–10

Lu Y, Shan S, Li H, Shi J, Zhang X, Li Z (2018) Reversal effects of bound polyphenol from foxtail millet bran on multidrug resistance in human HCT-8/Fu colorectal cancer cell. J Agric Food Chem 66:5190–5199

Luceri C, Caderni G, Sanna A, Dolara P (2002) Red wine and black tea polyphenols modulate the expression of cycloxygenase-2, inducible nitric oxide synthase and glutathione-related enzymes in azoxymethane-induced f344 rat colon tumors. J Nutr 132:1376–1379

Ma J, Duan W, Han S, Lei J, Xu Q, Chen X, Jiang Z, Nan L, Li J, Chen K (2015) Ginkgolic acid suppresses the development of pancreatic cancer by inhibiting pathways driving lipogenesis. Oncotarget 6:20993

Macheix JJ, Sapis JC, Fleuriet A, Lee C (1991) Phenolic compounds and polyphenoloxidase in relation to browning in grapes and wines. Crit Rev Food Sci Nutr 30:441–486

Maher DM, Bell MC, O'donnell EA, Gupta BK, Jaggi M, Chauhan SC (2011) Curcumin suppresses human papillomavirus oncoproteins, restores p53, rb, and ptpn13 proteins and inhibits benzo [a] pyrene-induced upregulation of HPV E7. Mol Carcinog 50:47–57

Maitan-Alfenas GP, de ALLG, de Almeida MN, Visser EM, de Rezende ST, Guimaraes VM (2014) Hydrolysis of soybean isoflavones by Debaryomyces hansenii UFV-1 immobilised cells and free beta-glucosidase. Food Chem 146:429–436

Mazewski C, Liang K, Gonzalez de Mejia E (2018) Comparison of the effect of chemical composition of anthocyanin-rich plant extracts on colon cancer cell proliferation and their potential mechanism of action using in vitro, in silico, and biochemical assays. Food Chem 242:378–388

Milner JA, McDonald SS, Anderson DE, Greenwald P (2001) Molecular targets for nutrients involved with cancer prevention. Nutr Cancer 41:1–16

Montales MT, Simmen RC, Ferreira ES, Neves VA, Simmen FA (2015) Metformin and soybean-derived bioactive molecules attenuate the expansion of stem cell-like epithelial subpopulation and confer apoptotic sensitivity in human colon cancer cells. Genes Nutr 10:49

Moon D-O, Kim M-O, Choi YH, Kim G-Y (2008) β-Sitosterol induces G2/M arrest, endoreduplication, and apoptosis through the Bcl-2 and PI3K/Akt signaling pathways. Cancer Lett 264:181–191

Mosialos G (1993) Gilmore TD: v-Rel and c-Rel are differentially affected by mutations at a consensus protein kinase recognition sequence. Oncogene 8:721–730

Murad LD, Soares Nda C, Brand C, Monteiro MC, Teodoro AJ (2015) Effects of caffeic and 5-caffeoylquinic acids on cell viability and cellular uptake in human colon adenocarcinoma cells. Nutr Cancer 67:532–542

Muto S, Fujita K-i, Yamazaki Y, Kamataki T (2001) Inhibition by green tea catechins of metabolic activation of procarcinogens by human cytochrome P450. Mutat Res Fundam Mol Mech Mutagen 479:197–206

Neogi P, Lakner FJ, Medicherla S, Cheng J, Dey D, Gowri M, Nag B, Sharma SD, Pickford LB, Gross C (2003) Synthesis and structure-activity relationship studies of cinnamic acid-based novel thiazolidinedione antihyperglycemic agents. Bioorg Med Chem 11:4059–4067

O'Keefe SJ, Li JV, Lahti L, Ou J, Carbonero F, Mohammed K, Posma JM, Kinross J, Wahl E, Ruder E (2015) Fat, fibre and cancer risk in African Americans and rural Africans. Nat Commun 6:ncomms7342

Owczarek K, Hrabec E, Fichna J, Sosnowska D, Koziolkiewicz M, Szymanski J, Lewandowska U (2017) Flavanols from Japanese quince (Chaenomeles japonica) fruit suppress expression of cyclooxygenase-2, metalloproteinase-9, and nuclear factor-kappaB in human colon cancer cells. Acta Biochim Pol 64:567–576

Palozza P, Colangelo M, Simone R, Catalano A, Boninsegna A, Lanza P, Monego G, Ranelletti FO (2010) Lycopene induces cell growth inhibition by altering mevalonate pathway and Ras signaling in cancer cell lines. Carcinogenesis 31:1813–1821

Pan MH, Chang YH, Badmaev V, Nagabhushanam K, Ho CT (2007) Pterostilbene induces apoptosis and cell cycle arrest in human gastric carcinoma cells. J Agric Food Chem 55:7777–7785

Pan MH, Chang YH, Tsai ML, Lai CS, Ho SY, Badmaev V, Ho CT (2008) Pterostilbene suppressed lipopolysaccharide-induced up-expression of iNOS and COX-2 in murine macrophages. J Agric Food Chem 56:7502–7509

Pan MH, Chiou YS, Chen WJ, Wang JM, Badmaev V, Ho CT (2009) Pterostilbene inhibited tumor invasion via suppressing multiple signal transduction pathways in human hepatocellular carcinoma cells. Carcinogenesis 30:1234–1242

Pandey KB, Rizvi SI (2009) Plant polyphenols as dietary antioxidants in human health and disease. Oxidative Med Cell Longev 2:270–278

Papi A, Farabegoli F, Iori R, Orlandi M, De Nicola GR, Bagatta M, Angelino D, Gennari L, Ninfali P (2013) Vitexin-2-O-xyloside, raphasatin and (−)-epigallocatechin-3-gallate synergistically affect cell growth and apoptosis of colon cancer cells. Food Chem 138:1521–1530

Patel BB, Gupta D, Elliott AA, Sengupta V, Yu Y, Majumdar AP (2010) Curcumin targets FOLFOX-surviving colon cancer cells via inhibition of EGFRs and IGF-1R. Anticancer Res 30:319–325

Paul S, Rimando AM, Lee HJ, Ji Y, Reddy BS, Suh N (2009) Anti-inflammatory action of pterostilbene is mediated through the p38 mitogen-activated protein kinase pathway in colon cancer cells. Cancer Prev Res (Phila) 2:650–657

Paul S, DeCastro AJ, Lee HJ, Smolarek AK, So JY, Simi B, Wang CX, Zhou R, Rimando AM, Suh N (2010) Dietary intake of pterostilbene, a constituent of blueberries, inhibits the beta-catenin/p65 downstream signaling pathway and colon carcinogenesis in rats. Carcinogenesis 31:1272–1278

Peng Y, Zhao B, Kang Q, Liu J, Chen C, Li B, Xie Y, Wu Q (2018) Colorectal cancer preventive effect of combined administration of phenolic acids and supercritical extracts from Angelica sinensis. Zhongguo Zhong yao za zhi= Zhongguo zhongyao zazhi= China J Chinese Materia Medica 43:1235–1240

Petrick JL, Steck SE, Bradshaw PT, Trivers KF, Abrahamson PE, Engel LS, He K, Chow WH, Mayne ST, Risch HA, Vaughan TL, Gammon MD (2015) Dietary intake of flavonoids and oesophageal and gastric cancer: incidence and survival in the United States of America (USA). Br J Cancer 112:1291–1300

Pontiki E, Hadjipavlou-Litina D, Litinas K, Geromichalos G (2014) Novel cinnamic acid derivatives as antioxidant and anticancer agents: design, synthesis and modeling studies. Molecules (Basel Switzerland) 19:9655–9674

Priego S, Feddi F, Ferrer P, Mena S, Benlloch M, Ortega A, Carretero J, Obrador E, Asensi M, Estrela JM (2008) Natural polyphenols facilitate elimination of HT-29 colorectal cancer xenografts by chemoradiotherapy: a Bcl-2- and superoxide dismutase 2-dependent mechanism. Mol Cancer Ther 7:3330–3342

Prietsch RF, Monte LG, da Silva FA, Beira FT, Del Pino FA, Campos VF, Collares T, Pinto LS, Spanevello RM, Gamaro GD, Braganhol E (2014) Genistein induces apoptosis and autophagy in human breast MCF-7 cells by modulating the expression of proapoptotic factors and oxidative stress enzymes. Mol Cell Biochem 390:235–242

Qi W, Weber CR, Wasland K, Savkovic SD (2011) Genistein inhibits proliferation of colon cancer cells by attenuating a negative effect of epidermal growth factor on tumor suppressor FOXO3 activity. BMC Cancer 11:219

Qiao L, Zheng J, Jin X, Wei G, Wang G, Sun X, Li X (2017) Ginkgolic acid inhibits the invasiveness of colon cancer cells through AMPK activation. Oncol Lett 14:5831–5838

Qin J, Chen JX, Zhu Z, Teng JA (2015) Genistein inhibits human colorectal cancer growth and suppresses miR-95, Akt and SGK1. Cell Physiol Biochem 35:2069–2077

Qu Q, Qu J, Guo Y, Zhou BT, Zhou HH (2014) Luteolin potentiates the sensitivity of colorectal cancer cell lines to oxaliplatin through the PPARgamma/OCTN2 pathway. Anti-Cancer Drugs 25:1016–1027

Ramos S (2008) Cancer chemoprevention and chemotherapy: dietary polyphenols and signalling pathways. Mol Nutr Food Res 52:507–526

Ramos S, Rodríguez-Ramiro I, Martín MA, Goya L, Bravo L (2011) Dietary flavanols exert different effects on antioxidant defenses and apoptosis/proliferation in Caco-2 and SW480 colon cancer cells. Toxicol In Vitro 25:1771–1781

Rao V (2012) Phytochemicals-a global perspective of their role in nutrition and health. BoD–Books on Demand, Norderstedt

Reddivari L, Charepalli V, Radhakrishnan S, Vadde R, Elias RJ, Lambert JD, Vanamala JKP (2016) Grape compounds suppress colon cancer stem cells in vitro and in a rodent model of colon carcinogenesis. BMC Complement Altern Med 16:278–278

Rimando AM, Suh N (2008) Biological/chemopreventive activity of stilbenes and their effect on colon cancer. Planta Med 74:1635–1643

Rimando AM, Cuendet M, Desmarchelier C, Mehta RG, Pezzuto JM, Duke SO (2002) Cancer chemopreventive and antioxidant activities of pterostilbene, a naturally occurring analogue of resveratrol. J Agric Food Chem 50:3453–3457

Rimando AM, Kalt W, Magee JB, Dewey J, Ballington JR (2004) Resveratrol, pterostilbene, and piceatannol in vaccinium berries. J Agric Food Chem 52:4713–4719

Roy N, Narayanankutty A, Nazeem PA, Valsalan R, Babu TD, Mathew D (2016) Plant phenolics ferulic acid and P-coumaric acid inhibit colorectal cancer cell proliferation through EGFR down-regulation. Asian Pac J Cancer Prev: APJCP 17:4019–4023

Sebastian RS, Wilkinson Enns C, Goldman JD, Moshfegh AJ (2017) Dietary flavonoid intake is inversely associated with cardiovascular disease risk as assessed by body mass index and waist circumference among adults in the United States. Nutrients 9:827

Shao H, Jing K, Mahmoud E, Huang H, Fang X, Yu C (2013) Apigenin sensitizes colon cancer cells to antitumor activity of ABT-263. Mol Cancer Ther 12:2640–2650

Sharma RA, McLelland HR, Hill KA, Ireson CR, Euden SA, Manson MM, Pirmohamed M, Marnett LJ, Gescher AJ, Steward WP (2001) Pharmacodynamic and pharmacokinetic study of oral Curcuma extract in patients with colorectal cancer. Clin Cancer Res 7:1894–1900

Sharma SH, Rajamanickam V, Nagarajan S (2018) Antiproliferative effect of p-Coumaric acid targets UPR activation by downregulating Grp78 in colon cancer. Chem Biol Interact 291:16–28

Shehzad A, Wahid F, Lee YS (2010) Curcumin in cancer chemoprevention: molecular targets, pharmacokinetics, bioavailability, and clinical trials. Arch Pharm 343:489–499

Shilpa P, Kaveri K, Salimath BP (2015) Anti-metastatic action of anacardic acid targets VEGF-induced signalling pathways in epithelial to mesenchymal transition. Drug Discov Ther 9:53–65

Simental-Mendia LE, Caraglia M, Majeed M, Sahebkar A (2017) Impact of curcumin on the regulation of microRNAs in colorectal cancer. Expert Rev Gastroenterol Hepatol 11:99–101

Simopoulos AP, Milner JA (2010) Personalized nutrition. Karger Medical and Scientific Publishers, Basel

Song HM, Park GH, Eo HJ, Lee JW, Kim MK, Lee JR, Lee MH, Koo JS, Jeong JB (2015) Anti-proliferative effect of Naringenin through p38-dependent downregulation of Cyclin D1 in human colorectal cancer cells. Biomol Ther 23:339–344

Suh N, Paul S, Hao X, Simi B, Xiao H, Rimando AM, Reddy BS (2007) Pterostilbene, an active constituent of blueberries, suppresses aberrant crypt foci formation in the azoxymethane-induced colon carcinogenesis model in rats. Clin Cancer Res 13:350–355

Surh Y-J (2003) Cancer chemoprevention with dietary phytochemicals. Nat Rev Cancer 3:768

Tili E, Michaille JJ (2011) Resveratrol, microRNAs, inflammation, and cancer. J Nucleic Acids 2011:102431

Tili E, Michaille JJ, Alder H, Volinia S, Delmas D, Latruffe N, Croce CM (2010) Resveratrol modulates the levels of microRNAs targeting genes encoding tumor-suppressors and effectors of TGFbeta signaling pathway in SW480 cells. Biochem Pharmacol 80:2057–2065

Tomás-Barberán FA, Clifford MN (2000) Dietary hydroxybenzoic acid derivatives–nature, occurrence and dietary burden. J Sci Food Agric 80:1024–1032

Tong W, Wang Q, Sun D, Suo J (2016) Curcumin suppresses colon cancer cell invasion via AMPK-induced inhibition of NF-kappaB, uPA activator and MMP9. Oncol Lett 12:4139–4146

Tse G, Eslick GD (2016) Soy and isoflavone consumption and risk of gastrointestinal cancer: a systematic review and meta-analysis. Eur J Nutr 55:63–73

Velioglu Y, Mazza G, Gao L, Oomah B (1998) Antioxidant activity and total phenolics in selected fruits, vegetables, and grain products. J Agric Food Chem 46:4113–4117

Wang X, Hao M-W, Dong K, Lin F, Ren J-H, Zhang H-Z (2009) Apoptosis induction effects of EGCG in laryngeal squamous cell carcinoma cells through telomerase repression. Arch Pharm Res 32:1263–1269

Wang C, Xie J, Guo J, Manning HC, Gore JC, Guo N (2012) Evaluation of CD44 and CD133 as cancer stem cell markers for colorectal cancer. Oncol Rep 28:1301–1308

Wargovich MJ, Chen CD, Harris C, Yang E, Velasco M (1995) Inhibition of aberrant crypt growth by non-steroidal anti-inflammatory agents and differentiation agents in the rat colon. Int J Cancer 60:515–519

Wenzel U, Kuntz S, Brendel MD, Daniel H (2000) Dietary flavone is a potent apoptosis inducer in human colon carcinoma cells. Cancer Res 60:3823–3831

Willett CG, Chang DT, Czito BG, Meyer J, Wo J (2013) Cancer genome atlas network. Comprehensive molecular characterization of human colon and rectal cancer. Nature 2012.(5). Int J Radiat Oncol Biol Phys 86

Wolter F, Akoglu B, Clausnitzer A, Stein J (2001) Downregulation of the cyclin D1/Cdk4 complex occurs during resveratrol-induced cell cycle arrest in colon cancer cell lines. J Nutr 131:2197–2203

Yan GR, Xiao CL, He GW, Yin XF, Chen NP, Cao Y, He QY (2010) Global phosphoproteomic effects of natural tyrosine kinase inhibitor, genistein, on signaling pathways. Proteomics 10:976–986

Yang G, Qiu J, Wang D, Tao Y, Song Y, Wang H, Tang J, Wang X, Sun YU, Yang Z, Hoffman RM (2018) Traditional Chinese medicine curcumin sensitizes human colon cancer to radiation by altering the expression of DNA repair-related genes. Anticancer Res 38:131–136

Yao H, Wang H, Zhang Z, Jiang BH, Luo J, Shi X (2008) Sulforaphane inhibited expression of hypoxia inducible factor-1α in human tongue squamous cancer cells and prostate cancer cells. Int J Cancer 123:1255–1261

Yeo SK, Liong MT (2010) Angiotensin I-converting enzyme inhibitory activity and bioconversion of isoflavones by probiotics in soymilk supplemented with prebiotics. Int J Food Sci Nutr 61:161–181

Yeom SJ, Kim BN, Kim YS, Oh DK (2012) Hydrolysis of isoflavone glycosides by a thermostable beta-glucosidase from Pyrococcus furiosus. J Agric Food Chem 60:1535–1541

Zhang C, He LJ, Ye HZ, Liu DF, Zhu YB, Miao DD, Zhang SP, Chen YY, Jia YW, Shen J, Liu XP (2018) Nrf2 is a key factor in the reversal effect of curcumin on multidrug resistance in the HCT8/5Fu human colorectal cancer cell line. Mol Med Rep 18:5409–5416

Zheng Q, Hirose Y, Yoshimi N, Murakami A, Koshimizu K, Ohigashi H, Sakata K, Matsumoto Y, Sayama Y, Mori H (2002) Further investigation of the modifying effect of various chemopreventive agents on apoptosis and cell proliferation in human colon cancer cells. J Cancer Res Clin Oncol 128:539–546

Zhou C, Li X, Du W, Feng Y, Kong X, Li Y, Xiao L, Zhang P (2010) Antitumor effects of ginkgolic acid in human cancer cell occur via cell cycle arrest and decrease the Bcl-2/Bax ratio to induce apoptosis. Chemotherapy 56:393–402

Zhou P, Wang C, Hu Z, Chen W, Qi W, Li A (2017) Genistein induces apoptosis of colon cancer cells by reversal of epithelial-to-mesenchymal via a Notch1/NF-kappaB/slug/E-cadherin pathway. BMC Cancer 17:813

Zhu B, Shang B, Li Y, Zhen Y (2016) Inhibition of histone deacetylases by trans-cinnamic acid and its antitumor effect against colon cancer xenografts in athymic mice. Mol Med Rep 13:4159–4166

Zhu J, Ren J, Tang L (2018) Genistein inhibits invasion and migration of colon cancer cells by recovering WIF1 expression. Mol Med Rep 17:7265–7273

Zuo Q, Wu R, Xiao X, Yang C, Yang Y, Wang C, Lin L, Kong AN (2018) The dietary flavone luteolin epigenetically activates the Nrf2 pathway and blocks cell transformation in human colorectal cancer HCT116 cells. J Cell Biochem 119:9573–9582

Chapter 2
Adiponectin Signaling in Colorectal Cancer

Gowru Srivani, Begum Dariya, Ganji Purnachandra Nagaraju, and Afroz Alam

Abstract Colorectal cancer (CRC) is an obesity-correlated malignancy. Obesity leads to dysregulation of adipocytokines. The adiponectin is an adipokine, secreted by adipose tissue, plays an essential role in energy metabolism, anti-inflammatory and insulin sensitizing properties. Besides this it also possess anti-neoplastic properties including promotion of apoptosis and inhibition of differentiation and progression of tumor cells. Adiponectin in serum/plasma level always depend on physical activity, diet and inheritance. High concentration of adiponectin are observed to play a crucial role in inhibiting tumor growth, which is also evidenced from the EPIC study where lower adiponectin circulation are at higher risk of cancer including CRC. Thus these results suggest an epidemiological link existing between obesity and cancer in association with adiponectin, playing its crucial role. Even though there are several studies reported in regard of CRC, the results with anti-cancerous property of adiponectin are still inconsistent. Adiponectin has numerous systemic effects that regulates homeostasis and constantly exists in high concentration. However, from the epidemological studies it was reported that low concentration of adiponectin are always at higher risk for CRC. The present article includes molecular evidence with a prominence on the regulation of adiponectin signalling pathways and their crosstalk with other signalling pathways in relation to CRC progression and provides several therapeutic future perspectives for the crucial role of adiponectin, aiming it as a target for the prevention of CRC.

Keywords CRC · Adiponectin · Obesity · Inflammation · Signalling pathways

---

G. Srivani · B. Dariya · A. Alam (✉)
Department of Bioscience and Biotechnology, Banasthali University, Vanasthali, Rajasthan, India

G. P. Nagaraju
Department of Hematology and Medical Oncology, Winship Cancer Institute, Emory University, Atlanta, GA, USA

G. S. R. Raju, L.V.K.S. Bhaskar (eds.), *Theranostics Approaches to Gastric and Colon Cancer*, Diagnostics and Therapeutic Advances in GI Malignancies, https://doi.org/10.1007/978-981-15-2017-4_2

Abbreviations

5-FU	Fluorouracil
Acrp30	Adipocyte complement related protein 30
ADP 355	Adenosine di phosphate
Akt	Protein kinase B
AMPK	Adenosine monophosphate Activated protein-kinase
BMI	Body mass index
CRC	Colorectal cancer
DFS	Disease free survival
EGFR	Epidermal growth factor receptor
EPIC	European prospective investigation into cancer and nutrition study
ERK	Extracellular signal—regulated kinase
FAS	Fatty acid synthase
GBP28	Gelatin binding protein
IGF-1	Insulin—like growth factor
IGFBP-1	Insulin like growth factor binding protein-1
IL	Interleukin
mTOR	Mammalian target of rapamycin
NF-κB	Nuclear factor kappa B cells
OS	Overall survival
PCI-1	Plasminogen activator inhibitor-1
PI3K	Phosphadidylinositol-3-kinase
PPARγ	Peroxisome proliferator activated receptor γ
SIRT1	Sirtulin (Silent mating type information regulation 2 homolog) 1
STAT3	Signal transducer and activator of transcription
TNF-α	Tumour necrosis factor
VEGF	Vascular endothelial growth factor

2.1 Introduction

Colorectal cancer (CRC) increasing in prevalence and has an increased mortality rate. It is the third most common cancer and ranking as the fourth leading cause of mortality (Jemal et al. 2011). Its initiation is associated with multifaceted interactions between host and environmental factors. These interactions stimulate the acquisition of genetic and epigenetic mutations. Thus contributes the inactivation of tumour suppresser genes and abnormal activation of proto oncogenes. The resulting deregulation of various intracellular signalling pathways, including growth factor (EGFR, VEGF) mediated signalling pathways, Notch, Wnt signalling, PI3K-Akt-mTOR and p53 pathways. These promote the oncogenic characteristics, such as abnormal cell proliferation, invasion, apoptosis, angiogenesis, improved metastasis and multi drug resistance (Malih and Najafi 2015; Wu et al. 2013). Predominately the most significant predisposing factor for the development and progression of CRC is diet, obesity and inheritance (Haggar and Boushey 2009; Malih and Najafi 2015). Obesity is characterized as abnormal elevated levels of fat deposition in the adipose tissue (Divella et al. 2016). The total incidence rate estimated in US and Europe for obesity-related colon cancer reports for 14–35% (Calle and Kaaks 2004). In addition, several studies revealed that obesity showed 7–60% of increased risk of CRC than the non-obese individuals (Dai et al. 2007; Ma et al. 2013). Adipose tissue is the long standing energy reservoir in the body. It not only stores the energy but also act as largest endocrine organ in the body. The energy is stored in the form of lipids to maintain physiologic homeostasis, metabolism, and immunity, bone remodelling and endocrine balance (Desruisseaux et al. 2007; Kelesidis et al. 2006; MacDougald and Burant 2007).

Adipose tissue secretes various bio-active factors that functions like hormones called adipocytes or adipocytokines possess several biological functions and are therefore classified as endocrine organ recently. Adipocytokines contain resistin, leptin, ghrelin, tumor necrosis factor alpha (TNF-α), plasminogen activator inhibitor-1 (PCI-1) and adiponectin (Otani et al. 2017). Adiponectin is a pleiotropic adipocytokine, functions in both paracrine and endocrine way. Several studies demonstrated that adiponectin have anti-obesity, anti-diabetic, anti-angiogenic, anti-inflammatory and anti-atherosclerotic properties (Barb et al. 2006). It play a vital role in the development and progression of various obesity associated diseases (Barb et al. 2006; Gavrila et al. 2003). The serum levels of adiponectin is decreased in obesity and paradoxically enhanced in non-obese individuals (Arita et al. 1999b). Furthermore, low serum concentrations of adiponectin extensively related with an increased risk of many types of cancers including CRC (Otake et al. 2010; Wei et al. 2005a). Supporting this, many epidemiological studies revealed, reduced adiponectin concentration are with highest risk of endometrial cancer (Dossus et al. 2007) and breast cancer (Körner et al. 2007) in women, colorectal (Otake et al. 2010; Wei et al. 2005a) and prostate cancer (Li et al. 2010) in men. These epidemiological results strongly elucidates that adiponectin play a significant role in the occurrence of malignancies. In this report, we summarized the mechanism and mediated signalling pathways associated between the adiponectin and CRC.

2.2 Biology of Adiponectin and Its Receptors

Adiponectin, a fat-derived plasma protein and insulin-sensitizing adipokine, was discovered in 1995. It is an abundantly secreted adipokine by adipocytes. It is also known as AdipoQ, GBP28 (gelatin binding protein), Acrp30 (adipocyte complement related protein 30) and apM1 gene product. The protein weighs about 30 kDa and its gene location is on the chrosomsome 3, 3q27. It includes a C-terminal globular domain and a N-terminal collagenous single peptide domain with variable and collagen like domains in the middle of the N and C-terminal domain, made of 244 amino oacids and it is the most abundant adipokine in human body. They exist as globular and full length adiponectin circulating in plasma or serum. Earlier to the secretion, it endures wide amendments post-transitionally in endoplasmic reticulum (ER), that contribute to multimerization and potentiality. The secretion of adiponectin oligomers are regulated by the molecular chaperons in ER, including ER oxidoreductase 1-Lalpha (Ero1-Lalpha) and ER protein of 44 kDa (Sundaram et al. 2013). Its posttranslational and pre-secretion processing produces various isoforms in the cells and plasma including trimers (LMW, ~67 kDa), hexamers (MMW, ~136 kDa) and high molecular weight multimers (HMW, >300 kDa) (Kadowaki and Yamauchi 2005; Simpson and Whitehead 2010; Tsao et al. 2003). The HMW-adiponectin is the leading form present in the plasma and are physiologically most relevant with greater pro-inflammatory activity, secreted more in females than males, interestingly they never gets interconverted during their circulation (Lara-Castro et al. 2006; Pajvani et al. 2003). The LMW forms have greater anti-inflammatory properties. The concentration of adiponectin is maintained in the blood, as it undergoes O-glycosylation. The plasma adiponectin levels are determined to be high in healthy individuals, however their lower levels are considerably associated with higher risk for various types of malignancies including CRC. Furthermore the serum level of adiponectin is also inversely proportional to the BMI (Arita et al. 1999a). Thus suggests that adiponectin plays a potential role in suppressing carcinogenesis. It is transcriptionally controlled positively by E47, peroxisome proliferator activated receptor γ (PPARγ), C/EBP, sterol regulatory element binding protein and negatively controlled by Id3 protein (Rosen et al. 2009). Moreover its biological functions are pleiotropic ranging from playing a key role in energy homeostasis, anti-apoptosis, anti-inflammatory to anti-angiogenic (Deng and Scherer 2010; Landskroner-Eiger et al. 2009). Adiponectin plays a major role in condensing phagocytosis and inhibiting myelomonocytic progenitor cell proliferation (Yokota et al. 2000). In addition adiponectin also induces the anti-inflammatory cytokine secretion including IL-10 and IL-1RA in monocyte derived macrophages, dendritic cells and inhibits the activation of NK cells and T cells mediated by IL-2 extending its anti-inflammatory potentiality (Kim et al. 2006; Wilk et al. 2011; Wolf et al. 2004).

2.3 Adeponectin Receptors

The adiponectin receptors for the first time identified by Yamauchi et al. (2003) using expression cloning. The adiponectin receptors are AdipoR1, AdipoR2 and T-cadherin. The receptors AdipoR1/R2 are ubiquitously expressed isoforms and shows different affinities, they have seven transmembrane domains that has opposite transmembrane topology to the G-protein coupled receptors. They act as receptors for both full length and globular adiponectin though has much affinity between AdipoR1with globular, expressed mainly in skeletal muscles and endothelial cells and for AdipoR2 with both globular and full length adiponectin (Yamauchi et al. 2003) expressed highly in liver cells. The recent crystal structure study of human have provided an insight of both AdipoR1/R2 that has a large cavity enclosed with seven transmembrane helices and preserved with three histidine residues coordinated to a zinc ion, these Zn ion binding motifs are engaged in stimulating adiponectin to activate AMPK and PPAR-α (Tanabe et al. 2015). The multimers of the adiponectin exhibits diverse biological functions binding to their respective receptors. The trimer binds to AdipoR1, whereas hexamer binds to AdipoR2. The complex of adiponectin and AdipoR1 activates AMPK thus promotes lipid oxidation where as AdipoR2 increases peroxisome-proliferator-activated receptor (PPAR) ligand activity by reducing steatosis, enhancing insulin sensitivity through the activation of AMPK. Moreover the over expression of AdipoR1 in the liver stimulate AMPK activation thereby repress de novo lipogenesis, hepatic gluconeogenesis and promotes fatty acid oxidation (Yamauchi et al. 2007). The removal of AdipoR1 weakens the adiponectin mediated activation of AMPK and SIRT1 (Iwabu et al. 2010) and the functionality of AdipoR2 is still not elucidated properly. Their expression is observed in various cancer cell lines including endometrium, stomach, breast and colon. The T-cadherin cells on the other hand is also called as adiponectin-binding protein are located on the cell surfaces of epithelial, smooth muscle and endothelial cells (Hug et al. 2004). Both the hexamer and HMW multimer acts as ligand for T-cadherin (Hug et al. 2004; Yamauchi et al. 2003). However, T-cadherin never binds to the globular form or trimer of adiponectin due to the lack of cytoplasmic sequence and thus does not perform intracellular signaling. Its function is not fully elucidated, but later found to regulate vascular function in breast cancer mice model (Hebbard et al. 2008). Thus the potentiality of adiponectin depends not only on its circulation in the blood but also with its receptors for tissue specific expression and therefore determining the functionality of the receptors involved in the signaling pathways of adiponectin during the metastasis and tumor growth could be a novel therapeutic strategy.

2.4 Obesity ↔ Adiponectin ↔ CRC

It is suggested from various reviews that obesity is associated with CRC and almost 7–60% are with greater risk for this malignancy when compared with the non-obese people. It was also hypothesized that insulin and adipokines are the regulators for the correlation of obesity and CRC. The meta-analysis performed by Larsson and Wolk (2007) observed that the risk for CRC and obesity are related with the site of cancer and sex. High BMI that reflects high waist-hip ratio and waist circumference are at high risk for CRC. The epidemiological studies also revealed that abdominal obesity and complete obese are at higher risk, however abdominal obesity are much prone to the disease (Pischon et al. 2006; Vainio et al. 2002). Furthermore the accumulation of fat at visceral part of the body shows higher incidence rate for CRC (Kang et al. 2010; Nam et al. 2010). In addition to this the biological and genetically alterations in the obese people including sex steroids, insluin growth factors (IGF), insulin and adipokines affects the association of obesity and the CRC risk. IGF binding protein-

1 concentration is inversely associated with adipocity and thus decrease in IGFBP-1in one way promotes the cell proliferation. As IGFBP-1 fails to bind to the IGF-1 leading to the more bioavailability of free IGF-1 and thus involved with the CRC risk (Wei et al. 2005b). Furthermore, researchers determined obesity is also associated with DNA methylation due to adiposity and that was identified with 187 gene loci methylated at their CpG sites that are engaged in lipid and lipoprotein metabolism and pathways involved in inflammation (Wahl et al. 2017). In addition the genes regulated by obesity also includes intronic sequence harboring miRNA that exert oncogenic and anti-tumoral potentiality basing on their functionality. The adipokine regulated miRs have been characterized for their tumor biological functions that include invasion, angiogenesis, apoptosis and proliferation. The profile that includes the obesity linked malignancies are renal, colon cancer and esophageal adenocarcinoma including oncogenic miRs like miR-21, miR-155,miR-31, miR-185 and miR-96 are induced by adipokines, however adiponectin causing CRC in this regard is in fact controversial (Jasinski-Bergner and Kielstein 2019). It was also revealed from human studies that adiponectin is inversely proportional to plasma insulin and is eventually reduced in obese and type II diabetic patients that has developed an insulin resistance state. Thus mediating the insulin sensitivity in one way, the adiponectin affect the biological pathways that develops obesity associated cancers including CRC and is inversely associated with the rapid growth of tumor. The known factor in obesity is that it induces PI3K/Akt signaling pathway controlling its downstream signaling pathways, eventually develops carcinogenesis. The activation of PI3K/Akt pathway blocks the tumor suppressor protein, P53 that induces apoptosis by stimulating the oncogenic protein Mdm2 (Oren et al. 2002), thus in another way P53 mutation in obese individuals also plays a crucial role in obesity related CRC. Additionally PI3K/Akt pathway also controls fatty acid synthase (FAS) ligand system and low expression of FAS was observed in colon cancer (O'Connell et al. 1998), however no evidence found in relation with FAS related PI3K/Akt controlling obesity related colon cancer. The adiponectin exerts the tumor suppressing potentiality via the adenosine monophosphate-activated protein kinase (AMPK) activation and indeed inhibits various signaling pathways including PI3K/Akt pathways involved in tumorigenesis. Thus exhaustive knowledge about the downstream target molecules of obesity and their role in tumorigenesis would aid the development of novel approach for the prevention of obesity related CRC.

2.5 Evidence for the Association Between the Adiponectin and CRC

Adipnectin is usually expressed in high levels of about 30 μg/ml in healthy individual, however its concentration decline in CRC patients to about 15.9 μg/ml. There is an inverse relation between the serum concentrations of the adiponectin and CRC.

An et al. (2012) performed a meta analysis on the 2632 cases of CRC and 2753 controls. This analysis showed that significant lower levels of adiponectin in CRC patients, especially in men ($P < 0.001$) than in controls (Non-CRC) (An et al. 2012). Similarly in another study showed that plasma adiponectin concentrations inversely linked with greater risk of CRC (Wei et al. 2005a). A meta analysis study addressed that adenoma and CRC demonstrated noticeably lower levels of adiponectin than healthy controls (Xu et al. 2011). In addition, plasma levels of adiponectin decreased in CRC patients compared with healthy individuals. Adiponectin receptors (Adipo R1 and AdipoR2) significantly expressed in CRC tissues and adenocarcinoma cells, could arbitrates its effects on apoptosis and cell proliferation. Thus results suggesting that adiponectin and their receptors associated in the CRC development and progression (Byeon et al. 2010). As the position of CRC is low and located deep in the pelvic cavity, the recurrence rate and infection rate raises in the postoperative stage of a radical surgery (Tariq and Ghias 2016). The pulmonary infection in CRC patients occurs due to their malnutrition, anemia and abdominal infection (abdominal fat). He et al. (2019) determined that the levels of anti-inflammatory factors IL-4 and IL-10 were high and there is a decline of adiponectin levels in the infection group of CRC patients when compared to the non-infectious group to strengthen the anti-inflammatory effect. IL-4 and IL-10 induces B-cells, T-cells, macrophages and inhibits TNF, IL-6 and IL-1 secretion to promote immune response in the body, whereas adiponectin is consumed in the body due to exacerbate response of inflammation accelerating immune response resulting with the low level of adiponectin. Similarly the levels of IL-4, IL-10 (positive correlation) and adiponectin (negative correlation) are also correlated with the stage of CRC. Thus the serum levels of IL-4, IL-10 and adiponectin is worth to be promoted clinically to determine the presence of postoperative infection of CRC. Moreover adiponectin is also involved in controlling many genes involved in anti-inflammatory properties. It inhibits NF-κB pathway via downregulating the IκB phosphorylation that plays a vital role in cancer cell proliferation and metastasis. Canhoroz et al. (2014) determined if the expression of adiponectin is associated in clinical and histopathologic features. They have analyzed overall survival (OS) and disease free survival (DFS) associated with the expression of adiponectin using Kaplan-Meier test and determined OS for the group of CRC patients with adiponectin expression showed 42.9% recurrence, whereas without adiponectin expression showed 34.4% recurrence, similarly the mean DFS for the expression of adiponectin was 65 months and 67 months without adiponectin expression. Thus revealed from the survival curves that adiponectin has negative effect in DFS but was no statistically significant difference was observed with adiponectin and OS. They suggested that adiponectin promotes proliferation in lower concentration and may affect the colon cancer cells, indeed taken as a marker for poor prognosis.

2.6 Adiponectin Mediated Signalling in CRC

The adiponectin elicits various downstream signalling events to exert its tumor suppressing effect that facilitate colorectal carcinogenesis, through its receptors express both in normal as well in the colon cancer tissue. It was also evident that a mice lacking adiponectin gene and receptors had an increase occurrence of high fat diet stimulated colorectal polyps, compared with active adiponectin mice. Thus adiponectin gene defect is correlated with the occurrence of CRC. In the whole process of signalling cascade AMPK plays a crucial central role, that is activated by adiponectin and inhibiting PI3K/Akt, mTOR, JAK/STAT, MAPK and Wnt-GSK3β pathways (Fig. 2.1). Vetvik et al. (2014) proposed that the globular adiponectin and AMPK pathway functions in an autocrine manner in CRC altering the cellular oxidative capacity in the tumour. However, there are no intrinsic protein kinase detected in any of these receptors. The adiponectin receptors hence undergo certain conformational changes and pair their intracellular domain with the signalling molecule through their extracellular adiponectin binding protein. Mao et al. (2006) identified APPL1 as the intracellular binding protein partner of both the receptors

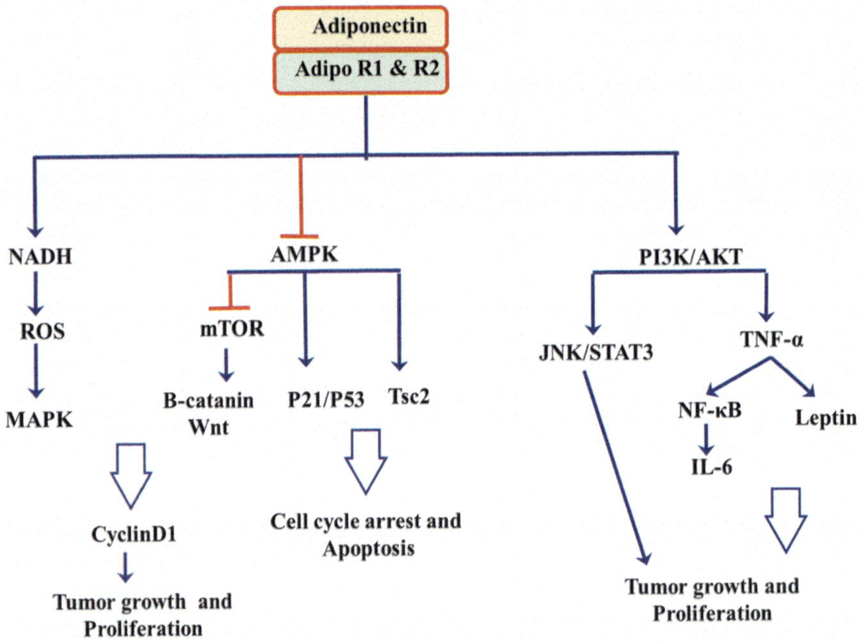

Fig. 2.1 Adiponectin mediated signaling in CRC (Nagaraju et al. 2015): Adiponectin reduces the tumor growth and proliferation through the suppression of adiponectin receptors and AMPK mediated signaling pathway, PI3K/AKT signaling pathway. As resulted, inhibition of tumor survival through the induces the blockage of cell cycle progression, apoptosis and reducing the activity of JNK/STAT3, NF-κB, mTOR, Wnt and MAPK signalling pathways

using yeast two-hybrid technology. It directly binds to the intracellular domain of both AdipoR1 and AdipoR2, mediating the adiponectin action for regulating energy metabolism as well insulin sensitivity. It also plays a crucial role in mediating AMPK activation in lipid metabolism to promote fatty acid oxidation as well mediates adiponectin induced activation of p38 MAPK pathway. However its over expression leads to enhanced stimulatory action of adiponectin in glucose metabolism in muscles and potentiates insulin stimulated AKT signalling and thereby inhibits hepatic gluconeogenesis (Cheng et al. 2009), in addition to this it was revealed from human genetic studies that the point mutation in Appl1 coding region is correlated with the distribution of body fat and high prevalence rate for diabetes (Barbieri et al. 2013). Similarly APPL2 which is an isoform of APPL1 directly binds to AdipoR1/R2 preventing the interaction of APPL1 to the receptors or heterodimerizes with APPL1 to impair adiponection functionality (Miaczynska et al. 2004). Thus APPL2 blocks the signalling of adiponectin competing with APPL1. Further, adiponectin can also modulate the dissociation of the APPL1 and APPL2 that is also triggered by metformin (Wang et al. 2009). Thus the opposing roles of both the protein APPL1 and APPL2 regulate PI3K/Akt/NF-κB pathways in macrophages (Mao et al. 2014). AMPK is required for adiponectin action. The activated AMPK impede the cellular signalling through mTOR and suppresses the promotion of tumor cell adhesion, migration and eventually causes carcinogenesis, and adiponectin is arbitrated by the inhibition of mTOR (Sugiyama et al. 2009). In addition, the activated AMPK, suppresses mTOR and induces cell cycle arrest at G1/S phase in CRC cell growth. Besides this AMPK downregulates the expression of fatty acid synthase (FAS), thus preventing fatty acid synthesis, which is another way correlated with the development of cancers like CRC (Zhan et al. 2008), breast and prostate (Brusselmans et al. 2005; Chajes et al. 2006; Luo et al. 2010). As, AMPK/mTOR are the downstream target of PI3K/Akt signalling pathways, that promotes insulin and growth factor induced signalling results in cellular growth and proliferation is activated in various cancers including CRC. However, adiponection inhibits this pathway thereby suppressing tumour growth. The PI3K/Akt pathway that promotes the activation of transcription growth factors, NF-κB elicit survival signalling pathways (Wang et al. 1998). In the cytoplasm, NF-κB complexes with IkB, which is however degraded by Akt by phosphorylating IkB, eventually translocates NF-κB into the nucleus, that is engaged in transcribing the oncogenes (Chandrasekar et al. 2003; Plummer et al. 1999). The activated NF-κB develops resistance to apoptosis and induces angiogenesis and migration. NF-κB activated in obese patients is clearly elucidated (Ferrante Jr. 2007). There is no clear evidence if Akt mediates this process in developing cancer, however a high fat diet feeding caused the activation of Akt in prostate that consequently potentiates the activation of NF-κB (Shankar et al. 2012). TNF-α also induces the activation of NF-κB in obese patients (Nagaraju et al. 2013). Moreover, pro-survival effects due to TNF-α in cancer cells and IL-6 upregulating the expression of JAK/STAT3 was also observed. The JAK/STAT3 pathway is activated by the phosphorylation and induces cell differentiation and proliferation. Adiponectin however, decreases the phosphorylation of STAT3, thus inhibiting tumour development (Fenton and Birmingham

2010; Miyazaki et al. 2005; Sharma et al. 2010). In addition to this adiponectin also inhibits the phosphorylation of GSK3β, that is activated by Wnt signaling pathway and inactivates its ability to remove β-catenin. Wnt signalling pathway induces various cancers by promoting the accumulation of β-catenin in the nucleus and thus increases the expression of cyclin-D1 that promotes cell cycles to occur. C-Jun N-terminal kinase (JNK) is a potential oncogene present in the intestine and is activated due to obesity that plays a potent role in developing insulin resistance. Elevated levels of JNK are observed in adipose tissues, liver and muscles (Endo et al. 2009). Increased expression of JNK is observed in CRC and plays a crucial role in progression of tumor (Fang and Richardson 2005). It was even theoretically determined that the low level of adiponectin fail to control the activity of JNK, that is even supported by proteomic analysis of Cox-2, pSTAT3 and pAMPK (Saxena et al. 1822; Tilg and Kaser 2009). Thus it is evident that lack of adiponectin in the body contributes to CRC and it plays a crucial role in obesity related tumorigenesis. Moreover, adiponectin also inhibits the leptin-stimulated immortalized colon epithelial cell proliferation (Fenton et al. 2008). Thus adipnectin regulates the insulin senstivity, inflammatory responses and acting directly on the tumor cells at the cellular level.

2.7 Epidemiological Studies

Obesity is positively correlated with CRC. Although there are numerous studies describing the adiponectin circulation concentration levels and their risk for CRC, the results for the epidemiological studies are still inconsistent. Various epidemological studies in association with genetic variants of adiponectin (ADIPOQ) gene and adiponectin receptor (ADIPOR) gene polymorphism in relation with the risk of CRC have studied, but the result were incomplete and inconsistent. Initially a meta analysis study was performed that included six control studies with three polymorphism in the ADIPOQ gene related with the risk of CRC, however the association was observed only with Asian population and not with Caucasian populations (Yang et al. 2015). Further, from the pooled analysis, that also included genetic markers, polymorphism in the ADIPOQ gene have been identified in genomide wide association studies for adiponectin concentration were not related with risk of CRC (Song et al. 2015). Later, Tan et al. (2017) in a systematic way conducted a meta analysis to assess the association taking all published, updated case control studies from PubMed, Elseiver, China knowledge infrastructure and Chongqing VIP. They determined that ADIPOQ variants (rs2241766 T/G, rs1501299 G/T) and ADIPOR variant (rs1342387 G/A) population are specifically correlated with the risk of CRC that is mediated by insulin resistance. This study is however with certain limitation including lack of literature related to the polymorphism of few ADIPOQ gene, therefore cannot be pooled for estimation and eligible studies and participants very few for stratified analysis. Similarly Zhou et al. (2017) investigated the association between AdipoR2 gene polymorphism, its expression at protein level and risk for CRC. They identified that the variant AdipoR2, rs10773989 polymorphism is related with the degree of tumor infiltration, however the another variant form rs1044471 is associated with degree of differentiation and Dukes' staging. Thus the findings indicated that single nucleotide polymorphism of AdipoR2 and their protein expression level were vital for CRC progression. These findings also support the assumption that, profuse amount of AdipoR inhibit the carcinogenic effect by promoting ADIPOQ suppressing effect for the CRC growth, whereas low AdipoR expression levels promote CRC progression acting against ADIPOQ effect (Byeon et al. 2010). Song et al. (2013) performed a prospective nested case-control study to evaluate the association of plasma adiponectin and soluble leptin receptor for CRC risk. They determined that plasma adiponectin is inversely associated with the risk of CRC in men and not in women. They also found that HMW form of adiponectin is more closely related to CRC risk than the other forms of adiponectin. In addition European prospective investigation into cancer and nutrition study (EPIC) also suggested that non-HMW adiponectin are with reduced risk for CRC (Aleksandrova et al. 2012). However, plasma soluble leptin receptor did not appear to be related with CRC risk. Similar meta-analysis done on 13 epidemological studies showed sex differences in both case control and prospective studies, where a weak inverse association was observed between colorectal cancer and adiponectin in men but not in women (Xu et al. 2011). Thus many more prospective studies are essential to confirm all the observations and further investigation is warranted to determine the potential role of receptors of adiponectin as a therapeutic target for the survival of CRC.

2.8 In Vitro and In Vivo Experiments

Fenton et al. (2010) in their experiment treated adiponectin to MC 38 murine colon carcinoma cells, they observed no inhibition of insulin induced cell prolferation, but suppressing IL-6 induced cell proliferation was observed via negatively controlling phosphorylation of STAT3 and is activation. Adiponectin inhibiting the proliferation of colon cancer cell lines-HCT116, LoVo and HT29 via cell cycle arrest at G1/S phase was also observed (Kim et al. 2010). The expression of adiponectin receptors was observed in the CRC cell lines HT29, HCT116 and LoVo cells, in addition the proliferation assay also showed that adiponectin inhibits CRC cell growth via activating AMPK and downregulating mTOR pathway (Sugiyama et al. 2009). Hiyoshi et al. (2012) also observed the expression of AdipoR1 and AdipoR2 in human CRC decreased in cancer tissue and thus down regulated when compared with the normal healthy colon cells, in addition to this they also determined from the analysis of clinico-pathological parameters correlating adiponectin receptor expression and revealed that low level AdipoR2 mRNA expression was observed in the lymph node metastasis. Tae et al. (2014) examined the serum adiponectin concentrations, mRNA and protein expression of adiponectin and their receptors AdipoR in tissue samples from the patients with CRC, advanced adenoma and a normal colon.

They determined that both the expression of adiponectin and its receptors are observed mutually in all the samples. Moreover the decrease in the expression of AdipoRs in the colonic epithelium accompanied with increase in the adiponectin expression in the mesenchyma, In addition to this they also observed that increase in adiponectin, downregulates the mRNA expression of COX-2 and upregulates T-cadherin mRNA expression in HCT-116 cell lines. Thus, suggested a novel insight for understanding the relationship between adiponectin and carcinogenesis. Otani et al. (2010) administered adiponectin exogenously in C57BL/6 J-Apc(Min/+) mice infected with intestinal polyps as well possess point mutation in the Apc gene. They observed that adiponectin decreased the size of polyp as well inhibited the intestinal adenoma growth in the mice. Further, Mutoh et al. (2011) used a Min mice, that is deficient of adiponectin, the mice was detected with more number of polyps compared to the wild type adeponectin mice. Similarly C57BL/6 J mice that is also deficient for adiponectin, also treated with azoxymethane that induces CRC, thus the incidence for CRC was observed in higher number when compared to the adiponectin wild type mice. Moon et al. (2013) for the first time found that administration of recombinant adiponectin inhibits the tumor growth of implanted colon cancer cells in mice. They observed better effect in adipnectin deficiency with metabolic dysfunction and western diet induced obesity. They also suggested that adiponectin directly controls the malignant properties including proliferation, colony formation, adhesion and invasion, regulates pathways including AMPK-S6, cell cycles by controlling the expression of p21/p53/p27/cyclins and inflammatory pathways that include IL12, STAT3-VEGF in human and mouse colon cancer cell lines in a LKB1-dependent pathway. Further, research work in relation with chronic inflammation associated colon cancer is carried on. In this regard, a chronic colonic inflammation model mice was taken and observed that adiponectin prevents the goblet cell apoptosis, as well increases the epithelium formation, that eventually reduces chronic inflammation in colon cell (Saxena et al. 1822). Thus, adiponectin may play a vital role in controlling chronic inflammation associated CRC. The adiponectin expression is very much essential during the chronic phase of inflammation rather than in the acute phase, as its deficiency could lead to tumorigenesis in colon.

2.9 Therapeutic Role of Adiponectin

It is evident from various research studies about the potentiality for high concentration levels of adiponectin having therapeutic benefit against obesity related malignancies like CRC that have very low availability of this protein. Synthesizing and processing adiponectin as a viable drug for human use is extremely a difficult task. Therefore advanced research efforts are aimed at identifying pathways and related factors that enhances the circulation of adiponectin levels to combat obesity associated cancer. Otvos et al. (2011) developed a novel adiponectin based short peptide called ADP 355, that imitate adiponectin functionally in a dose dependent controlled proliferation observed in the cancer cell lines that had numerous adiponectin receptors. It also enhanced adiponectin signaling pathways including AMPK, Akt, STAT3 and ERK1/2 as well suppressed the cancer tumor xenograft growth by 31% (Delort et al. 2012). Further, an orally administered synthetic small molecule AdipoR (AdipoRon) agonists was discovered by Okada-Iwabu et al. (2013) for the first time, that with much affinity binds and activates both the adiponectin receptors. Their survey on this synthetic molecule include improved insulin resistance, lower plasma glucose, increased fatty acid oxidation and decrease in proinflammatory cytokine expression (TNF-α) are observed. They showed similar effects of adiponectin, activating PPAR-α and AMPK pathways in liver and muscles. 'AdipoRon' was also prepared by Zhang et al. (2015) on post-ischemic mycordial apoptosis mouse models. The oral administration resulted in enhanced cardiac function in wild-type mice and attenuated in post-ischemic cardiac injury. In addition, it also induced apoptotic cell death in adiponectin knockout mice or cardiomyocyte-specific AMPK dominant negative. Thus this adiponectin receptor agonist, 'AdipoRon', could be a therapy choice for obesity related CRC chemoprevention (Malih and Najafi 2015). A class of non-thiazolidinedione PPAR ligands called SPPARMINT131 are detected to enhance the HMW adiponectin concentration (Higgins and Depaoli 2010). Similarly, few pharmacological agents that with much potentiality enhances the adiponectin circulation and its signaling pathways, includes selective PPAR (SPPARM) agonist and apolipoprotein peptide that mimics L-4F via its receptors (Fowler et al. 2011). Further, therapy that includes chemodrugs inhibiting risk factors as well the pathways leading to obesity associated malignancy are useful as modulator to chemotherapy. Moon and Mantzoros (2013) observed that IL1β controls cell proliferation, apoptosis and cell cycle regulating genes and eventually inducing CRC. They observed that these IL1β are attenuated by adiponectin with or without the combination of metformin by upregulating AMPK and down regulating STAT3 pathways in human and colon cancer cell lines. Thus adiponectin alone or in combination of metformin could be an useful agent as chemoprevention for IL1β-induced colon carcinogenesis. Slomian et al. (2014) determined an association between the chemotherapy (used 5-FU, oxaliplatin and irinotecan) inducing plasma adipokines production in CRC patients. They observed that the chemo drug positively induces cytokine production-increasing the production of adiponectin (47%) and decreasing the production of visfatin and resistin. The restoring of adiponectin in the body is essential for better prognosis. There was a positive correlation associated with TNF-α, IGF-1 and leptin concentrations with BMI and negative correlation associated with adiponectin and BMI, thus the circulating levels of TNF-α, IGF-1 and adiponectin are associated with risk for colorectal adenoma and can be taken as a better biomarkers for the detection of CRC (Ashktorab et al. 2018). A group of studies reported hypermethylation of T-cadherin at its promoter region were detected in CRC (Toyooka et al. 2002) and adenomas with almost 83% poorly differentiated CRC were detected (Kawashima et al. 2017). In addition to this Scarpa et al. (2016) also reported T-cadherin promoter in its hypermethylation status in non-neoplastic mucosa as marker of ulcerative colitis associated CRC. Novel strategies that promotes the expression of the adiponectin receptors to adiponectin and designing agonist would provide new therapeutic approach and novel modalities could provide a therapeutic strategy for obesity correlated CRC.

2.10 Conclusion

Despite of vigorous research work carried for the understanding the anti-proliferative and tumor-inhibiting nature of adiponectin, the results are still conflicting. Worldwide progress made in the area of research to develop solid benefits for the obese individual who are at increased risk for cancer. Further, research that implicate obesity associated biomarkers are essential to quantify the individual mediating role for targeted pharmacological and lifestyle intrusion targeting obesity associated cancer prevention. Developing certain procedures for the administration of analogues would be a therapeutic strategy in preventing CRC. Thus standardization for maintaining adiponectin levels and associated assays are essential to make adiponecin as a potential diagnostic target for obesity related cancer and thorough understanding of adiponectin and its downstream signalling pathways are also essential for developing novel drugs in the therapy.

References

Aleksandrova K, Boeing H, Jenab M, Bueno-de-Mesquita HB, Jansen E, van Duijnhoven FJ, Fedirko V, Rinaldi S, Romieu I, Riboli E, Romaguera D, Westphal S, Overvad K, Tjonneland A, Boutron-Ruault MC, Clavel-Chapelon F, Kaaks R, Lukanova A, Trichopoulou A, Lagiou P, Trichopoulos D, Agnoli C, Mattiello A, Saieva C, Vineis P, Tumino R, Peeters PH, Arguelles M, Bonet C, Sanchez MJ, Dorronsoro M, Huerta JM, Barricarte A, Palmqvist R, Hallmans G, Khaw KT, Wareham N, Allen NE, Crowe FL, Pischon T (2012) Total and high-molecular weight adiponectin and risk of colorectal cancer: the European prospective investigation into cancer and nutrition study. Carcinogenesis 33:1211–1218

An W, Bai Y, Deng S-X, Gao J, Ben Q-W, Cai Q-C, Zhang H-G, Li Z-S (2012) Adiponectin levels in patients with colorectal cancer and adenoma: a meta-analysis. Eur J Cancer Prev 21:126–133

Arita Y, Kihara S, Ouchi N, Takahashi M, Maeda K, Miyagawa J-i, Hotta K, Shimomura I, Nakamura T, Miyaoka K (1999a) Paradoxical decrease of an adipose-specific protein, adiponectin, in obesity. Biochem Biophys Res Commun 257:79–83

Arita Y, Kihara S, Ouchi N, Takahashi M, Maeda K, Miyagawa J-i, Hotta K, Shimomura I, Nakamura T, Miyaoka K, Kuriyama H, Nishida M, Yamashita S, Okubo K, Matsubara K, Muraguchi M, Ohmoto Y, Funahashi T, Matsuzawa Y (1999b) Paradoxical decrease of an adipose-specific protein, adiponectin, in obesity. Biochem Biophys Res Commun 257:79–83

Ashktorab H, Soleimani A, Nichols A, Sodhi K, Laiyemo AO, Nunlee-Bland G, Nouraie SM, Brim H (2018) Adiponectin, Leptin, IGF-1, and tumor necrosis factor alpha as potential serum biomarkers for non-invasive diagnosis of colorectal adenoma in African Americans. Front Endocrinol (Lausanne) 9:77–77

Barb D, Pazaitou-Panayiotou K, Mantzoros CS (2006) Adiponectin: a link between obesity and cancer. Expert Opin Investig Drugs 15:917–931

Barbieri M, Esposito A, Angellotti E, Rizzo MR, Marfella R, Paolisso G (2013) Association of genetic variation in adaptor protein APPL1/APPL2 loci with non-alcoholic fatty liver disease. PLoS One 8:e71391–e71391

Brusselmans K, De Schrijver E, Verhoeven G, Swinnen JV (2005) RNA interference-mediated silencing of the acetyl-CoA-carboxylase-alpha gene induces growth inhibition and apoptosis of prostate cancer cells. Cancer Res 65:6719–6725

Byeon JS, Jeong JY, Kim MJ, Lee SM, Nam WH, Myung SJ, Kim JG, Yang SK, Kim JH, Suh DJ (2010) Adiponectin and adiponectin receptor in relation to colorectal cancer progression. Int J Cancer 127:2758–2767

Calle EE, Kaaks R (2004) Overweight, obesity and cancer: epidemiological evidence and proposed mechanisms. Nat Rev Cancer 4:579

Canhoroz M, Kanat O, Saraydaroglu O, Buluc E, Avci N, Cubukcu E, Olmez OF, Manavoglu O (2014) Clinical significance of adiponectin expression in colon cancer patients. J Cancer Res Ther 10:347–353

Chajes V, Cambot M, Moreau K, Lenoir GM, Joulin V (2006) Acetyl-CoA carboxylase alpha is essential to breast cancer cell survival. Cancer Res 66:5287–5294

Chandrasekar B, Mummidi S, Perla RP, Bysani S, Dulin NO, Liu F, Melby PC (2003) Fractalkine (CX3CL1) stimulated by nuclear factor kappaB (NF-kappaB)-dependent inflammatory signals induces aortic smooth muscle cell proliferation through an autocrine pathway. Biochem J 373:547–558

Cheng KK, Iglesias MA, Lam KS, Wang Y, Sweeney G, Zhu W, Vanhoutte PM, Kraegen EW, Xu A (2009) APPL1 potentiates insulin-mediated inhibition of hepatic glucose production and alleviates diabetes via Akt activation in mice. Cell Metab 9:417–427

Dai Z, Xu Y-C, Niu L (2007) Obesity and colorectal cancer risk: a meta-analysis of cohort studies. World J Gastroenterol 13:4199–4206

Delort L, Jardé T, Dubois V, Vasson M-P, Caldefie-Chézet F (2012) New insights into anticarcinogenic properties of adiponectin: a potential therapeutic approach in breast cancer? Vitam Horm 90:397–417

Deng Y, Scherer PE (2010) Adipokines as novel biomarkers and regulators of the metabolic syndrome. Ann N Y Acad Sci 1212:E1–E19

Desruisseaux MS, Nagajyothi, Trujillo ME, Tanowitz HB, Scherer PE (2007) Adipocyte, adipose tissue, and infectious disease. Infect Immun 75:1066–1078

Divella R, De Luca R, Abbate I, Naglieri E, Daniele A (2016) Obesity and cancer: the role of adipose tissue and adipo-cytokines-induced chronic inflammation. J Cancer 7:2346

Dossus L, Jenab M, Slimani N, Rinaldi S, Kaaks R, Cust AE, Overvad K, Fo C-C, Mesrine S, Joulin V, Linseisen J, Rohrmann S, Boeing H, Pischon T, Trichopoulou A, Trichopoulos D, Benetou V, Palli D, Berrino F, Tumino R, Sacerdote C, Mattiello A, Quirós JR, Mendez MA, Sánchez M-J, Larrañaga N, Tormo MJ, Ardanaz E, Bueno-de-Mesquita HB, van Gils CH, Peeters PHM, Khaw K-T, Bingham S, Allen N, Key T, Riboli E, Friedenreich C, Bonnet F, Laville M, Lukanova A, Lundin E, Olsen A, Tjønneland A (2007) Plasma adiponectin levels and endometrial cancer risk in pre- and postmenopausal women. J Clin Endocrinol Metabol 92:255–263

Endo H, Hosono K, Fujisawa T, Takahashi H, Sugiyama M, Yoneda K, Nozaki Y, Fujita K, Yoneda M, Inamori M, Wada K, Nakagama H, Nakajima A (2009) Involvement of JNK pathway in the promotion of the early stage of colorectal carcinogenesis under high-fat dietary condition. Gut 58:1637–1643

Fang JY, Richardson BC (2005) The MAPK signalling pathways and colorectal cancer. Lancet Oncol 6:322–327

Fenton JI, Birmingham JM (2010) Adipokine regulation of colon cancer: adiponectin attenuates interleukin-6-induced colon carcinoma cell proliferation via STAT-3. Mol Carcinog 49:700–709

Fenton JI, Birmingham JM, Hursting SD, Hord NG (2008) Adiponectin blocks multiple signaling cascades associated with leptin-induced cell proliferation in Apc min/+ colon epithelial cells. Int J Cancer 122:2437–2445

Ferrante AW Jr (2007) Obesity-induced inflammation: a metabolic dialogue in the language of inflammation. J Intern Med 262:408–414
Fowler JA, Lwin ST, Drake MT, Edwards JR, Kyle RA, Mundy GR, Edwards CM (2011) Host-derived adiponectin is tumor-suppressive and a novel therapeutic target for multiple myeloma and the associated bone disease. Blood 118:5872–5882
Gavrila A, Chan JL, Yiannakouris N, Kontogianni M, Miller LC, Orlova C, Mantzoros CS (2003) Serum adiponectin levels are inversely associated with overall and central fat distribution but are not directly regulated by acute fasting or leptin administration in humans: cross-sectional and interventional studies. J Clin Endocrinol Metabol 88:4823–4831
Haggar FA, Boushey RP (2009) Colorectal cancer epidemiology: incidence, mortality, survival, and risk factors. Clin Colon Rectal Surg 22:191–197
He J, Wang Z, Zhang S (2019) Correlation analysis of IL-4, IL-10 and APN levels with postoperative infection of colorectal cancer. Oncol Lett 17:1603–1608
Hebbard LW, Garlatti M, Young LJ, Cardiff RD, Oshima RG, Ranscht B (2008) T-cadherin supports angiogenesis and adiponectin association with the vasculature in a mouse mammary tumor model. Cancer Res 68:1407–1416
Higgins LS, Depaoli AM (2010) Selective peroxisome proliferator-activated receptor gamma (PPARgamma) modulation as a strategy for safer therapeutic PPARgamma activation. Am J Clin Nutr 91:267s–272s
Hiyoshi M, Tsuno NH, Otani K, Kawai K, Nishikawa T, Shuno Y, Sasaki K, Hongo K, Kaneko M, Sunami E, Takahashi K, Nagawa H, Kitayama J (2012) Adiponectin receptor 2 is negatively associated with lymph node metastasis of colorectal cancer. Oncol Lett 3:756–760
Hug C, Wang J, Ahmad NS, Bogan JS, Tsao T-S, Lodish HF (2004) T-cadherin is a receptor for hexameric and high-molecular-weight forms of Acrp30/adiponectin. Proc Natl Acad Sci 101:10308–10313
Iwabu M, Yamauchi T, Okada-Iwabu M, Sato K, Nakagawa T, Funata M, Yamaguchi M, Namiki S, Nakayama R, Tabata M, Ogata H, Kubota N, Takamoto I, Hayashi YK, Yamauchi N, Waki H, Fukayama M, Nishino I, Tokuyama K, Ueki K, Oike Y, Ishii S, Hirose K, Shimizu T, Touhara K, Kadowaki T (2010) Adiponectin and AdipoR1 regulate PGC-1alpha and mitochondria by Ca(2+) and AMPK/SIRT1. Nature 464:1313–1319
Jasinski-Bergner S, Kielstein H (2019) Adipokines regulate the expression of tumor-relevant microRNAS. Obes Facts 12:211–225
Jemal A, Bray F, Center MM, Ferlay J, Ward E, Forman D (2011) Global cancer statistics. CA Cancer J Clin 61:69–90
Kadowaki T, Yamauchi T (2005) Adiponectin and adiponectin receptors. Endocr Rev 26:439–451
Kang HW, Kim D, Kim HJ, Kim CH, Kim YS, Park MJ, Kim JS, Cho SH, Sung MW, Jung HC, Lee HS, Song IS (2010) Visceral obesity and insulin resistance as risk factors for colorectal adenoma: a cross-sectional, case-control study. Am J Gastroenterol 105:178–187
Kawashima K, Maeda K, Saigo C, Kito Y, Yoshida K, Takeuchi T (2017) Adiponectin and intelectin-1: important adipokine players in obesity-related colorectal carcinogenesis. Int J Mol Sci 18:866
Kelesidis I, Kelesidis T, Mantzoros CS (2006) Adiponectin and cancer: a systematic review. Br J Cancer 94:1221–1225
Kim AY, Lee YS, Kim KH, Lee JH, Lee HK, Jang SH, Kim SE, Lee GY, Lee JW, Jung SA, Chung HY, Jeong S, Kim JB (2010) Adiponectin represses colon cancer cell proliferation via AdipoR1- and -R2-mediated AMPK activation. Mol Endocrinol 24:1441–1452
Körner A, Kiess W, Kaprara A, Pazaitou-Panayiotou K, Neuwirth A, Williams CJ, Kelesidis I, Bullen J, Mitsiades N, Kelesidis T, Mantzoros CS, Tseleni S (2007) Total and high-molecular-weight adiponectin in breast cancer: in vitro and in vivo studies. J Clin Endocrinol Metabol 92:1041–1048
Kim K-y, Kim JK, Han SH, Lim J-S, Kim KI, Cho DH, Lee M-S, Lee J-H, Yoon D-Y, Yoon SR (2006) Adiponectin is a negative regulator of NK cell cytotoxicity. J Immunol 176:5958–5964

Landskroner-Eiger S, Qian B, Muise ES, Nawrocki AR, Berger JP, Fine EJ, Koba W, Deng Y, Pollard JW, Scherer PE (2009) Proangiogenic contribution of adiponectin toward mammary tumor growth in vivo. Clin Cancer Res 15:3265–3276

Lara-Castro C, Luo N, Wallace P, Klein RL, Garvey WT (2006) Adiponectin multimeric complexes and the metabolic syndrome trait cluster. Diabetes 55:249–259

Larsson SC, Wolk A (2007) Obesity and colon and rectal cancer risk: a meta-analysis of prospective studies. Am J Clin Nutr 86:556–565

Li H, Stampfer MJ, Mucci L, Rifai N, Qiu W, Kurth T, Ma J (2010) A 25-year prospective study of plasma adiponectin and leptin concentrations and prostate cancer risk and survival. Clin Chem 56:34–43

Luo Z, Zang M, Guo W (2010) AMPK as a metabolic tumor suppressor: control of metabolism and cell growth. Future Oncol 6:457–470

Ma Y, Yang Y, Wang F, Zhang P, Shi C, Zou Y, Qin H (2013) Obesity and risk of colorectal cancer: a systematic review of prospective studies. PLoS One 8:e53916–e53916

MacDougald OA, Burant CF (2007) The rapidly expanding family of adipokines. Cell Metab 6:159–161

Malih S, Najafi R (2015) AdipoRon: a possible drug for colorectal cancer prevention? Tumour Biol 36:6673–6675

Mao X, Kikani CK, Riojas RA, Langlais P, Wang L, Ramos FJ, Fang Q, Christ-Roberts CY, Hong JY, Kim RY, Liu F, Dong LQ (2006) APPL1 binds to adiponectin receptors and mediates adiponectin signalling and function. Nat Cell Biol 8:516–523

Mao L, Lin W, Nie T, Hui X, Gao X, Li K, Ding M, Tang X, Li P, Wang Y, Xu A, Liu P, Wu D (2014) Absence of Appl2 sensitizes endotoxin shock through activation of PI3K/Akt pathway. Cell Biosci 4:60

Miaczynska M, Christoforidis S, Giner A, Shevchenko A, Uttenweiler-Joseph S, Habermann B, Wilm M, Parton RG, Zerial M (2004) APPL proteins link Rab5 to nuclear signal transduction via an endosomal compartment. Cell 116:445–456

Miyazaki T, Bub JD, Uzuki M, Iwamoto Y (2005) Adiponectin activates c-Jun NH2-terminal kinase and inhibits signal transducer and activator of transcription 3. Biochem Biophys Res Commun 333:79–87

Moon HS, Mantzoros CS (2013) Adiponectin and metformin additively attenuate IL1beta-induced malignant potential of colon cancer. Endocr Relat Cancer 20:849–859

Moon HS, Liu X, Nagel JM, Chamberland JP, Diakopoulos KN, Brinkoetter MT, Hatziapostolou M, Wu Y, Robson SC, Iliopoulos D, Mantzoros CS (2013) Salutary effects of adiponectin on colon cancer: in vivo and in vitro studies in mice. Gut 62:561–570

Mutoh M, Teraoka N, Takasu S, Takahashi M, Onuma K, Yamamoto M, Kubota N, Iseki T, Kadowaki T, Sugimura T, Wakabayashi K (2011) Loss of adiponectin promotes intestinal carcinogenesis in min and wild-type mice. Gastroenterology 140:2000–2008

Nagaraju GP, Zhu S, Wen J, Farris AB, Adsay VN, Diaz R, Snyder JP, Mamoru S, El-Rayes BF (2013) Novel synthetic curcumin analogues EF31 and UBS109 are potent DNA hypomethylating agents in pancreatic cancer. Cancer Lett 341:195–203

Nagaraju GP, Aliya S, Alese OB (2015) Role of adiponectin in obesity related gastrointestinal carcinogenesis. Cytokine Growth Factor Rev 26:83–93

Nam SY, Kim BC, Han KS, Ryu KH, Park BJ, Kim HB, Nam BH (2010) Abdominal visceral adipose tissue predicts risk of colorectal adenoma in both sexes. Clin Gastroenterol Hepatol 8:443–450

O'Connell J, Bennett MW, O'Sullivan GC, Roche D, Kelly J, Collins JK, Shanahan F (1998) Fas ligand expression in primary colon adenocarcinomas: evidence that the Fas counterattack is a prevalent mechanism of immune evasion in human colon cancer. J Pathol 186:240–246

Okada-Iwabu M, Yamauchi T, Iwabu M, Honma T, Hamagami K, Matsuda K, Yamaguchi M, Tanabe H, Kimura-Someya T, Shirouzu M, Ogata H, Tokuyama K, Ueki K, Nagano T, Tanaka A, Yokoyama S, Kadowaki T (2013) A small-molecule AdipoR agonist for type 2 diabetes and short life in obesity. Nature 503:493–499

Oren M, Damalas A, Gottlieb T, Michael D, Taplick J, Leal JF, Maya R, Moas M, Seger R, Taya Y, Ben-Ze'Ev A (2002) Regulation of p53: intricate loops and delicate balances. Ann N Y Acad Sci 973:374–383

Otake S, Takeda H, Fujishima S, Fukui T, Orii T, Sato T, Sasaki Y, Nishise S, Kawata S (2010) Decreased levels of plasma adiponectin associated with increased risk of colorectal cancer. World J Gastroenterol: WJG 16:1252

Otani K, Kitayama J, Yasuda K, Nio Y, Iwabu M, Okudaira S, Aoki J, Yamauchi T, Kadowaki T, Nagawa H (2010) Adiponectin suppresses tumorigenesis in Apc(min)(/+) mice. Cancer Lett 288:177–182

Otani K, Ishihara S, Yamaguchi H, Murono K, Yasuda K, Nishikawa T, Tanaka T, Kiyomatsu T, Hata K, Kawai K (2017) Adiponectin and colorectal cancer. Surg Today 47:151–158

Otvos L Jr, Haspinger E, La Russa F, Maspero F, Graziano P, Kovalszky I, Lovas S, Nama K, Hoffmann R, Knappe D, Cassone M, Wade J, Surmacz E (2011) Design and development of a peptide-based adiponectin receptor agonist for cancer treatment. BMC Biotechnol 11:90

Pajvani UB, Du X, Combs TP, Berg AH, Rajala MW, Schulthess T, Engel J, Brownlee M, Scherer PE (2003) Structure-function studies of the adipocyte-secreted hormone Acrp30/adiponectin implications for metabolic regulation and bioactivity. J Biol Chem 278:9073–9085

Pischon T, Lahmann PH, Boeing H, Friedenreich C, Norat T, Tjonneland A, Halkjaer J, Overvad K, Clavel-Chapelon F, Boutron-Ruault MC, Guernec G, Bergmann MM, Linseisen J, Becker N, Trichopoulou A, Trichopoulos D, Sieri S, Palli D, Tumino R, Vineis P, Panico S, Peeters PH, Bueno-de-Mesquita HB, Boshuizen HC, Van Guelpen B, Palmqvist R, Berglund G, Gonzalez CA, Dorronsoro M, Barricarte A, Navarro C, Martinez C, Quiros JR, Roddam A, Allen N, Bingham S, Khaw KT, Ferrari P, Kaaks R, Slimani N, Riboli E (2006) Body size and risk of colon and rectal cancer in the European prospective investigation into Cancer and nutrition (EPIC). J Natl Cancer Inst 98:920–931

Plummer SM, Holloway KA, Manson MM, Munks RJ, Kaptein A, Farrow S, Howells L (1999) Inhibition of cyclo-oxygenase 2 expression in colon cells by the chemopreventive agent curcumin involves inhibition of NF-kappaB activation via the NIK/IKK signalling complex. Oncogene 18:6013–6020

Rosen E, Eguchi J, Xu Z (2009) Transcriptional targets in adipocyte biology. Expert Opin Ther Targets 13:975–986

Saxena A, Chumanevich A, Fletcher E, Larsen B, Lattwein K, Kaur K, Fayad R (1822) Adiponectin deficiency: role in chronic inflammation induced colon cancer. Biochim Biophys Acta 2012:527–536

Scarpa M, Scarpa M, Castagliuolo I, Erroi F, Kotsafti A, Basato S, Brun P, D'Inca R, Rugge M, Angriman I, Castoro C (2016) Aberrant gene methylation in non-neoplastic mucosa as a predictive marker of ulcerative colitis-associated CRC. Oncotarget 7:10322–10331

Shankar E, Vykhovanets EV, Vykhovanets OV, Maclennan GT, Singh R, Bhaskaran N, Shukla S, Gupta S (2012) High-fat diet activates pro-inflammatory response in the prostate through association of Stat-3 and NF-κB. Prostate 72:233–243

Sharma D, Wang J, Fu PP, Sharma S, Nagalingam A, Mells J, Handy J, Page AJ, Cohen C, Anania FA, Saxena NK (2010) Adiponectin antagonizes the oncogenic actions of leptin in hepatocellular carcinogenesis. Hepatology 52:1713–1722

Simpson F, Whitehead JP (2010) Adiponectin—it's all about the modifications. Int J Biochem Cell Biol 42:785–788

Slomian G, Swietochowska E, Malinowska-Borowska J, Kasperczyk S, Rogalska A, Nowak P (2014) Association between chemotherapy and plasma adipokines in patients with colorectal cancer. Pharmacol Rep 66:902–907

Song M, Zhang X, Wu K, Ogino S, Fuchs CS, Giovannucci EL, Chan AT (2013) Plasma adiponectin and soluble leptin receptor and risk of colorectal cancer: a prospective study. Cancer Prev Res 6:875–885

Song M, Gong J, Giovannucci EL, Berndt SI, Brenner H, Chang-Claude J, Curtis KR, Harrison TA, Hoffmeister M, Hsu L, Jiao S, Le Marchand L, Potter JD, Schoen RE, Seminara D, Slattery ML,

White E, Wu K, Ogino S, Fuchs CS, Hunter DJ, Tworoger SS, Hu FB, Rimm E, Jensen M, Peters U, Chan AT (2015) Genetic variants of adiponectin and risk of colorectal cancer. Int J Cancer 137:154–164

Sugiyama M, Takahashi H, Hosono K, Endo H, Kato S, Yoneda K, Nozaki Y, Fujita K, Yoneda M, Wada K, Nakagama H, Nakajima A (2009) Adiponectin inhibits colorectal cancer cell growth through the AMPK/mTOR pathway. Int J Oncol 34:339–344

Sundaram S, Johnson AR, Makowski L (2013) Obesity, metabolism and the microenvironment: links to cancer. J Carcinog 12:19–19

Tae CH, Kim S-E, Jung S-A, Joo Y-H, Shim K-N, Jung H-K, Kim TH, Cho M-S, Kim KH, Kim JS (2014) Involvement of adiponectin in early stage of colorectal carcinogenesis. BMC Cancer 14:811–811

Tan X, Wang GB, Tang Y, Bai J, Ye L (2017) Association of ADIPOQ and ADIPOR variants with risk of colorectal cancer: a meta-analysis. J Huazhong Univ Sci Technol Med Sci [Hua zhong ke ji da xue xue bao Yi xue Ying De wen ban = Huazhong keji daxue xuebao Yixue Yingdewen ban] 37:161–171

Tanabe H, Fujii Y, Okada-Iwabu M, Iwabu M, Nakamura Y, Hosaka T, Motoyama K, Ikeda M, Wakiyama M, Terada T, Ohsawa N, Hato M, Ogasawara S, Hino T, Murata T, Iwata S, Hirata K, Kawano Y, Yamamoto M, Kimura-Someya T, Shirouzu M, Yamauchi T, Kadowaki T, Yokoyama S (2015) Crystal structures of the human adiponectin receptors. Nature 520:312–316

Tariq K, Ghias K (2016) Colorectal cancer carcinogenesis: a review of mechanisms. Cancer Biol Med 13:120–135

Tilg H, Kaser A (2009) Adiponectin and JNK: metabolic/inflammatory pathways affecting gastrointestinal carcinogenesis. Gut 58:1576–1577

Toyooka S, Toyooka KO, Harada K, Miyajima K, Makarla P, Sathyanarayana UG, Yin J, Sato F, Shivapurkar N, Meltzer SJ, Gazdar AF (2002) Aberrant methylation of the CDH13 (H-cadherin) promoter region in colorectal cancers and adenomas. Cancer Res 62:3382–3386

Tsao T-S, Tomas E, Murrey HE, Hug C, Lee DH, Ruderman NB, Heuser JE, Lodish HF (2003) Role of disulfide bonds in Acrp30/adiponectin structure and signaling specificity different oligomers activate different signal transduction pathways. J Biol Chem 278:50810–50817

Vainio H, Kaaks R, Bianchini F (2002) Weight control and physical activity in cancer prevention: international evaluation of the evidence. Eur J Cancer Prev 11(Suppl 2):S94–S100

Vetvik KK, Sonerud T, Lindeberg M, Luders T, Storkson RH, Jonsdottir K, Frengen E, Pietilainen KH, Bukholm I (2014) Globular adiponectin and its downstream target genes are up-regulated locally in human colorectal tumors: ex vivo and in vitro studies. Metab Clin Exp 63:672–681

Wahl S, Drong A, Lehne B, Loh M, Scott WR, Kunze S, Tsai PC, Ried JS, Zhang W, Yang Y, Tan S, Fiorito G, Franke L, Guarrera S, Kasela S, Kriebel J, Richmond RC, Adamo M, Afzal U, Ala-Korpela M, Albetti B, Ammerpohl O, Apperley JF, Beekman M, Bertazzi PA, Black SL, Blancher C, Bonder MJ, Brosch M, Carstensen-Kirberg M, de Craen AJ, de Lusignan S, Dehghan A, Elkalaawy M, Fischer K, Franco OH, Gaunt TR, Hampe J, Hashemi M, Isaacs A, Jenkinson A, Jha S, Kato N, Krogh V, Laffan M, Meisinger C, Meitinger T, Mok ZY, Motta V, Ng HK, Nikolakopoulou Z, Nteliopoulos G, Panico S, Pervjakova N, Prokisch H, Rathmann W, Roden M, Rota F, Rozario MA, Sandling JK, Schafmayer C, Schramm K, Siebert R, Slagboom PE, Soininen P, Stolk L, Strauch K, Tai ES, Tarantini L, Thorand B, Tigchelaar EF, Tumino R, Uitterlinden AG, van Duijn C, van Meurs JB, Vineis P, Wickremasinghe AR, Wijmenga C, Yang TP, Yuan W, Zhernakova A, Batterham RL, Smith GD, Deloukas P, Heijmans BT, Herder C, Hofman A, Lindgren CM, Milani L, van der Harst P, Peters A, Illig T, Relton CL, Waldenberger M, Jarvelin MR, Bollati V, Soong R, Spector TD, Scott J, McCarthy MI, Elliott P, Bell JT, Matullo G, Gieger C, Kooner JS, Grallert H, Chambers JC (2017) Epigenome-wide association study of body mass index, and the adverse outcomes of adiposity. Nature 541:81–86

Wang X, Martindale JL, Liu Y, Holbrook NJ (1998) The cellular response to oxidative stress: influences of mitogen-activated protein kinase signalling pathways on cell survival. Biochem J 333(Pt 2):291–300

Wang C, Xin X, Xiang R, Ramos FJ, Liu M, Lee HJ, Chen H, Mao X, Kikani CK, Liu F, Dong LQ (2009) Yin-Yang regulation of adiponectin signaling by APPL isoforms in muscle cells. J Biol Chem 284:31608–31615

Wei EK, Giovannucci E, Fuchs CS, Willett WC, Mantzoros CS (2005a) Low plasma adiponectin levels and risk of colorectal cancer in men: a prospective study. J Natl Cancer Inst 97:1688–1694

Wei EK, Ma J, Pollak MN, Rifai N, Fuchs CS, Hankinson SE, Giovannucci E (2005b) A prospective study of C-peptide, insulin-like growth factor-I, insulin-like growth factor binding protein-1, and the risk of colorectal cancer in women. Cancer Epidemiol Biomarkers Prev 14:850–855

Wilk S, Scheibenbogen C, Bauer S, Jenke A, Rother M, Guerreiro M, Kudernatsch R, Goerner N, Poller W, Elligsen-Merkel D (2011) Adiponectin is a negative regulator of antigen-activated T cells. Eur J Immunol 41:2323–2332

Wolf AM, Wolf D, Rumpold H, Enrich B, Tilg H (2004) Adiponectin induces the anti-inflammatory cytokines IL-10 and IL-1RA in human leukocytes. Biochem Biophys Res Commun 323:630–635

Wu WK, Wang XJ, Cheng AS, Luo MX, Ng SS, To KF, Chan FK, Cho CH, Sung JJ, Yu J (2013) Dysregulation and crosstalk of cellular signaling pathways in colon carcinogenesis. Crit Rev Oncol Hematol 86:251–277

Xu XT, Xu Q, Tong JL, Zhu MM, Huang ML, Ran ZH, Xiao SD (2011) Meta-analysis: circulating adiponectin levels and risk of colorectal cancer and adenoma. J Dig Dis 12:234–244

Yamauchi T, Kamon J, Ito Y, Tsuchida A, Yokomizo T, Kita S, Sugiyama T, Miyagishi M, Hara K, Tsunoda M (2003) Cloning of adiponectin receptors that mediate antidiabetic metabolic effects. Nature 423:762

Yamauchi T, Nio Y, Maki T, Kobayashi M, Takazawa T, Iwabu M, Okada-Iwabu M, Kawamoto S, Kubota N, Kubota T, Ito Y, Kamon J, Tsuchida A, Kumagai K, Kozono H, Hada Y, Ogata H, Tokuyama K, Tsunoda M, Ide T, Murakami K, Awazawa M, Takamoto I, Froguel P, Hara K, Tobe K, Nagai R, Ueki K, Kadowaki T (2007) Targeted disruption of AdipoR1 and AdipoR2 causes abrogation of adiponectin binding and metabolic actions. Nat Med 13:332–339

Yang X, Li J, Cai W, Yang Q, Lu Z, Yu J, Yu H, Zhang N, Sun D, Qu Y, Guo H, Wen F, Ji Y (2015) Adiponectin gene polymorphisms are associated with increased risk of colorectal Cancer. Med Sci Monit 21:2595–2606

Yokota T, Oritani K, Takahashi I, Ishikawa J, Matsuyama A, Ouchi N, Kihara S, Funahashi T, Tenner AJ, Tomiyama Y (2000) Adiponectin, a new member of the family of soluble defense collagens, negatively regulates the growth of myelomonocytic progenitors and the functions of macrophages. Blood 96:1723–1732

Zhan Y, Ginanni N, Tota MR, Wu M, Bays NW, Richon VM, Kohl NE, Bachman ES, Strack PR, Krauss S (2008) Control of cell growth and survival by enzymes of the fatty acid synthesis pathway in HCT-116 colon cancer cells. Clin Cancer Res 14:5735–5742

Zhang Y, Zhao J, Li R, Lau WB, Yuan YX, Liang B, Li R, Gao EH, Koch WJ, Ma XL, Wang YJ (2015) AdipoRon, the first orally active adiponectin receptor activator, attenuates postischemic myocardial apoptosis through both AMPK-mediated and AMPK-independent signalings. Am J Phys Endocrinol Metab 309:E275–E282

Zhou L, Zhang HF, Ning W, Song X, Liu X, Liu JX (2017) Associations of adiponectin receptor 2 (AdipoR2) gene polymorphisms and AdipoR2 protein expression levels with the risk of colorectal cancer: a case-control study. Mol Med Rep 16:3983–3993

Chapter 3
Role of Matrix Metalloproteinases in Colorectal Cancer

Neha Merchant, Gayathri Chalikonda, and Ganji Purnachandra Nagaraju

Abstract Colorectal cancer or CRC is amongst the top three leading causes of cancer related fatalities in the US. Advanced stages of CRC have very restricted treatment options. Matrix metalloproteinases (MMPs) are vital to maintain extracellular homeostasis, however, they play an important in CRC invasion and progression. Activities of MMP1, MMP2, MMP3, MMP7, MMP9, MMP13, and MT1-MMP are overexpressed in CRC and associated with worse results, whereas activities of MMP12 is known to be protective. Therefore, MMPs have become an important therapeutic target. Earlier clinical trials by broad-spectrum inhibitors of MMP have proven to be unsuccessful due to off-target toxicity as well as lack of efficiency. Currently, safe and selective inhibitors have transformed its therapeutic ability to target MMPs.

Keywords Colorectal cancer · Growth · Metastasis · Matrix metalloproteinases · Therapy

Abbreviations

CRC	Colorectal cancer
MMP	Matrix metalloproteinases
TIMP	Tissue Inhibitors of Metalloproteinase
mRNA	microRNA

N. Merchant · G. Chalikonda · G. P. Nagaraju (✉)
Department of Hematology and Medical Oncology, Winship Cancer Institute, Emory University, Atlanta, GA, USA
e-mail: pganji@emory.edu

G. S. R. Raju, L.V.K.S. Bhaskar (eds.), *Theranostics Approaches to Gastric and Colon Cancer*, Diagnostics and Therapeutic Advances in GI Malignancies,
https://doi.org/10.1007/978-981-15-2017-4_3

3.1 Introduction

Colorectal Cancer or CRC is amongst the top three cancers that cause the highest number of fatalities in males and females in the United States. Pathogenesis of CRC is an intricate and complex process stating from the neoplastic alteration of non-cancerous cells, invasion of tissue, vascular extravasation and intravasation, and finally seeding in other organs such as the liver. The extracellular matrix forms the basic structural framework for supporting cells, regulates intercellular and intracellular signaling, plays a vital role in cell differentiation, invasion, and adhesion. The extracellular matrix is comprised of proteins that play important role in cell migration, proliferation and metastasis. Various proteases regulate the extracellular matrix degradation and remodeling. Matrix metalloproteinases are one such group of enzymes that as known as key intermediaries in the extracellular degradation process (Hua et al. 2011). MMPs are a union of enzymes that degrade the extracellular matrix that are expressed in different stages of colorectal cancer or CRC. MMPs are associated with survival and diagnosis. Several preclinical studies have revealed effectiveness of MMP inhibitors at various stages of tumor progression in CRC cases such as inhibition of preliminary tumor growth by MMP inhibitors. MMPs are a family of 25 zinc-dependent enzymes—endopeptidases that can degrade the extracellular matrix and its components. MMPs are characterized largely on the basis of their basic features such as stromelysins, matrilysins, gelatinases, membrane-type, and collagenases. The domain structure of all MMPs include a catalytic domain, C terminal domain that looks like a hemopexin, a pro-peptide, and a hinge region which is link between the catalytic domain and the hemopexin (Hu et al. 2007). MMPs are synthesized either as a membrane-related or as a secreted inactive zymogen that has to be proteolytically managed to an active state. The management encompasses of cysteine residue elimination, thereby stopping its interaction with zinc ions from the active site that eventually leads to MMP activation. MMPs control shedding, a cell surface growth factor responsible for modulating the proteolytic release of various proteins like chemokines, adhesion molecules as well as growth factors. In various cancers, including CRC, elevated MMP expression level is highly associated with angiogenesis, metastasis and invasion with low survival. Recent investigations have also revealed MMPs to have tumor shielding properties (Van Der Jagt et al. 2010).

3.2 Role of Matrix Metalloproteinases in Colorectal Cancer

3.2.1 Role of MMP1 in CRC

MMP1 is a collagenase that targets the main components of the intestinal stroma including collagen type I–III and it is responsible for extracellular matrix degradation. The expression of MMP1 is associated with advanced stages of CRC as well as poor prognosis (Sunami et al. 2000). An investigation involving human specimens revealed a high correlation between elevated MMP1 expression levels and metastasis and invasion (Shiozawa et al. 2000). Another investigation revealed discrete MMP1 expression levels in primary CRC as compared to metastatic CRC. Primary CRC tumors involving lymph node showed a strong MMP1 activity whereas metastatic CRC revealed a fairly declined MMP1 activity (Bendardaf et al. 2007). Studies have also revealed that MMP1 expressions were not over-expressed in distant metastasis, which suggests the role of MMP1 during early stages of invasion, while weakening as metastases progresses. Elevated CRC susceptibility is related with MMP1 gene polymorphisms. The promotor region of MMP1 revealed a guanine removal/insertion, where an allele consisted of only one guanine nucleotide and a second allele consisted of a pair of guanine nucleotides (Decock et al. 2008). The allele with pair of guanine nucleotides showed elevated levels of transcriptional expressions as compared the allele with single guanine nucleotide. It was also linked with elevated metastasis and vulnerability to CRC.

3.2.2 Role of MMP13 in CRC

MMP13, also a collagenase, is structurally homologous to MMP1 is known to effectively degrade type II collagen in the extra cellular matrix. Investigations have shown that elevated immunohistochemical staining of MMP13 is associated with low survival rate in CRC cases (Leeman et al. 2002). The role of MMP13 as a potential biomarker for CRC showed that elevated expression of MMP13 is correlated with advanced stages of CRC and heightened risk for post-operative decline (Huang et al. 2010).

3.2.3 Role of MMP2 in CRC

Another sub group of enzymes within the MMP family are the gelatinases that consist of MMP2 and MMP9 that play a role in degrading type IV collagen of the extracellular matrix and gelatin. These gelatinases also share proteolytic movement against other ECM molecules. Worsened outcomes in CRC prognosis and elevated expression levels of MMP2 and MMP9 are often correlated. According to a study, elevated levels of MMP2 expressions were detected in CRC patients with positive lymph node in comparison to those without such metastasis (Langenskiöld et al. 2005). Numerous investigations have studied effect of serum MMPs as biomarkers in CRC invasion. CEA and CA19–9 are commonly used markers in clinical settings. It was revealed in an investigation (Dragutinović et al. 2011) that the protein levels of MMP2 and MMP9 were relatively expressed at a much-elevated rate in CRC patients.

CRC cells need integrins and MMPs for adhesion and proteolysis, respectively. The interaction of MMP2 and β1 integrins was investigated in a study in CRC cells. The investigations showed MMP2 up regulations in invasive CRC by degrading β1 integrins and eventually leading to enhanced motility and reduced cell adhesion (Kryczka et al. 2012). Gelatinases are activated via various signaling pathways. SMAD proteins play a major role in regulating cell cycle, apoptosis, differentiation as well as TGF-β signaling. These proteins bind with the receptor modulated counterparts and suppresses CRC cell migration via MMP9 regulation. Elevated levels of c-Jun synthesis is linked with over-expressed levels of p38 gamma MAPK in CRC cells that leads to increased transcription of MMP9 as well as MMP9 linked invasion (Loesch et al. 2010).

3.2.4 Role of MMP9 in CRC

MMP9 expression is diminished by TGF-β receptor kinase inhibitors and the CRC metastasis is blocked into the liver (Papageorgis et al. 2011; Zhang et al. 2010). MMP9 expression is elevated in the inflamed intestines of patients with inflammatory bowel syndrome. This disease is frequently related with being progressed to colitis-associated colon cancer or CAC. The MMP9-null mice showed higher vulnerability to CAC as compared to WT mice. It also displayed elevated Notch-1 stimulation, which is a vital transcription factor in order to determine the epithelial lineage and also reduced proto-oncogene, β-catenin expression that plays a chief role in modulating gene transcription as well as Wnt signaling (Garg et al. 2010). Even as a pro-inflammatory mediator in colitis, MMP9 has its strong presence as a protector as well as a tumor suppressor for CAC patients (Fig. 3.1).

3.2.5 Co-expression of MMP3 and MMP9 in CRC

Studies have reported that MMP3 and MMP9 co-expression is associated with CRC tumors. A study hypothesized that MMP9 co-expressed with uPA in CRC is

Fig. 3.1 MMPs induced inflammation, colon cancer growth and metastasis

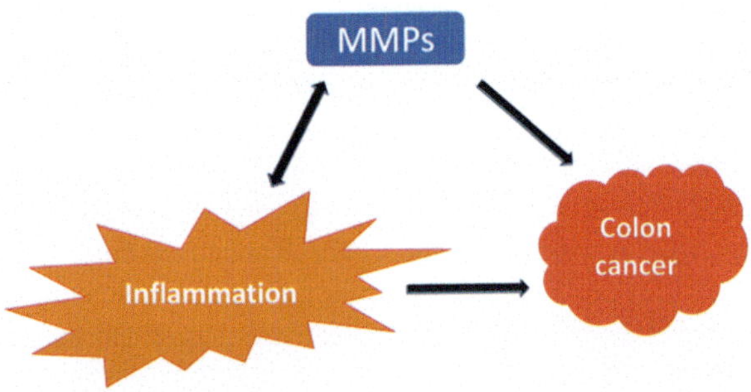

responsible for plasminogen activation to plasmin (Inuzuka et al. 2000). Plasmin stimulates proMMP3 to MMP3, eventually activating proMMP9 and CRC progression.

3.2.6 Role of MMP7 in CRC

More than 80% of CRC cases reveal MMP7 overexpression (Brabletz et al. 1999). Reduced survival rate in advanced stages of CRC and tumor progression is strongly correlated with MMP7 serum levels (Maurel et al. 2007). An investigation conducted using adenomatous polyposis coli (Apc)+/Min-FCCC mice and a fluorescent MMP7 probe, approximately 90% of CRC tumors are successfully identified, which suggests the use of such probe in detecting early stages of malignancy and intervention (Clapper et al. 2011). CRC invasion is promoted by MMP7 through the ECM proteolytic cleavage, which also stimulates other MMPs such as proMMP9 as well as proMMP2 in order to promote tumor invasion (Ii et al. 2006). The ECM proteolytic cleavage alters non-ECM proteins, which results in its stimulation, degradation as well as proteolytic shedding. MMP7 modulates cell proliferation as well as apoptosis via proHB-EGF proteolytic shedding to activate EGFR signaling and produce the mature HB-EGF (Ii et al. 2006). Moreover, stimulation of muscarinic receptors leads to elevated levels of MMP7 in human CRC cell lines (Xie et al. 2009). In a study including SCID mice, treatment with human CRC cell lines, which overexpressed MMP7 elevated tumor metastasis as well as invasion (Adachi et al. 1999). Furthermore, in multiple intestinal neoplasia mice models, MMP7 knockout significantly decreased CRC multiplicity and polyp size (Wilson et al. 1997), therefore further accentuating the role played by MMP7 in carcinogenesis.

3.2.7 Role of MMP13 in CRC

According to immunohistochemical staining of MMP13 performed on CRC biopsy specimens, it was identified that more than 90% of cases exhibited MMP13 activity and tumor cell cytoplasm had localized immunoreactivity (Leeman et al. 2002). Elevated levels of MMP13 expression was revealed in CRC cells lines as compared to noncancerous tissues. Increased MMP13 expression was linked with poor survival. A possible mechanism of action of MMP13 is thought to be based on its ability to stimulate proMMP9. proMMP13 is activated by plasmin, MMP2, as well as MT1-MMP (Knäuper et al. 1996).

3.2.8 *Role of MT1-MMP in CRC*

Increasing CRC stage was associated with elevated MT1MMP expression levels, according to an investigation (Sardinha et al. 2000). Immunoperoxidase staining was performed on adenomas, metastatic tumor, and carcinoma in situ to evaluate MT1-MMP role in colon tumor to invasive CRC evolution (Malhotra et al. 2002). Substantial MT1-MMP progression was shown in adenomas transformation to carcinoma in situ and invasive CRC. A study revealed that more than 90% of RC cases exhibited elevated MT1-MMP expression levels (Takahashi et al. 2005), which suggests that MT1-MMP is an important target of the catenin/Tcf complex involving the Wnt signaling pathway.

3.2.9 *Role of MMP12 in CRC*

MMP12 is an independent metalloelastase that is not a part of any subfamilies and is well-known for degrading various substrates. Even though MMP12 is primarily expressed in macrophages, various investigations have revealed the protective role of MMP12 in CRC (Decock et al. 2011; Overall and Kleifeld 2006). MMP12 was known to diminish VEGF expression and elevate, which is an endogenous angiogenesis inhibitor (angiostatin) (Shi et al. 2006). Many investigations have suggested the role of MMP12 as a tumor suppressor and aids in enhanced survival (Xu et al. 2008; Yang et al. 2001).

3.3 Role of Tissue Inhibitors of Metalloproteinases in Colorectal Cancer

TIMPs or tissue inhibitors of MMPs are a family of protease inhibitors, which occur naturally and are specific to MMPs. TIMPs aid in minimizing degradation of the ECM. MMP-TIMP complexes inhibit the proteolytic activities of the MMP. TIMPs also participate in various physiological processes such as angiogenesis, migration, proliferation, invasion, and apoptosis (Bourboulia and Stetler-Stevenson 2010). TIMPs are known to play dual roles: as modulators of MMP actions and operating independently of MMPs. According to a study, TIMP-1 showed resistance against toxicity incurred via TNF-α and IL-2 in human CRC cells. It was also responsible for clonogenicity as well as tumor advancement during initial stage of disease progression (Kim et al. 2012). However, abnormal TIMP-1 glycosylation was observed during the final stages of tumor advancement accompanied by lost TIMP-1 inhibition of collagenases, eventually encouraging additional invasive tumors. TIMP-1 selectively inhibits MMP1, MMP3, MMP7, and MMP9 (Bourboulia and Stetler-Stevenson 2010). Elevated levels of TIMP-1 were associated with CRC patients and worse results (Lee et al. 2011). A study was conducted to examine probable biomarkers for initial detection of CRC tumors, which involved identifying CRC patients with high specificity as well as sensitivity also had elevated right sided CRC predictive value (Holten-Andersen et al. 2002). MMP2 and MMP9 are inhibited by TIMP-2 serve as an adaptor protein for pro-MMP2 stimulation (Lu et al. 2004). An investigation conducted on Korean patients with CRC, TIMP-2 was correlated to generic polymorphism with elevated risk of worse prognosis and metastasis (Park et al. 2011). CD133 is a presumed stem cell as well as a tumor stem cell marker. CRC cell with siRNA knockdown of CD133 displayed TIMP-2 down regulation and diminished invasiveness (Zhang et al. 2013). TIMP-3 inhibits MMPs activity as well as it inhibits ADAMs, which are a family of peptidase proteins. In CRC cell lines, TIMP-3 is known to suppress neoplasia by induction of apoptosis, which is an action presumed to be mediated by TNF-α receptors stabilization. In stroma surrounding CRC, TIMP-3 expression was decreased (Powe et al. 1997). A study conducted examined the use of TIMP-3 as a therapeutic agent for CRC in which adenovirus facilitated TIMP-3 transduction arrested tumor cell growth as well as induced apoptosis. According to in vitro studies, elevated levels of TIMP-3 leads diminished adhesion, invasion, and migration of CRC cell lines. Whereas, in vivo investigations reveal that transduction of TIMP-3 decreases liver metastasis as well as tumor advancement (Lin et al. 2012). The catalytic actions of MMP2 is inhibited via TIMP-4. An investigation involving multivariate analysis showed that TIMP-3 stromal cytoplasmic staining is the only CRC biomarker that holds some prognostic value and therefore highlights its role as a promising CRC marker (Hilska et al. 2007).

3.4 MMPs as Therapeutic Target

Given present understanding about MMPs as well as its association with abrupt tumor growth, MMP inhibitors were tested and developed in clinical settings. During the 1990s, MMP inhibitors that were tested to target tumor revealed promising outcomes in phase I and phase II trials. As expected, MMP inhibitors exhibited inhibitory properties on primary as well as metastatic CRC. Whereas, several phase III trials showed no promising advantages and its use was also related with toxicity. The failure of synthetic MMP inhibitors due to lack of selectivity. These inhibitors were speculated to have inhibited MMP activity, which were not over expressed in any specific cancer or inhibited MMPs activities with tumor suppressing properties (Coussens et al. 2002). An investigation that highlights another reason of MMP inhibitor's therapeutic failure involved a synthetic inhibitor MMI270's effect on CRC metastasis rat model (Ogata et al. 2006). It was studied that MMI270 competitively bound to zinc in active state of various MMPs such as MMP1, MMP2, MMP3, MMP9, and MMP13, subsequently leading to their inhibition. Initial MMI270 administration after eliminating primary CRC, lead to diminished lung metastasis as compared to delayed administration. Due to its reduced selectivity and in order to elevate its substrate selectivity, novel structure-based MMP inhibitors are developed that allow for the three-dimensional enzymatic conformation active site instead of the earlier substrate based MMP inhibitors. Additional MMP inhibitors are SB-3CT that covalently binds with MMP2 active site in order to achieve the pro peptide state of the enzyme (Roy et al. 2009). Currently, a group of MMP inhibitors are being investigated that are chemically modified tetracyclines or CMTs. Although CMTs are deficient in intrinsic antibiotic activity, their mechanism of action is presumed to be binding to the calcium or zinc ions, or via MMP transcriptional regulation (Roy et al. 2009). Alternative technique to target MMPs is via microRNA or miRNAs, which work in transcriptional as well as post-transcriptional regulation of gene. These biological modulators block the MMP secretion and MMP activity, thereby highlighting their probable role as a tumor suppressor (Van Der Jagt et al. 2010). One such instance was renowned that included using antisense oligonucleotides, which targeted MMP7 microRNA, prevented human CRC cell invasion through MMP7 activity inhibition (Miyazaki et al. 1999). Using antisense oligonucleotides towards MMP7 microRNA in human CRC cell lines xenograft models revealed blockade of basement membrane infiltration as well as liver metastases suppression (Hasegawa et al. 1998). A recent group of investigations that were designed to investigate microRNA-34a role in human CRC were designed to highlight the developing role of microRNA targeting in CRC. microRNA-34a is known to act as a p53 transcriptional target, a gene which is vital in cell cycle regulation and also acts as a tumor suppressor. Transfection of microRNA-34a in human CRC cell lines, it was revealed that MMP1 and MMP9 expression was reduced significantly (Wu et al. 2011) and eventually inhibits human CRC cell invasion and migration.

Current investigations are undergoing to investigate the best possible ways to target MMPs in CRC, as previously investigated therapeutic inhibition of MMPs have encountered several difficulties. Present extent of investigation includes MMP's role as a prognostic marker in the treatment of CRC, its utilization to monitor therapeutic responses, as well as to evaluate individuals who are expected to respond best towards a particular therapeutic regimen. In the light of current understanding about the functions of MMP in different pathological conditions, the subsequent goal of all future investigations should be towards developing an efficient treatment strategy by selectively targeting MMPs.

References

Adachi Y, Yamamoto H, Itoh F, Hinoda Y, Okada Y, Imai K (1999) Contribution of matrilysin (MMP-7) to the metastatic pathway of human colorectal cancers. Gut 45:252–258

Bendardaf R, Buhmeida A, Ristamäki R, Syrjänen K, Pyrhönen S (2007) MMP-1 (collagenase-1) expression in primary colorectal cancer and its metastases. Scand J Gastroenterol 42:1473–1478

Bourboulia D, Stetler-Stevenson WG (2010) Matrix metalloproteinases (MMPs) and tissue inhibitors of metalloproteinases (TIMPs): positive and negative regulators in tumor cell adhesion. Semin Cancer Biol 20:161–168

Brabletz T, Jung A, Dag S, Hlubek F, Kirchner T (1999) β-Catenin regulates the expression of the matrix metalloproteinase-7 in human colorectal cancer. Am J Pathol 155:1033–1038

Clapper ML, Hensley HH, Chang W-CL, Devarajan K, Nguyen MT, Cooper HS (2011) Detection of colorectal adenomas using a bioactivatable probe specific for matrix metalloproteinase activity. Neoplasia 13:685–691

Coussens LM, Fingleton B, Matrisian LM (2002) Matrix metalloproteinase inhibitors and cancer—trials and tribulations. Science 295:2387–2392

Decock J, Paridaens R, Ye S (2008) Genetic polymorphisms of matrix metalloproteinases in lung, breast and colorectal cancer. Clin Genet 73:197–211

Decock J, Thirkettle S, Wagstaff L, Edwards DR (2011) Matrix metalloproteinases: protective roles in cancer. J Cell Mol Med 15:1254–1265

Dragutinović VV, Radonjić NV, Petronijević ND, Tatić SB, Dimitrijević IB, Radovanović NS, Krivokapić ZV (2011) Matrix metalloproteinase-2 (MMP-2) and-9 (MMP-9) in preoperative serum as independent prognostic markers in patients with colorectal cancer. Mol Cell Biochem 355:173–178

Garg P, Sarma D, Jeppsson S, Patel NR, Gewirtz AT, Merlin D, Sitaraman SV (2010) Matrix metalloproteinase-9 functions as a tumor suppressor in colitis-associated cancer. Cancer Res 70:792–801

Hasegawa S, Koshikawa N, Momiyama N, Moriyama K, Ichikawa Y, Ishikawa T, Mitsuhashi M, Shimada H, Miyazaki K (1998) Matrilysin-specific antisense oligonucleotide inhibits liver metastasis of human colon cancer cells in a nude mouse model. Int J Cancer 76:812–816

Hilska M, Roberts PJ, Collan YU, Laine VJO, Kössi J, Hirsimäki P, Rahkonen O, Laato M (2007) Prognostic significance of matrix metalloproteinases-1,-2,-7 and-13 and tissue inhibitors of metalloproteinases-1,-2,-3 and-4 in colorectal cancer. Int J Cancer 121:714–723

Holten-Andersen MN, Christensen IJ, Nielsen HJ, Stephens RW, Jensen V, Nielsen OH, Sørensen S, Overgaard J, Lilja H, Harris A (2002) Total levels of tissue inhibitor of metalloproteinases 1 in plasma yield high diagnostic sensitivity and specificity in patients with colon cancer. Clin Cancer Res 8:156–164

Hu J, Van den Steen PE, Sang Q-XA, Opdenakker G (2007) Matrix metalloproteinase inhibitors as therapy for inflammatory and vascular diseases. Nat Rev Drug Discov 6:480

Hua H, Li M, Luo T, Yin Y, Jiang Y (2011) Matrix metalloproteinases in tumorigenesis: an evolving paradigm. Cell Mol Life Sci 68:3853–3868

Huang M-Y, Chang H-J, Chung F-Y, Yang M-J, Yang Y-H, Wang J-Y, Lin S-R (2010) MMP13 is a potential prognostic marker for colorectal cancer. Oncol Rep 24:1241–1247

Ii M, Yamamoto H, Adachi Y, Maruyama Y, Shinomura Y (2006) Role of matrix metalloproteinase-7 (matrilysin) in human cancer invasion, apoptosis, growth, and angiogenesis. Exp Biol Med 231:20–27

Inuzuka K, Ogata Y, Nagase H, Shirouzu K (2000) Significance of coexpression of urokinase-type plasminogen activator, and matrix metalloproteinase 3 (stromelysin) and 9 (gelatinase B) in colorectal carcinoma. J Surg Res 93:211–218

Kim Y-S, Ahn YH, Song KJ, Kang JG, Lee JH, Jeon SK, Kim H-C, Yoo JS, Ko J-H (2012) Overexpression and β-1, 6-N-Acetylglucosaminylation-initiated aberrant glycosylation of TIMP-1 A "DOUBLE WHAMMY" STRATEGY IN COLON CANCER PROGRESSION. J Biol Chem 287:32467–32478

Knäuper V, Will H, López-Otin C, Smith B, Atkinson SJ, Stanton H, Hembry RM, Murphy G (1996) Cellular mechanisms for human procollagenase-3 (MMP-13) activation evidence that MT1-MMP (MMP-14) and gelatinase a (MMP-2) are able to generate active enzyme. J Biol Chem 271:17124–17131

Kryczka J, Stasiak M, Dziki L, Mik M, Dziki A, Cierniewski CS (2012) Matrix metalloproteinase-2 cleavage of the β1 integrin ectodomain facilitates colon cancer cell motility. J Biol Chem 287:36556–36566

Langenskiöld M, Holmdahl L, Falk P, Ivarsson M-L (2005) Increased plasma MMP-2 protein expression in lymph node-positive patients with colorectal cancer. Int J Color Dis 20:245–252

Lee J-H, Choi J-W, Kim Y-S (2011) Plasma or serum TIMP-1 is a predictor of survival outcomes in colorectal cancer: a meta-analysis. J Gastrointestin Liver Dis 20:287–291

Leeman M, McKay J, Murray GI (2002) Matrix metalloproteinase 13 activity is associated with poor prognosis in colorectal cancer. J Clin Pathol 55:758–762

Lin H, Zhang Y, Wang H, Xu D, Meng X, Shao Y, Lin C, Ye Y, Qian H, Wang S (2012) Tissue inhibitor of metalloproteinases-3 transfer suppresses malignant behaviors of colorectal cancer cells. Cancer Gene Ther 19:845

Loesch M, Zhi H-Y, Hou S-W, Qi X-M, Li R-S, Basir Z, Iftner T, Cuenda A, Chen G (2010) p38γ MAPK cooperates with c-Jun in trans-activating matrix metalloproteinase 9. J Biol Chem 285:15149–15158

Lu KV, Jong KA, Rajasekaran AK, Cloughesy TF, Mischel PS (2004) Upregulation of tissue inhibitor of metalloproteinases (TIMP)-2 promotes matrix metalloproteinase (MMP)-2 activation and cell invasion in a human glioblastoma cell line. Lab Investig 84:8

Malhotra S, Newman E, Eisenberg D, Scholes J, Wieczorek R, Mignatti P, Shamamian P (2002) Increased membrane type 1 matrix metalloproteinase expression from adenoma to colon cancer. Dis Colon Rectum 45:537–543

Maurel J, Nadal C, Garcia-Albeniz X, Gallego R, Carcereny E, Almendro V, Mármol M, Gallardo E, Maria Augé J, Longarón R (2007) Serum matrix metalloproteinase 7 levels identifies poor prognosis advanced colorectal cancer patients. Int J Cancer 121:1066–1071

Miyazaki K, Koshikawa N, Hasegawa S, Momiyama N, Nagashima Y, Moriyama K, Ichikawa Y, Ishikawa T, Mitsuhashi M, Shimada H (1999) Matrilysin as a target for chemotherapy for colon cancer: use of antisense oligonucleotides as antimetastatic agents. Cancer Chemother Pharmacol 43:S52–S55

Ogata Y, Matono K, Nakajima M, Sasatomi T, Mizobe T, Nagase H, Shirouzu K (2006) Efficacy of the MMP inhibitor MMI270 against lung metastasis following removal of orthotopically transplanted human colon cancer in rat. Int J Cancer 118:215–221

Overall CM, Kleifeld O (2006) Validating matrix metalloproteinases as drug targets and anti-targets for cancer therapy. Nat Rev Cancer 6:227

Papageorgis P, Cheng K, Ozturk S, Gong Y, Lambert AW, Abdolmaleky HM, Zhou J-R, Thiagalingam S (2011) Smad4 inactivation promotes malignancy and drug resistance of colon cancer. Cancer Res 71:998–1008

Park KS, Kim SJ, Kim KH, Kim JC (2011) Clinical characteristics of TIMP2, MMP2, and MMP9 gene polymorphisms in colorectal cancer. J Gastroenterol Hepatol 26:391–397

Powe D, Brough J, Carter G, Bailey E, Stetler-Stevenson W, Turner D, Hewitt R (1997) TIMP-3 mRNA expression is regionally increased in moderately and poorly differentiated colorectal adenocarcinoma. Br J Cancer 75:1678

Roy R, Yang J, Moses MA (2009) Matrix metalloproteinases as novel biomarkers and potential therapeutic targets in human cancer. J Clin Oncol 27:5287

Sardinha TC, Nogueras JJ, Xiong H, Weiss EG, Wexner SD, Abramson S (2000) Membrane-type 1 matrix metalloproteinase mRNA expression in colorectal cancer. Dis Colon Rectum 43:389–395

Shi H, Xu JM, Hu NZ, Wang XL, Mei Q, Song YL (2006) Transfection of mouse macrophage metalloelastase gene into murine CT-26 colon cancer cells suppresses orthotopic tumor growth, angiogenesis and vascular endothelial growth factor expression. Cancer Lett 233:139–150

Shiozawa J, Ito M, Nakayama T, Nakashima M, Kohno S, Sekine I (2000) Expression of matrix metalloproteinase-1 in human colorectal carcinoma. Mod Pathol 13:925

Sunami E, Tsuno N, Osada T, Saito S, Kitayama J, Tomozawa S, Tsuruo T, Shibata Y, Muto T, Nagawa H (2000) MMP-1 is a prognostic marker for hematogenous metastasis of colorectal cancer. Oncologist 5:108–114

Takahashi M, Nakamura Y, Obama K, Furukawa Y (2005) Identification of SP5 as a downstream gene of the β-catenin/Tcf pathway and its enhanced expression in human colon cancer. Int J Oncol 27:1483–1487

Van Der Jagt MF, Wobbes T, Strobbe LJ, Sweep FC, Span PN (2010) Metalloproteinases and their regulators in colorectal cancer. J Surg Oncol 101:259–269

Wilson CL, Heppner KJ, Labosky PA, Hogan BL, Matrisian LM (1997) Intestinal tumorigenesis is suppressed in mice lacking the metalloproteinase matrilysin. Proc Natl Acad Sci 94:1402–1407

Wu J, Wu G, Lv L, Ren Y-F, Zhang X-J, Xue Y-F, Li G, Lu X, Sun Z, Tang K-F (2011) MicroRNA-34a inhibits migration and invasion of colon cancer cells via targeting to Fra-1. Carcinogenesis 33:519–528

Xie G, Cheng K, Shant J, Raufman J-P (2009) Acetylcholine-induced activation of M3 muscarinic receptors stimulates robust matrix metalloproteinase gene expression in human colon cancer cells. Am J Physiol Gastrointest Liver Physiol 296:G755–G763

Xu Z, Shi H, Li Q, Mei Q, Bao J, Shen Y, Xu J (2008) Mouse macrophage metalloelastase generates angiostatin from plasminogen and suppresses tumor angiogenesis in murine colon cancer. Oncol Rep 20:81–88

Yang W, Arii S, Gorrin-Rivas MJ, Mori A, Onodera H, Imamura M (2001) Human macrophage metalloelastase gene expression in colorectal carcinoma and its clinicopathologic significance. Cancer 91:1277–1283

Zhang B, Halder SK, Kashikar ND, Cho YJ, Datta A, Gorden DL, Datta PK (2010) Antimetastatic role of Smad4 signaling in colorectal cancer. Gastroenterology 138:969–980. e963

Zhang M, Liu Y, Feng H, Bian X, Zhao W, Yang Z, Gu B, Li Z, Liu Y (2013) CD133 affects the invasive ability of HCT116 cells by regulating TIMP-2. Am J Pathol 182:565–576

Chapter 4
The Role of Hypoxia-Inducible Factor 1-Alpha in Colorectal Cancer

Saimila Momin and Ganji Purnachandra Nagaraju

Abstract Colorectal cancer is fourth common form of cancer occurring on a global scale, with a strong prevalence in the Western world and industrialized nations. Even though there are many factors that contribute towards to the advancement of colorectal cancer, currently, investigations have shown a strong correlation between hypoxia-inducible factor and colorectal cancer. Hypoxia-inducible factor is a heterodimeric transcription factor, which encodes one of its regulatory subunits known as hypoxia-inducible factor 1-alpha (HIF-1-alpha). HIF-1-alpha is a protein which plays a key role in the regulation and modulation of gene expression and cellular response related to hypoxia. Hypoxia results when there is an inadequate delivery of oxygen to the cells. Studies have confirmed that the over-expression of HIF-1-alpha is strongly correlated to gastrointestinal cancers, including colorectal cancer. Specifically, the over-expression of HIF-1-alpha and PAS domain protein 1 serves as strong prognostic indicators to assess the advancement of the disease. Studying the correlation between HIF-1-alpha and colorectal cancers or in general cancers is an attractive point of study because targeting HIF-1-alpha, by disrupting its cellular pathways, can potentially prevent tumor growth development. In this paper, we will examine the role of HIF-1-alpha in colorectal cancer and how its expression and regulation could possibly lead to therapeutic options for treatment of colorectal cancer.

Keywords Colorectal cancer · Hypoxia-inducible factor 1-alpha · Signalling pathways · Therapy

S. Momin · G. P. Nagaraju (✉)
Department of Hematology and Medical Oncology, Winship Cancer Institute, Emory University, Atlanta, GA, USA
e-mail: pganji@emory.edu

G. S. R. Raju, L.V.K.S. Bhaskar (eds.), *Theranostics Approaches to Gastric and Colon Cancer*, Diagnostics and Therapeutic Advances in GI Malignancies, https://doi.org/10.1007/978-981-15-2017-4_4

4.1 Introduction

4.1.1 *Overview of Colorectal Cancer*

Colorectal cancer is described as tumorigenesis of the colon or the rectum and can start in either areas. Colorectal cancer, depending on whether is starts in the colon or rectum, are also known as just colon cancer or rectal cancer, respectively. Currently it is one of the fourth leading causes of death in both men and women (Bray et al. 2018). Understanding the factors and molecular mechanisms that are strongly correlated to colorectal cancer will aid us in understanding the development of this cancer and lead to strong improvements and developments of therapeutic treatments.

In the gastrointestinal system, the colon and rectum play a role in removing wastes from the body. The first part of the colon is known as the ascending colon with a cecum, which is a pouch-like part that takes the remaining, undigested food from the small intestine. The next parts of the colon, respectively, are called the transverse colon, descending colon, and the sigmoid colon, which connects to the rectum. The colon serves as an important part of the digestive system; in which, its primary goal is to reabsorb water and electrolytes and remove wastes in the form of stool. The colon and rectum both play an important role in the overall digestive system and disruption to these organs can lead to deadly problems. The growth of polyps in the colon can lead to colorectal cancer, which can grow to other parts of the body, rapidly bordering on fatality (Bray et al. 2018; De Leon and Di Gregorio 2001). Development of polyps generally lead to tumorigenesis.

Generally, these cancers advance from a small group of benign, cancerous cells, which are described as neoplastic polyps (Bray et al. 2018). Polyps can start to proliferate abnormally and uncontrollably and spread to other areas of the body. More specifically, colorectal tumor cells rise from adenomatous polyps, which are described as groups of epithelial dysplasia (De Leon and Di Gregorio 2001). In other words, normal cells undergo a transition and become abnormal cells and at this point these abnormal cells may become cancerous. In this case with adenomatous polyps, they have uncontrollable cell division (De Leon and Di Gregorio 2001). This can then lead to the advancement of colorectal cancer.

Symptoms for this cancer include but are not limited to hematochezia, unusually shaped stools, changes in bowel movements, including diarrhea and constipation, abdominal pain in the lower quadrants, fatigue, and weight loss (Augenlicht 2019; Bray et al. 2018; De Leon and Di Gregorio 2001). The pathophysiology of colorectal cancer is still under study; however, there are certain risk factors that are strongly correlated to this cancer. Some factors include but are not limited to type of dietary factors, genetics, family history of other hereditary cancers, alcohol, tobacco smoking, age, and geographical factors (Haggar and Boushey 2009). Other studies have shown that factors such as pelvic irradiation, cholecystectomy, inflammatory bowel disease, and deficiency of calcium and vitamin D (Rawla et al. 2019). Interestingly, when the body has little to no calcium in the mucosa, bile acids are unable to successfully bind to the calcium (Rawla et al. 2019). Without the binding, bile acids can lead to neoplastic changes, by modifying DNA and cellular proliferating (Rawla et al. 2019). Additionally, other factors that lead to colorectal cancer include Western diet, foods that are rich in fat, animal protein, diets that are low in fiber, vegetables, and fruits (Rawla et al. 2019). All these factors play a major role in the pathophysiology of colorectal cancer.

However, even though there are strong correlations between these factors and colorectal cancer, the exact cause and development of this type of cancer is still unknown. The study of the relation between HIF-1-alpha and colorectal cancer will definitely aid in understanding the pathophysiology of colorectal cancer and can potentially bring us a step closer to finding therapeutic treatments for those affected with colorectal cancer.

4.1.2 Basic Pathophysiology of Colorectal Cancer

Colorectal cancer generally begins as polyps on the inner lining of the colon or rectum. Polyps are described as abnormal tissue growth in the shape of flat bumps (Augenlicht 2019). Additionally, polyps can be divided into two types, adenomatous polyps or also known as adenomas and hyper plastic polyps and inflammatory polyps, depending on whether they will become cancerous or not over time (Augenlicht 2019; De Leon and Di Gregorio 2001). Adenomatous polyps or adenomas are more likely to transition into cancer. On the other hand, hyper plastic polyps and inflammatory polyps are generally more common and are not likely to transition into cancer. Other factors that contribute towards colorectal cancer in the context of polyps include the size and quantity of polyps found (De Leon and Di Gregorio 2001). If polyps are larger than 1 cm and if multiple polyps are found, it is most likely pre- cancerous (De Leon and Di Gregorio 2001). Additionally, if dysplasia occurs especially after the polyps are removed, can also contribute to leading to colorectal cancer (De Leon and Di Gregorio 2001). Generally, the formation and development of polyps occurs on the innermost layer or the mucosa of the walls of the colon and rectum. The cancer cells, over time, develop and begin to grow and migrate to the other layers of the colon and rectum. Later, these cancer cells can develop into the lymph vessels and the lymph can carry these cells to lymph nodes and other parts of the body including blood vessels.

4.1.3 HIF-1-Alpha

Hypoxia-inducible factor 1 is a heterodimeric transcription factor with a helix-loop-helix structure that consists of two important subunits, an alpha and a beta subunit. The HIF1A gene encodes the hypoxia-inducible factor 1-alpha or HIF-1-alpha subunit. HIF-1-alpha unit is a sensitive, oxygen dependent subunit, which is primarily responsible for the delivery of oxygen to the all the cells in the body through angiogenesis (Masoud and Li 2015). Angiogenesis is described as the development of new blood vessels from pre-existing blood vessels. Under normal conditions, HIF-1-alpha is hydroxylated at specific proline residue sites, where it interacts with the von Hippel-Lindau tumor suppressor protein (pVHL), leading to ubiquitination and degradation of proteasome (Imamura et al. 2009). These two processes allow for the proper formation and generation of proteins essential for the body's normal function.

However, under hypoxic conditions, HIF-1-alpha is unable to undergo hydroxylation at the proline residues because these sites are inhibited. And due to this, there is a cascade effect, in which the HIF-1-alpha is unable to interact with pVHL and therefore both ubiquitination and proteasomal degradation are unable to occur (Imamura et al. 2009; Masoud and Li 2015). Instead, under hypoxic conditions, the amount of HIF-1alpha increases and interacts with HIF-1-beta, specifically dimerization occurs between the two subunits leading to the binding to hypoxia response elements in promoter regions of DNA (Imamura et al. 2009; Masoud and Li 2015). Overall, this leads to the activation of specific genes that lead to the overall promotion of tumorigenesis.

4.2 Role of HIF-1-Alpha in Colorectal Cancer

HIF-1-alpha plays either an oncogenic or tumor-suppressive role in a variety of cancers. In colorectal cancer, HIF-1-alpha plays an unfavorable role and therefore functions as an oncogene, resulting in tumor angiogenesis, which aids in the development and advancement of colorectal cancer. HIF-1-alpha is triggered by both hypoxia independent or dependent factors (Baba et al. 2010). HIF-1-alpha is able to program itself to promote tumor angiogenesis through specific cellular mechanisms and pro-angiogenic factors, such as vascular endothelial growth factor or VEGF, which is a protein that induces the formation of blood vessels (Baba et al. 2010). Overall, HIF-1-alpha is able to activate genes that are designed for the growth and development of tumors. Over- expression of HIF-1-alpha in cancers is strongly correlated with patient mortality in a variety of cancers, including colorectal cancer. In one study, they observed the role of HIF-1-alpha in colon cancer cells and xenografts and noted that HIF-1-alpha promoted increased cellular proliferation and aerobic glycolysis (Baba et al. 2010; Dang et al. 2006). Additionally, the study also noted that HIF-1-alpha led to colon cancer cell invasion via regulation of proteins including MMP2, TFGA, and vimetin (Baba et al. 2010). Furthermore, other investigations have observed that HIF-1-alpha is able to up regulate prostaglandin-endoperoxide synthase 2 (PTGS2) or also known as cyclooxygenase-2 (COX-2) by binding to the hypoxia-responsive element (Baba et al. 2010; Kaidi et al. 2006). Up-regulation of PTGS2 promotes colorectal cancer cell survival and angiogenesis (Baba et al. 2010; Kaidi et al. 2006). Overall, the over-expression of HIF-1- alpha can result in the advancement of colorectal cancer by inhibiting interaction with pVHL, ultimately preventing ubiquitination and degradation of proteasome.

4.3 Current Investigations

Cellular or tissue hypoxia is a common phenomenon in tumorigenesis. Here, the affected cells and tissue are not receiving adequate levels of oxygen to properly function. Cells in a hypoxic environment not only have an inadequate exchange of both nutrients and oxygen and insufficient tissue perfusion, but cellular metabolism and gene expression profiles of the cell can be altered. Investigating the association between the expression of HIF-1-alpha and colorectal cancer can aid us in identifying the cancer earlier to lead to a strong prognosis and potential therapeutic options.

In one study, they investigated the correlation between expression of the key regulatory subunits, HIF-1-alpha and endothelial PAS domain protein 1, of the HIF-1, and colorectal cancer. This investigation studied 731 cohort cases of colorectal cancers in which the data yielded that 142 tumors or about 19% of these studies, over-expressed HIF-1-alpha and 322 of these studies showed an over-expression of PAS domain protein 1 (Baba et al. 2010). Upon further statistical analysis, including Kaplan-Meier analysis, univariate Cox regression, and multivariate analysis, the study concludes that HIF-1-alpha expression is strongly associated with the poor prognosis in colorectal cancer, suggesting that expression of HIF-1-alpha can be a prognostic measure (Baba et al. 2010). Additionally, even though, there was a great deal of cases that showed over-expression of PAS domain protein 1, further analysis revealed that it was not associated with high mortality.

In another study, the expression of HIF-1-alpha and VEGF was observed in colorectal cancer tissues, since these two factors contribute heavily towards angiogenesis. The data of this study yielded the following: 54.93% of the patients had expression of HIF-1-alpha and 56.34% had expression of VEGF (Cao et al. 2009). Additionally, there was a strong association between HIF-1α and VEGF and tumor stage, lymph nodes, and liver metastases (Cao et al. 2009). Further statistical analysis led to the conclusion that expression of both HIF-1α and VEGF are two key players in tumorigenesis of colorectal cancer, which suggests that these two factors can be used as prognostic markers for advanced stages of colorectal cancer patients.

Using HIF-1-alpha as a potential prognostic marker would not only allow physicians to identify tumors, but also serve as a target for therapeutic treatment. However, according to a study conducted with Iranian patients, HIF-1-alpha is only a strong prognostic marker in advanced stages of colorectal cancer patients (Mansour et al. 2016). In this study, they took 75 cancer specimens, which were divided into four groups along with a control group; the group number was correlated to the stage of cancer (Mansour et al. 2016). This study used immunohistochemistry and reverse transcription-polymerase chain reaction to evaluate genetic expression at both the RNA and protein level (Mansour et al. 2016). Even though all groups showed an association between HIF-1-alpha and colorectal cancer; however,

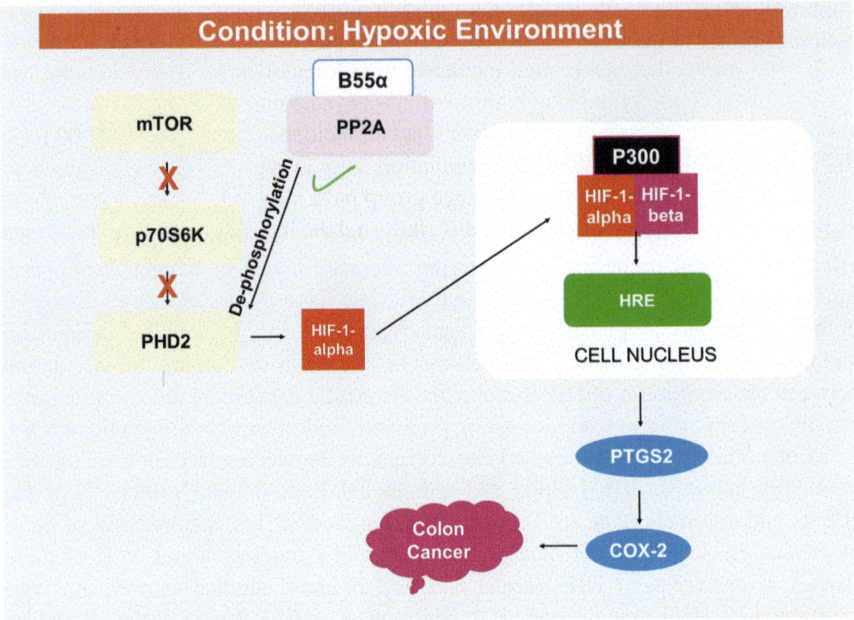

Fig. 4.1 During hypoxic conditions, the mTOR kinase is inhibited, which leads to the inactivation of p70S6K and results in the dephosphorylation of PHD2 by PP2A-B55α. The dephosphorylation of PHD2 by PP2A-B55α leads to the activation of HIF-1α and ultimately promotes cancer cell survival and tumor growth

the only group that reported an association that was statistically significant was group 4 or patients with stage 4 cancer (Mansour et al. 2016). The study strongly suggests that HIF-1-alpha gene expression is significant only in the late stages of colorectal cancer.

Lastly, in another study, investigators observed the cellular pathway involving the mammalian target of rapamycin or mTOR kinase to see what specific cellular factors in this pathway affect colorectal cancer cell survival. mTOR is a kinase that is primarily involved in the regulation of protein synthesis and cellular response to growth factors (Ioannou et al. 2015). During hypoxic conditions, mTOR is inhibited and ultimately allows for the activation of PP2A/ B55α. In this study, the mass spectrometry analysis revealed sites of phosphorylation; specifically, they concluded that PHD2 at Ser125 was phosphorylated by P70S6K (Ioannou et al. 2015). Additionally, they noted that dephosphorylation also occurred at PHD2 by PP2A/B55α (Ioannou et al. 2015). Overall, the inhibition of the mTOR pathway, allows P70S6K to dephosphorylate PHD2, which increases HIF-1α, which ultimately promotes tumor growth and angiogenesis (Fig. 4.1).

4.4 Conclusion

Multiple cellular mechanisms, factors, and proteins play a strong role in the development and advancement of cancers. The HIF-1-alpha under hypoxic conditions, for example, is one such cellular pathway that plays a key role in the development of colorectal cancer. Even though the pathophysiology and exact cause of colorectal cancer is not entirely understood, studying the various mechanisms that lead to tumorigenesis can allow for the generation of better therapeutic treatment options. Specifically, in the case of HIF-1-alpha, we can use HIF-1-alpha as a prognostic marker or biomarker. Additionally, we can learn how to manipulate mechanisms associated with HIF-1-alpha to prevent the dimerization of HIF-1-alpha and HIF-1-alpha. Further exploration and research in HIF-1-alpha is necessary to holistically understand the role of HIF-1-alpha in colorectal cancer. Potential future studies include but are not limited to: inhibition of the HIF-1-alpha pathway under hypoxic conditions and activation of the hydroxylation under hypoxic conditions.

Once we understand the role of HIF-1-alpha in detail, we can the learn how to manipulate and alter the pathway under hypoxic conditions, in order to promote activation of hydroxylation even under hypoxic conditions. This will prevent the development of tumors and angiogenesis. And in the future, the such changes can be implement and apply to live subjects to ultimately eliminate colorectal cancer. Understanding HIF-1-alpha's role and function in colorectal cancer and in other types of cancers can not only allow us to identify the cancer, but also allow us to find and generate treatments that target HIF-1-alpha to completely prevent the advancement of tumors.

References

Augenlicht LH (2019) Cell and molecular biology of colon cancer. CRC Press, Boca Raton

Baba Y, Nosho K, Shima K, Irahara N, Chan AT, Meyerhardt JA, Chung DC, Giovannucci EL, Fuchs CS, Ogino S (2010) HIF1A overexpression is associated with poor prognosis in a cohort of 731 colorectal cancers. Am J Pathol 176:2292–2301

Bray F, Ferlay J, Soerjomataram I, Siegel RL, Torre LA, Jemal A (2018) Global cancer statistics 2018: GLOBOCAN estimates of incidence and mortality worldwide for 36 cancers in 185 countries. CA Cancer J Clin 68:394–424

Cao D, Hou M, Guan Y-s, Jiang M, Yang Y, Gou H-f (2009) Expression of HIF-1alpha and VEGF in colorectal cancer: association with clinical outcomes and prognostic implications. BMC Cancer 9:432

Dang DT, Chen F, Gardner LB, Cummins JM, Rago C, Bunz F, Kantsevoy SV, Dang LH (2006) Hypoxia-inducible factor-1α promotes nonhypoxia-mediated proliferation in colon cancer cells and xenografts. Cancer Res 66:1684–1693

De Leon MP, Di Gregorio C (2001) Pathology of colorectal cancer. Dig Liver Dis 33:372–388

Haggar FA, Boushey RP (2009) Colorectal cancer epidemiology: incidence, mortality, survival, and risk factors. Clin Colon Rectal Surg 22:191–197

Imamura T, Kikuchi H, Herraiz MT, Park DY, Mizukami Y, Mino-Kenduson M, Lynch MP, Rueda BR, Benita Y, Xavier RJ (2009) HIF-1α and HIF-2α have divergent roles in colon cancer. Int J Cancer 124:763–771

Ioannou M, Paraskeva E, Baxevanidou K, Simos G, Papamichali R, Papacharalambous C, Samara M, Koukoulis G (2015) HIF-1α in colorectal carcinoma: review of the literature. J BUON 20:680–689

Kaidi A, Qualtrough D, Williams AC, Paraskeva C (2006) Direct transcriptional up-regulation of cyclooxygenase-2 by hypoxia-inducible factor (HIF)-1 promotes colorectal tumor cell survival and enhances HIF-1 transcriptional activity during hypoxia. Cancer Res 66:6683–6691

Mansour RN, Enderami SE, Ardeshirylajimi A, Fooladsaz K, Fathi M, Ganji SM (2016) Evaluation of hypoxia inducible factor-1 alpha gene expression in colorectal cancer stages of Iranian patients. J Cancer Res Ther 12:1313

Masoud GN, Li W (2015) HIF-1α pathway: role, regulation and intervention for cancer therapy. Acta Pharm Sin B 5:378–389

Rawla P, Sunkara T, Barsouk A (2019) Epidemiology of colorectal cancer: incidence, mortality, survival, and risk factors. Prz Gastroenterol 14(2):89–103

Chapter 5
Mechanisms and Pathways of Metabolic Reprogramming of Colorectal Cancer

A. Krishna Chaitanya, Seema Kumari, and Rama Rao Malla

Abstract Colorectal cancer is commonly developed solid cancers globally. Metabolic reprogramming is the most vital features of metastatic cancer cells which show high glycolysis due to Warburg effect. Among the various signaling pathway Wnt pathways plays and critical role in metastasis. Lipid, glucose and tryptophan metabolic remodeling contributes in severity of CRC. We have discussed about metabolic pathways and metabolic reprograming of CRC.

Keywords Colorectal cancer · Glycolysis · Metabolic reprogramming · Tryptophan · Wnt pathways

Abbreviations

APC	Adenomatous polyposis coli
ARH	Aromatic hydrocarbon receptor
CA	Carbohydrate antigen
CEA	Cancer antigen
CIMP	CpG Island Methylator Phenotype instability
CIN	Chromosomal instability
CRC	Colorectal cancer
EMT	Epithelial to mesenchymal transition

A. Krishna Chaitanya and Seema Kumari contributed equally with all other contributors.

A. Krishna Chaitanya
Department of Biochemistry and Bioinformatics, GIS, GITAM (Deemed to be University), Visakhapatnam, Andhra Pradesh, India

S. Kumari · R. R. Malla (✉)
Cancer Biology Lab, Department of Biochemistry and Bioinformatics, GIS, GITAM (Deemed to be University), Visakhapatnam, Andhra Pradesh, India

Department of Biochemistry and Bioinformatics, GIS, GITAM (Deemed to be University), Visakhapatnam, Andhra Pradesh, India

G. S. R. Raju, L.V.K.S. Bhaskar (eds.), *Theranostics Approaches to Gastric and Colon Cancer*, Diagnostics and Therapeutic Advances in GI Malignancies, https://doi.org/10.1007/978-981-15-2017-4_5

FA	Fatty acid
HDL	High density lipoprotein
IDO1	Indoleamine 2, 3-dioxygenase 1
KRAS	Kirsten rat sarcoma viral oncogene homolog
LDL	Low density lipoprotein
MLH1	MutL homolog 1
MSI	Microsatellite instability
PC	Phosphatidylcholine
PDK1	Pyruvate dehydrogenase kinase 1
TC	Total cholesterol
TGF-β	Transforming growth factor beta
TGs	Triacylglycerols
Wnt	Wingless-related integration site
5HT	5-hydroxytryptamine
KYNA	Kynurenine
COX	Cyclooxygenase
NSAID	Nonsteroidal antiinflammatory drugs

5.1 Highlights

- Wnt pathway in colorectal cancer
- Metabolism alteration in colorectal cancer
- Metabolic toxicity in colon cancer

5.2 Colorectal Cancer (CRC)

CRC is most commonly developing solid cancers globally. Inspite of progress made in diagnosis and treatment in surgical and oncological management, death due to CRCs is about 600,000 per year account for 50% of gastrointestinal cancer. The carcinogenesis of CRCs is associated with adenoma-carcinoma, alternative and serrated pathway. The 'traditional' pathway is associated with mutation of adenomatous polyposis coli (APC), heterozygosity loss as well as chromosomal variability. The 'alternative' pathway involves both KRAS and APC mutations with highly heterogeneous, not well characterized, than traditional and serrated pathways. The 'serrated' pathway progresses from CpG methylation, specifically BRAF mutations and late development of MSI. Colorectal adenomas show lesion and link between the normal mucosa and the cancer cells. They signify critical land scope for CRC and serve as risk factors as well as biomarkers.

5.3 CRC and Genetic Instability

CRCs can be induced by 3 kinds of genetic instabilities i.e., a) Chromosomal instability (CIN), b) Microsatellite instability (MSI) and CpG Island Methylator Phenotype instability (CIMP).

a) Chromosomal Instability (CIN)
In irregular CRCs, chromosomal variability includes insertion, inversion, deletion and rearrangement of the chromosome called as CINs. This results in different gene appearance, either by insertion or deletion, via structural change by rearrangements that results certainty of gene controlled by different promoter. The most commonly used methods for detection of the copy number of tumor are insitu hybridization, flowcytometry, and genome hybridization.

b) Microsatellite Instability (MSI)
The hereditary mutations in mismatch repair (MMR) genes causes microsatellite instability (MSI) or Lynch syndrome. However, methylation of CpG islands is common consequences from epigenetic mutations in MLH1 of MSI. Frequent mutations of CpG initiated in tumors with BRAF mutations, and tumors with numerous genes are differentially expressed.

c) CpG Island Methylator Phenotype (CIMP)
CIMP is another type of genomic instability determines of CRC, related to methylation status of promoters of genes.

5.4 Metabolic Reprogramming in Cancer

In addition to genetic factors, nutrient and environmental factors also disturb the signaling of cancer cells. Mutation in important oncogenes changes the cellular functions to establish additional environment for survival of cancer cells. It provides by high glucose, maintain high substrate levels for the metabolism of one-carbon metabolite and also facilitates excessive transport of acetyl-CoA. This causes abnormal epigenetic mutations in chromatin and DNA leading to cancer development (American Cancer Society 2017).

5.5 Metabolic Remodeling and Wnt Signaling in Colon Cancer

Metastatic cancers show metabolic reprogramming for high glycolysis called Warburg effect. This promotes high intake of glucose and glutamine for building of nucleotide and amino acid in growing cells (Deye et al. 2016; Yin et al. 2015). In addition to genetic and epigenetic variations, reprogrammed tumor microenvironment, redox state, and inflammation have deep impact on cancer cell metabolism. Considering this fact, metabolic reprogramming could be a novel approaches for cancer therapy (Satoh et al. 2017).

Wnt signaling contributes in tissue growth and conservation, nonetheless abnormal Wnt signaling is concerned with development of CRC. Overexpressed Wnt signaling causes constitutive activation of β-catenin, which in turn activates Wnt target genes through lymphoid relish factor/T-cell factor (Zheng et al. 2018). Wnt signaling contributes to induction of EMT, neovascularization and metastasis. Glycolysis in cancer cells is driven by hypoxia as well as PI3K/AKT pathway and crosstalk with metabolic pathways. Limited data is available about the direct association of Wnt with metabolic reprogramming (Pate et al. 2014).

Abnormal Wnt/β-catenin pathway contribute a significantly in progression of CRC with over 91% of mutations in APC or β-catenin. Colorectal tumors prompt high Wnt3 and Evi/Wls/GPR177 expression and binding of APC with β-catenin, which signifies that mutant APC or β-catenin be contingent on Wnt ligands (Bakopoulou et al. 2015; Liu et al. 2009). Mechanisms through which Wnt signaling stimulates proliferation of metastasis are promising in controlling cell cycle of CRCs. Pyruvate dehydrogenase kinase 1 (PDK1) is specific target for PDK1-mediated pyruvate flux to mitochondrial respiration and Wnt-inhibited glycolysis and angiogenesis in the tumor (Hunter et al. 2015).

5.6 Detection/Diagnosis

The traditional "adenomas-carcinoma" dogma defines CRC as a complex process. Mutation related to CRC are APC, TP53, KRAS, which are highly intricate with Wnt, TGF-b, and p53 pathways. APC mutations caused CRC in more than 51% patients. At present CEA199, CEA, CA742, CEA242, and CA125 are in use for clinical diagnosis (Zhang et al. 2016).

5.7 Lipid Metabolism

Hallmark of CRC is altered lipid metabolism. Lipids are a key group of metabolites with variance composition of fatty acid. Lipidomic investigations in thoughtful, diagnosis and treatment of cancer generate an chance to innovate targeted therapies, prognostic or biomarkers for screening (Zajac et al. 2014). Lipid levels are estimated by lipid profiling of serum (Simi et al. 2017). Mass spectrometry used to study thorough perception lipid molecules. Lipidomic, delivers the lipid dysregulation in cancer. A vast majority of studies were conducted to relate lipids and cancer. Considering the link between the cancer and lipidome gives an insight on delivery methods, pathogenesis and identification of novel biomarkers (Jansson and Lindberg 2012).

Lipid profiling in CRC may provide an improved understanding the aberrant metabolism in cancer and pave new therapeutic perception. Major pathways associated with biogenesis of FAs are promising targets for prevention of tumor growth. There are some FAs which noteworthy involved in proinflammation and anti-inflammation. Perhaps, proinflammatory molecules are key associators of besity and CRC (Brown et al. 2000). Furthermore, obesity is related to lipidome variations could be predisposed to cancer (Ballor et al. 1996). Deviations of FAs, may cause change in tumor microenvironment, which could be well-thought-out biomarkers of CRC (Helms et al. 2014). Yet, clinically valuable lipid biomarkers needs reliable methodology. Additional limitation of earlier studies is incomplete sampling which may delay simplification of data onto the whole population of CRC patient (Agudelo et al. 2018). Yet, considering of lipid changes related with CRC might describe new approach in the diagnosis and treatment of CRC.

5.8 Tryptophan Metabolism

Tryptophan is an essential amino acid which regulates defence system, CNS, GIT, nervous system and intestinal microflora. High-grade malignancy disease like CRC, associated with diverse factors and most common is inflammation, dysbacteriosis, and metabolic aberrancy which supports inflammation due to interrupts tryptophan metabolism (Lamas et al. 2016). Precisely, IDO1 can be a possible target of CRC (Ridler 2016). Moreover, the decrease of tryptophan is related to high mortality.

5-HT and KYNA are derivatives of tryptophan, which plays a vital role in controlling functions of immune system (Marsland 2016). During the progress of CRC, inflammatory reactions, change in intestinal microenvironment due to damaged intestinal barrier and its absorption increases, leading to an outflow of endotoxin. Arrangement of the intestinal microflora fluctuates the proportion of Enterococci and Clostridium. TRP metabolites inhibit inflammation and also repair the intestinal lining a links with beneficial microbes in the intestinal tract, which can inhibit onset of CRC (Lamas et al. 2017).

Cytotoxic therapies basis of important side effects for CRC patients with no cure in advanced stage. Immunotherapies are a new approach to connect the immune system and inflammatory response (Lamas et al. 2016; Kumari et al. 2015). Kynurenine pathway is a vital target for immune-based cancer therapy. Indoleamine 2,3 dioxygenase 1 (IDO1) is the widely explored ezymes which is overexpressed in CRC. It is important factors in CRC pathogenesis, transcriptional pathways of neoplastic development and luminal microbiota (Kumari and Malla 2015). TRP metabolism has a role in preventing the growth of CRC, and revealing the association between the TRP metabolism and CRC for further studies.

5.9 Glucose Metabolism

Diabetes mellitus and CRC are key reasons death globally. There is a statistically important link between diabetes and CRC. Insulin is pro-proliferative, it inhibits apoptosis and enhance growth of CRC. Hyper-insulinaemia is recognized as risk assessment-insulin resistance (HOMA-IR) index using homeostasis model. Increased glucose, HOMA-IR, HbA1c and C peptide are amplified in CRC. It is well known that HOMA-IR is précised indicator for CRC. Furthermore, HbA1c and C peptide, are usually late markers of CRC. Additionally, insulin and IGF-1 which activate Ras proteins, mTOR and Wnt signalling pathway, causes increased sensitivity of CRC (Gagnière et al. 2016).

5.10 Exercise Metabolism

Exercise has a major impact on intratumoral metabolism of TCA cycle and electron transport chain. Many of the metabolites are differentially changed which are intricate in pathways that control and balance redox of cysteine metabolism, glycolysis, and β-oxidation of fatty acids (Klement and Kämmerer 2013).

5.11 Therapies Based on Pathways

Targeted therapy is a prevailing strategy based on patient molecular profile. The survival rate has been improved by Anti-EGFR-targeted. Growth Factor Receptor (GFR) signaling pathway typically involves dimerization of the EGF receptors by binding of growth factors (ligands), further activating MAPK and the PI3K-AKT which play pivotal role in regulation of genes, apoptosis and proliferation. EGF receptors are highly expressed in epithelial tumors, and can be inhibited by cetuximab. But in some cases, mutations in either KRAS or BRAF can cause the pathway to be active even in the presence of EGFR inhibitors.

COX-2 has a role in prostaglandin synthesis during inflammation which also has a role in CRC, particularly prostaglandin E2 (PGE2). Various other factors like including mitogens, promoters of tumor and chemical messengers are associated with transcription of COX-2 (Saydah et al. 2003). Warburg effect, provides immense benefits to cancer cells by enhancing energetics along with and biosynthesis (Fig. 5.1) (Rampal et al. 2014). Overexpressed glucose transporter (Glut), proto-oncogenes and p53 affects the metabolism, along with mutations increases a metabolic phenotype, which provides supports cancer progression (Gnagnarella et al. 2008). The oncogenic pathways like PI3K, HIFs, p53 and v-myc are significant for metabolism and metastasis (Kumari et al. 2017). HIF-1/2 are overexpressed in CRC (Seema Kumari et al. 2016). GRP78 heat shock protein, ER chaperone, are also

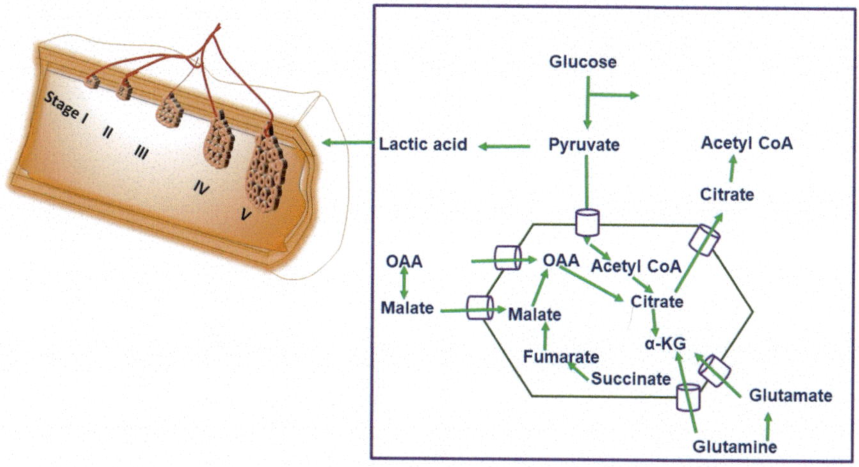

Fig. 5.1 In colorectal cancer cells, glucose consumption through glycolysis, involves break down into pyruvate, which enters into the tricarboxylic acid (TCA) cycle for maximum energy production. At a higher rate produce lactic acid by the adaptive metabolic shift, Warburg effect. This metabolic shift facilitates the uptake of glutamine to build essential metabolites required for proliferating cells

controlled by glucose metabolism. Proteomics and immunohistochemical analysis have revealed that GRP78 participates in the pathogenesis of CRC (Koteswari et al. 2018). Targeting GRP78 inhibit CRC through HIF-1A and VEGFR2 pathway (Malla et al. 2019). YAP, a downstream mediator of Hippo pathway has role in CRC (Badana et al. 2016). The Hippo pathway and AMP-activated protein kinase (AMPK) activation results in inactivation of YAP by phosphorylation and promotes cancer growth (Kumari et al. 2013). miRNA has an association with glucose metabolism.

5.12 Metabolic Toxicity in Colon Cancer

Recent developments in metabolo-genomics have given a better understanding the association of gut microbiota with various diseases including cancer. Gut microbiota has potential influence on host immune system as well as health condition (Bevara et al. 2017). Dysbiosis primes enhanced bacterial populations, which stimulate tumor formation and also contributes to epithelial carcinogenesis by altering metabolic properties such as bile acid and butyric acid and inflammatory process (Sears and Garrett 2014). The colonic microbiota such as enterotoxigenic Bacteroides fragilis (ETBF) may also promote CRC by promoting ng exaggerated immune responses via Th17 cells (Singh et al. 2014). In addition to carcinogenesis, gut microbiome also affect the toxicity of cancer immunotherapy (Borges-Canha et al. 2015). Some important compounds which have gained interest are azo-aromatic compounds, berberine, combinatorial libraries of mixed-ligand transition metal complexes of 2-butanone thiosemicarbazone (Gagnière et al. 2016; Irrazábal et al. 2014).

5.13 Polyphenols in Colon Cancer

Polyphenols are promising natural compounds with anti-cancer property and efficient. Most common natural compounds like flavonoids reported to have anticancer property (O'Keefe 2016). Flavonoids are derivative from medicinal plants including *Petroselinum crispum, Apium graveolens, Flemingia vestita, Phyllanthus emblica,* etc. (Belcheva et al. 2014). Natural flavonoids possess antioxidant, anti-cancerous actions on various pathways that induces cell cycle arrest and apoptosis as well as lowers the activity of nucleoside diphosphate kinase-B and suppress NF-kB pathway in diverse cancers (Turner et al. 2013).

5.14 Gut Metabolism in Colon Cancer

The occurrence of CRCs neoplastic serrated polyps is 14–29%. Serrated polyps display genetic (*BRAF* or *KRAS* mutations) as well as epigenetic (CpG island methylator phenotype (CIMP)) changes, which support the initiation and transformation of normal mucosa to polyps (Rowland 2009). Colon epithelium is exposed to elevated H_2S from gut microbes. H_2S is toxic to complex IV in the ETC (Gagnière et al. 2016). Sulfide quinone oxidoreductase (SQR), which catalyzes oxidation of sulfides and coupled to complex III, is poisoned by H_2S. The H_2S-induced stress response are mapped using redox-sensitive centers of carbon metabolism.

5.15 Metabolic Targets of Colon Cancer

The metabolic difference induces GLUT-1 and -3, and glutamine metabolism causing cancer cell growth. Glutor targets GLUT-1, -2, and -3, and inhibits glycolytic flux, which hinder cancer growth. The combination of Glutor and CB-839 showed synergistic of CRC growth. Such strategy is promising since it affects both metabolic plasticity and metabolic rescue reactions in CRC (Ou et al. 2013). Similarly, folate-dependent one-carbon (C1) metabolism is compartmentalized into the mitochondria and cytosol and supports cell progress through nucleotide and amino acid biosynthesis. Mitochondrial C1 metabolism, including SHMT 2, glycine, NAD(P)H, ATP, and C1 units for cytosolic biosynthetic reactions, and is concerned in the oncogenic phenotype across a wide range of cancers. Whereas multi-targeted inhibitors of cytosolic C1 metabolism such as pemetrexed are used and no anticancer drugs that specifically target mitochondrial C1 metabolism is available these are beneficial metabolic inhibitors.

5.16 Future Direction

Noteworthy advancement in understanding metabolism and signaling pathway associated with colorectal cancer gives a critical review on tumorigenesis. Further studies and developments may be required to drive future discovery and enhance therapeutic development in cancer research.

5.17 Conclusion

CRC common cancers globally, metabolic reprogramming is the most vital features of metastatic cancer cells which maintain extremely high glycolysis for the supply of energy regardless of oxygen availability, which is called Warburg effect. Among the various signaling pathway Wnt pathways plays and critical role in metastasis. Lipid, glucose and tryptophan metabolic remodeling contributes in severity of CRC. Understanding the altered signaling pathways associated to CRC could define new avenues for the developments of theranostics.

Acknowledgement The authors are thankful to GITAM (Deemed to be University) for providing necessary support.

References

Agudelo LZ, Ferreira DMS, Cervenka I, Bryzgalova G, Dadvar S, Jannig PR, Pettersson-Klein AT, Lakshmikanth T, Sustarsic EG, Porsmyr-Palmertz M, Correia JC, Izadi M, Martínez-Redondo V, Ueland PM, Midttun Ø, Gerhart-Hines Z, Brodin P, Pereira T, Berggren PO, Ruas JL (2018) Kynurenic acid and Gpr35 regulate adipose tissue energy homeostasis and inflammation. Cell Metab. https://doi.org/10.1016/j.cmet.2018.01.004

American Cancer Society (2017) Cancer facts and figures 2017. Genes Dev. https://doi.org/10.1101/gad.1593107

Badana A, Chintala M, Varikuti G, Pudi N, Kumari S, Kappala VR, Malla RR (2016) A lipid raft integrity is required for survival of triple negative breast cancer cells. J Breast Cancer 19:372–384

Bakopoulou A, Leyhausen G, Volk J, Papachristou E, Koidis P, Geurtsen W (2015) Wnt/β-catenin signaling regulates dental pulp stem cells' responses to pulp injury by resinous monomers. Dent Mater. https://doi.org/10.1016/j.dental.2015.02.004

Ballor DL, Harvey-Berino JR, Ades PA, Cryan J, Calles-Escandon J (1996) Contrasting effects of resistance and aerobic training on body composition and metabolism after diet-induced weight loss. Metabolism. https://doi.org/10.1016/S0026-0495(96)90050-5

Belcheva A, Irrazabal T, Robertson SJ, Streutker C, Maughan H, Rubino S, Moriyama EH, Copeland JK, Kumar S, Green B, Geddes K, Pezo RC, Navarre WW, Milosevic M, Wilson BC, Girardin SE, Wolever TMS, Edelmann W, Guttman DS, Philpott DJ, Martin A (2014) Gut microbial metabolism drives transformation of msh2-deficient colon epithelial cells. Cell. https://doi.org/10.1016/j.cell.2014.04.051

Bevara GB, Kumar ADN, Koteswramma K, Badana AK, Kumari S, Yarla NS, Malla RR (2017) C-glycosyl flavone from Urginea indica inhibits growth and dissemination of Ehrlich ascites carcinoma cells in mice. Anti Cancer Agents Med Chem 17. https://doi.org/10.2174/1871520617666170103101844

Borges-Canha M, Portela-Cidade JP, Dinis-Ribeiro M, Leite-Moreira AF, Pimentel-Nunes P (2015) Role of colonic microbiota in colorectal carcinogenesis: a systematic review. Rev Esp Enferm Dig 107(10):659–671

Brown RC, Cox CM, Goulding A (2000) High-carbohydrate versus high-fat diets: effect on body composition in trained cyclists. Med Sci Sports Exerc. https://doi.org/10.1097/00005768-200003000-00021

Deye N, Vincent F, Michel P, Ehrmann S, Da Silva D, Piagnerelli M et al (2016) Changes in cardiac arrest patientsâ€TM temperature management after the 2013 trial: results from an international survey. Annals of intensive, 61–66. Med Hypotheses 97:411–415. https://doi.org/10.1016/j.clae.2016.08.001

Gagnière J, Raisch J, Veziant J, Barnich N, Bonnet R, Buc E, Bringer MA, Pezet D, Bonnet M (2016) Gut microbiota imbalance and colorectal cancer. World J Gastroenterol. https://doi.org/10.3748/wjg.v22.i2.501

Gnagnarella P, Gandini S, La Vecchia C, Maisonneuve P (2008) Glycemic index, glycemic load, and cancer risk: a meta-analysis. Am J Clin Nutr 87(6):1793–1801

Helms ER, Zinn C, Rowlands DS, Brown SR (2014) A systematic review of dietary protein during caloric restriction in resistance trained lean athletes: a case for higher intakes. Int J Sport Nutr Exerc Metab. https://doi.org/10.1123/ijsnem.2013-0054

Hunter DJ, Bardet C, Mouraret S, Liu B, Singh G, Sadoine J, Dhamdhere G, Smith A, Tran XV, Joy A, Rooker S, Suzuki S, Vuorinen A, Miettinen S, Chaussain C, Helms JA (2015) Wnt acts as a Prosurvival signal to enhance dentin regeneration. J Bone Miner Res. https://doi.org/10.1002/jbmr.2444

Irrazábal T, Belcheva A, Girardin SE, Martin A, Philpott DJ (2014) The multifaceted role of the intestinal microbiota in colon cancer. Mol Cell. https://doi.org/10.1016/j.molcel.2014.03.039

Jansson A, Lindberg JE (2012) A forage-only diet alters the metabolic response of horses in training. Animal. https://doi.org/10.1017/S1751731112000948

Klement RJ, Kämmerer U (2013) Is there a role for carbohydrate restriction in the treatment and prevention of cancer? Clin Nutr Interface Between Metab Diet Dis. https://doi.org/10.1201/b16308

Koteswari LL, Kumari S, Kumar AB, Malla RR (2018) A comparative anticancer study on procyanidin C1 against receptor positive and receptor negative breast cancer. Nat Prod Res. https://doi.org/10.1080/14786419.2018.1557173

Kumari S, Malla R (2015) New insight on the role of plasminogen receptor in Cancer progression. Cancer Growth Metastasis 88:35–42. https://doi.org/10.4137/CGM.s27335

Kumari S, Kaladhar DSVGK, Sandeep Solmon K, Malla RR, Kishore G (2013) Anti-proliferative and metastatic protease inhibitory activities of protoberberines: an in silico and in vitro approaches. Process Biochem 48:1565–1571. https://doi.org/10.1016/j.procbio.2013.06.027

Kumari S, Gayatri DV, Badana A, Dasari VR, Rao Malla R (2015) CD151—a striking marker for Cancer therapy, biomark. Cancer:7–7. https://doi.org/10.4137/BiC.s21847

Kumari S, Badana AK, Mohan GM, Shailender Naik G, Malla RR (2017) Synergistic effects of coralyne and paclitaxel on cell migration and proliferation of breast cancer cells lines. Biomed Pharmacother. https://doi.org/10.1016/j.biopha.2017.04.027

Lamas B, Richard ML, Leducq V, Pham HP, Michel ML, Da Costa G, Bridonneau C, Jegou S, Hoffmann TW, Natividad JM, Brot L, Taleb S, Couturier-Maillard A, Nion-Larmurier I,

Merabtene F, Seksik P, Bourrier A, Cosnes J, Ryffel B, Beaugerie L, Launay JM, Langella P, Xavier RJ, Sokol H (2016) CARD9 impacts colitis by altering gut microbiota metabolism of tryptophan into aryl hydrocarbon receptor ligands. Nat Med. https://doi.org/10.1038/nm.4102

Lamas B, Richard ML, Sokol H (2017) Caspase recruitment domain 9, microbiota, and tryptophan metabolism: dangerous liaisons in inflammatory bowel diseases. Curr Opin Clin Nutr Metab Care. https://doi.org/10.1097/MCO.0000000000000382

Liu L, Ling J, Wei X, Wu L, Xiao Y (2009) Stem cell regulatory gene expression in human adult dental pulp and periodontal ligament cells undergoing Odontogenic/Osteogenic differentiation. J Endod. https://doi.org/10.1016/j.joen.2009.07.005

Malla RR, Kumari S, Gavara MM, Badana AK, Gugalavath S, Kumar DKG, Rokkam P (2019) A perspective on the diagnostics, prognostics, and therapeutics of microRNAs of triple-negative breast cancer. Biophys Rev. https://doi.org/10.1007/s12551-019-00503-8

Marsland BJ (2016) Regulating inflammation with microbial metabolites. Nat Med. https://doi.org/10.1038/nm.4117

O'Keefe SJD (2016) Diet, microorganisms and their metabolites, and colon cancer. Nat Rev Gastroenterol Hepatol. https://doi.org/10.1038/nrgastro.2016.165

Ou J, Carbonero F, Zoetendal EG, DeLany JP, Wang M, Newton K, Gaskins HR, O'Keefe SJD (2013) Diet, microbiota, and microbial metabolites in colon cancer risk in rural Africans and African Americans. Am J Clin Nutr. https://doi.org/10.3945/ajcn.112.056689

Pate KT, Stringari C, Sprowl-Tanio S, Wang K, TeSlaa T, Hoverter NP, McQuade MM, Garner C, Digman MA, Teitell MA, Edwards RA, Gratton E, Waterman ML (2014) Wnt signaling directs a metabolic program of glycolysis and angiogenesis in colon cancer. EMBO J. https://doi.org/10.15252/embj.201488598

Rampal S, Yang MH, Sung J, Son HJ, Choi YH, Lee JH, Kim YH, Chang DK, Rhee PL, Rhee JC, Guallar E, Cho J (2014) Association between markers of glucose metabolism and risk of colorectal adenoma. Gastroenterology. https://doi.org/10.1053/j.gastro.2014.03.006

Ridler C (2016) IBD: Dysbiosis underlies CARD9 risk alleles in colitis. Nat Rev Gastroenterol Hepatol. https://doi.org/10.1038/nrgastro.2016.82

Rowland I (2009) The role of the gastrointestinal microbiota in colorectal Cancer. Curr Pharm Des. https://doi.org/10.2174/138161209788168191

Satoh K, Yachida S, Sugimoto M, Oshima M, Nakagawa T, Akamoto S, Tabata S, Saitoh K, Kato K, Sato S, Igarashi K, Aizawa Y, Kajino-Sakamoto R, Kojima Y, Fujishita T, Enomoto A, Hirayama A, Ishikawa T, Taketo MM, Kushida Y, Haba R, Okano K, Tomita M, Suzuki Y, Fukuda S, Aoki M, Soga T (2017) Global metabolic reprogramming of colorectal cancer occurs at adenoma stage and is induced by MYC. Proc Natl Acad Sci. https://doi.org/10.1073/pnas.1710366114

Saydah SH, Platz EA, Rifai N, Pollak MN, Brancati FL, Helzlsouer KJ (2003) Association of markers of insulin and glucose control with subsequent colorectal cancer risk. Cancer Epidemiol Biomark Prev 12(5):412–418

Sears CL, Garrett WS (2014) Microbes, microbiota, and colon cancer. Cell Host Microbe. https://doi.org/10.1016/j.chom.2014.02.007

Seema Kumari PN, Anil Badana V, Gayatridevi MV, Malla RR (2016) Coralyne targets proteases involved in cancer progression: an in silico study. In: Computational intelligence techniques in health care. Springer, Singapore, pp 19–30

Simi B, Sempore B, Mayet MH, Favier RJ (2017) Additive effects of training and high-fat diet on energy metabolism during exercise. J Appl Physiol. https://doi.org/10.1152/jappl.1991.71.1.197

Singh N, Gurav A, Sivaprakasam S, Brady E, Padia R, Shi H, Thangaraju M, Prasad PD, Manicassamy S, Munn DH, Lee JR, Offermanns S, Ganapathy V (2014) Activation of Gpr109a, receptor for niacin and the commensal metabolite butyrate, suppresses colonic inflammation and carcinogenesis. Immunity. https://doi.org/10.1016/j.immuni.2013.12.007

Turner ND, Ritchie LE, Bresalier RS, Chapkin RS (2013) The microbiome and colorectal neoplasia: environmental modifiers of dysbiosis. Curr Gastroenterol Rep. https://doi.org/10.1007/s11894-013-0346-0

Yin X, Li J, Salmon B, Huang L, Lim WH, Liu B, Hunter DJ, Ransom RC, Singh G, Gillette M, Zou S, Helms JA (2015) Wnt signaling and its contribution to craniofacial tissue homeostasis. J Dent Res. https://doi.org/10.1177/0022034515599772

Zajac A, Poprzecki S, Maszczyk A, Czuba M, Michalczyk M, Zydek G (2014) The effects of a ketogenic diet on exercise metabolism and physical performance in off-road cyclists. Nutrients. https://doi.org/10.3390/nu6072493

Zhang J, Lu X, Feng G, Gu Z, Sun Y, Bao G, Xu G, Lu Y, Chen J, Xu L, Feng X, Cui Z (2016) Chitosan scaffolds induce human dental pulp stem cells to neural differentiation: potential roles for spinal cord injury therapy. Cell Tissue Res. https://doi.org/10.1007/s00441-016-2402-1

Zheng W, Li X, Liu D, Li J, Yang S, Gao Z, Wang Z, Yokota H, Zhang P (2018) Mechanical loading mitigates osteoarthritis symptoms by regulating endoplasmic reticulum stress and autophagy. FASEB J. https://doi.org/10.1096/fj.201801851r

Chapter 6
Targeting Metabolic Reprogramming of Colorectal Cancer

Seema Kumari and Rama Rao Malla

Abstract Cancer is characterized by an altered metabolism which facilitates their survival and proliferation. The mechanism of altered metabolism is called as metabolic reprogramming. We have emphasized on different aspects of metabolic reprogramming associated with colorectal cancer (CRC) and discussed about different targets which can be proven as novel therapeutic strategy against CRC. We have highlighted on the role of mitochondria and ROS in cancer progression and use of cancer stem cells, miRNA, autophagy as advanced strategy against cancer growth.

Keywords Metabolic reprogramming · Cancer stem cells · Autophagy · miRNA

Abbreviations

AMPK	5′ AMP-activated protein kinase
CAD	Carbamoyl-phosphate synthetase 2
CamKKB	Calcium/Calmodulin-Dependent Protein Kinase Kinase 2
CRC	Colorectal cancer
CSCs	Cancer stem cells
CTPS	Cytidine triphosphate synthetase
DARTS	Drug Affinity Responsive Target Stability
FAS	Fatty acid synthase
GLUT3	Glucose transporter 3
GPT2	Glutamate pyruvate transaminase 2
LDHA	Lactate dehydrogenase A
mROS	Mitochondria-derived reactive oxygen species
mTOR	Mammalian target of rapamycin
NADPH	Nicotinamide adenine dinucleotide phosphate

S. Kumari · R. R. Malla (✉)
Cancer Biology Lab, Department of Biochemistry and Bioinformatics, GIS, GITAM (Deemed to be University), Visakhapatnam, Andhra Pradesh, India

G. S. R. Raju, L.V.K.S. Bhaskar (eds.), *Theranostics Approaches to Gastric and Colon Cancer*, Diagnostics and Therapeutic Advances in GI Malignancies,
https://doi.org/10.1007/978-981-15-2017-4_6

PDH	Pyruvate dehydrogenase
PDHX	Pyruvate dehydrogenase protein X component
PI3K	Phosphoinositide 3-kinases
PKM2	Pyruvate kinase M2
RES	Resveratrol
SCFA	Short chain fatty acid
UMPS	Uridine monophosphate synthetase

Highlights
- Metabolic anomaly in colorectal cancer
- Bioenergetics and ROS in colorectal cancer
- Application of miRNA, cancer stem cells in CRC therapeutics

6.1 Metabolic Reprogramming

Cancer cells and its surrounding stromal cells with diverse genetic or epigenetic is called as intra-tumoral heterogeneity. Cancer cells microenvironment (ME) is different from normal cells as it has high adaptive responses to hypoxia and hyponutrient conditions called as metabolic reprogramming (Lee et al. 2015). Warburg effect is highly accepted metabolic reprogramming associated with metabolism of mitochondria and aerobic glycolysis. Cancer-associated fibroblasts (CAFs) exhibit unusual oxygen-independent metabolic pathway, glycolysis as well as autophagy compared to adjoined cancer cells. Further, diverse transporters of monocarboxylates cause heterogeneity in cellular metabolism leading to formation of lactate and its uptake. In CRC, heterogeneous metabolism promotes metabolic association, which leads to adaptation to change associated with nutrient microenvironment due to chemotherapy and leads to drug resistance (Nguyen et al. 2012). In general, under unfavorable conditions, the dynamic nature of metabolism contemplated as a robust character of cancer cells. Metabolic reprogramming helps in hyper adaptation of cancer cells, which can be used to target proliferation of cancer cells (Visvader and Lindeman 2012). Metabolic reprogramming has a vital role in tumorigenesis, targeting cancer bioenergetics is a very promising and rapidly growing research area for anti-cancer therapy by selectively and effectively inhibit metabolic enzymes that are important for cancer survival.

6.2 Colorectal Cancer and Metabolic Changes Associated with Butyrate

The underlying mechanisms that regulate cancer metabolic anomaly is poorly understood in many cancer (Munkley and Elliott 2016). Warburg effect is a common condition in cancer metastasis including colorectal cancer (CRC). From a clinical and therapeutic perception, detecting metabolic changes is of dominant for risk stratification and in targeting metabolic alterations with chemical agents such as metformin as a promising chemopreventive strategy (Vander Heiden et al. 2010). The short chain fatty acid, butyrate is abundant in the lumen of gastrointestinal tract. It serves as an energy source to induce normal colonocyte proliferation and inhibits histone deacetylase and proliferative capacity of cancerous analogous that experiencing the Warburg effect (Li et al. 2014). The combined metabolomics-proteomics approach has proven that butyrate-mediated arrest of proliferation and cellular metabolism. Study of metabolic profiling by metabolomics reported diminished levels of glycolytic intermediates and nucleotides as well as accumulation of pyruvate (Han et al. 2018). Upon supplementation of metabolic intermediates, the metabolism of cancer cells was affected and repressive effect of butyrate was modulated. Further, a proteomics approach showed evidence for the direct effect of butyrate on dephosphorylation of M2 isoform of a pyruvate kinase, PKM2. This phosphorylation causes tetramerization and activation of PKM2, which induces reprogramming mechanisms of metabolism and inhibits Warburg effect by favoring energetic metabolism (HUA et al. 2019). This mechanistic study strongly supports the anticancer activity of butyrate, which could be exploited as small molecules for unknown protein targets of cancer etiology.

6.3 MYC Induced Global Metabolic Reprogramming of Colorectal Cancer

It is widely accepted that tumor cells undergo metabolic changes to produce precursors of macromolecules. This reprogramming is caused mainly due to abnormal expression of MYC (Soong et al. 2000). Analyses of CRC and adjacent normal tissue samples using multi-omics revealed that metabolic transformation rise at the adenoma stage of CRC due to specific gene mutations (Kitamura et al. 2007). MYC modulated the expression of 121 metabolic and 39 transporter genes by inducing nearly 215 biochemical reactions (Walz et al. 2014). It can negatively modulates the expression of genes associated with biogenesis and maintenance of mitochondria, but positively regulates genes required for DNA and histone modifications (Dang 2012). Silencing of MYC in CRC restored the metabolism to normal and reduced the tumor growth. Further, MYC is a master regulator of transcription of LDH-A and CAD genes (Terunuma et al. 2014). MYC is a downstream mediator of the Wnt pathway and overexpressed by mutations in APC or β-catenin gene in most of the CRC tissues (Liu et al. 2008). In addition, MYC is deregulated by massive genetic anomalies and aberrant expression of transcription factors, abnormal activation of PI3K/AKT/mTOR pathway and receptor tyrosine kinases (Meyer and Penn 2008). These studies demonstrates that MYC-mediated metabolic reprogramming and enough nutrient supply at precancerous stage are indispensable for development and growth of CRC (Locasale 2013). Therefore, they can be implicated for prevention as well as treatment of MYC-deregulated CRCs. However, finding of small-molecule inhibitors for MYC is highly challenging due to lack of surface binding site as well as nuclear localization (Jager et al. 2007). Therefore, MYC is neither blocked by small molecule inhibitors nor neutralized by antibodies. Silencing of MYC suppressed the proliferation CRC cells as well as genes involved in pyrimidine biosynthesis. This provides the foundation for MYC-mediated regulation of genes controlling pyrimidine biosynthesis and its importance in targeting CRCs (Lebleu et al. 2014).

6.4 Targeting Mitochondria and ROS in CRC

Mitochondria is indispensable for maintenance of energy and regulation of mitochondria-derived reactive oxygen species (mROS). They interfere with function of tumor suppressors, induce DNA damage, and activate oncogenes and metastatic signaling mechanisms. The mitochondrial antioxidants SOD2, Grx2, GPrx, Trx and TrxR are key players of the cellular redox systems and tightly control the primary ROS signaling molecule, H_2O_2. The mROS also regulates the activity of enzymes, which decomposing H_2O_2 by controlling the incorporation of selenocysteine at epitranscriptomic levels (Liu et al. 2017). mROS inducers reduced the expression of Sp1, Sp3 and Sp4 transcription factors and their pro-oncogenic regulators, which are indispensable for controlling metastasis. The mROS also mediates activation of Sp-suppressors, ZBTB10 and ZBTB4 and modulation of miRNA-27a or miRNA-20a. This pathway made significant contribution to the anticancer property of ROS inducers and therefore, they could be exploited in preparation of drug formulations (Lin et al. 2018). Mitochondria based antiapoptotic proteins can be proven as as efficient target for cancer therapy. However, mitochondrial metabolism emerged as a target for cancer therapy as it is center for ROS generation. Therefore, therapeutics that induce mROS could be exploited for treatment of CRCs. Cancer cells are highly dependent on mitochondrial bioenergetics for survival, thus targeting mitochondria and ROS can be rationally approach in cancer therapy.

6.5 Resveratrol Targeting Pyruvate Dehydrogenase Complex and Reverses the Warburg Effect in Colon Cancer Cells

Resveratrol (RES), is a polyphenol with antioxidant, anti-inflammatory and anti-proliferative activity and also exhibit anticancer property, but it has limitation with rapid metabolism which reduces its bioavailability (Weinhouse et al. 1956). RES induces growth arrest in tumor cells by modifying lipid profile, reducing the glycolysis and PPP pathway and by increasing the oxidative phosphorylation and

ATP synthesis (Vazquez et al. 2016). RES modulates energetics of mitochondria by targeting pyruvate dehydrogenase (PDH). However, chelation of calcium or blockading of its transport prevents the RES-induced oxidative metabolism implying the role of calcium in metabolic shift (Laplante and Sabatini 2013). Targeting CamKKB or its downstream AMPK pathway partially inhibited RES-induced glucose metabolism. Further, this study suggests that RES enhances the oxidative ability of cancer cells via CamKKB/AMPK pathway (Kumari et al. 2018). RES causes reprograming of cancer cells to increase the oxidation of glucose, decrease HMP shunt, lactate formation and lipid biosynthesis from glucose. Other studies have demonstrated that RES showed antitumor activity by enhancing cell cycle regulators cyclin E and A21, 22 and also by reducing cyclin D19 and 21 (Kumari et al. 2017). RES reduced the aggressive and invasive capacity of tumor cells by shifting glycolysis phenotype to more oxygen dependent metabolism (Faber et al. 2006). Apart from that, cancer cells commonly exhibits alterations in lipid composition (Malla et al. 2018). RES decreased conversion of glucose to lipids by depleting NADPH, a cofactor of fatty acid synthase (FAS) (Malla et al. 2019a). The change in lipid composition promotes changes in membrane fluidity, signal transduction, and response to chemotherapeutic drugs. For example, RES increased the ceramides levels but decreased the sphingomyelins by upregulating acid sphingomyelinase in CRC model (Malla et al. 2019b). The PDH complex is important for metabolic adaptations of the CRC since inhibition of PDH activity or attenuation of mitochondrial function leads to metabolic deviations in cancer cells (Hardie et al. 2012, 2015). Thus, RES has important role in regulating cancer growth by targeting different metabolites which are altered as part of metabolic adaption of cancerous growth for survival and proliferation.

6.6 MicroRNA-26a Targeting PDHX in Colorectal Cancer Cells

miRNAs are reported to associate with energy metabolism. Interestingly, alterations in cellular metabolism linked to dysregulation of miRNA (Kumari et al. 2013). They strive cellular metabolism directly by regulating metabolic pathways and enzymes and indirectly by regulating the expression of genes (Fei et al. 2012). miRNA-195-5p is reported to reduce the uptake of glucose as well as glycolytic mechanism by reducing the expression of GLUT3 as well as hexokinase 2. miRNA-143 alter metabolism of glucose via AKT signaling pathway. miRNA-26a is highly expressed in colon cancers in hypoxic environment (Jiang et al. 2012; Gregersen et al. 2012). It play role in cellular differentiation, growth, metastasis and apoptosis (Hanahan and Weinberg 2011). The miRNA-26a has been also associated with energy metabolism via by upregulating GAPDH, phosphoglycerate kinase 1and phosphoglycerate mutase 1 in colon cancer (Dai and Grant 2010). In CRC, the pyruvate dehydrogenase, one of the component of pyruvate dehydrogenase complex is a direct target of miR-26a, which helps in processing of glucose metabolism (Hsu and Sabatini 2008). RNAi is an advanced technology, utilizing miRNAs as therapeutic against cancer. Indeed, several miRNAs are validated at preclinical stage, but further exploration is required to enter clinical trials for metastatic CRCs. Overall, the miRNA networks that associate with metabolic reprogramming may provide reasonable and promising target strategies for targeting CRCs.

6.7 Reprogramming of Glutamine Metabolism by Oncogenic PIK3CA Mutations in Colorectal Cancer

Cancer cells dependent on glutamine for growth which is used as marker to distinguish from noncancerous cells. In CRCs, the glutamine metabolism is reprogrammed to utilize more glutamine by PIK3CA mutations in glutamate pyruvate transaminase 2 (GPT2). PIK3CA mutant cells shifts considerable levels of glutamine into α-ketoglutarate for anaplerotic refilling of Kreb's cycle to generate more ATP. However, p110a mutant cells overexpress GPT2 gene via PDK1–RSK2–ATF4 signaling in a AKT-independent mechanism (Jones and Thompson 2009). The aminooxyacetate, the inhibitor of aminotransferases suppressed the growth of PIK3CA mutant CRCs in xenograft model. Thus, glutamine metabolism is a potential target for PIK3CA mutant CRC patients (Miura et al. 2008).

6.8 Targeting Lipid Metabolism as Therapeutic Strategy in Colorectal Cancer

The essential cellular mechanisms of CRCs are affected by nutrients. Therefore, nutrients are considered as key risk factors. Nonetheless, the correlation of nutrients with carcinogenic mechanisms of CRC is highly. For example, nutritional factors and genes related to lipid metabolism. Lipids control different cellular mechanisms including synthesis of ATP, organization of membrane and activation of pathways associated with cellular plasticity. Consequently, various metastatic events are disposed by reprogramming of lipid metabolism (Aguirre-Portolés et al. 2017). The progression of CRC to metastatic stage is promoted by network of abnormal acyl-CoA synthetases (ACSL) or stearoyl-CoA desaturases (SCD). Therefore, therapeutics that targeting lipid metabolic networks could be beneficial for improving clinical outcomes in CRCs (Ramos-Lopez et al. 2017). Subsequently, studies validated by quantitative-PCR, Western blotting and luciferase assay have confirmed with showing miR-544a, miR-142 and miR-19b-1 as key mediators of ACSL/SCD axis. Notably, under expression of miR-19b-1 expression associated with reduced survival of CRC patients (Nct 2013). Lipid metabolism has a crucial role cancer growth and metastasis. Thus, targeting novel molecular targets of lipid metabolism would improve the efficacy on cancer prevention and treatment.

6.9 Metabolism Based Therapeutic Strategies Targeting CRC Stem Cells

The heterogeneity of tumors is key drivers of disease progression and failure of therapy. Tumors consist of distinct subpopulations, which originate from cancer stem cells (CSCs) (Caino et al. 2015). CSCs promotes tumorigenesis as well as metastasis even at high chemotherapeutic doses by enhancing survival and driving relapse. Indeed, CSCs exhibits diverse metabolic landscapes across tumors as well as subclones within the tumors (Vaira et al. 2015). CSCs originate from either mutation in normal stem cell or cells differentiated by attaining stem-like properties (Vaira et al. 2015).

Diverse studies have reported aberrant expression of genes involved in pathways associated with maintenance of stem cells including c-MYC and developmental pathways such as Bmi-1, Hedgehog, Notch and Wnt. The developmental pathways are central for stem cell functions (Bhuvanalakshmi et al. 2015). Primarily stem cells depends on glycolysis but acquire metabolic plasticity and oxidative metabolism to activates maturation and differentiation of CSC lineage. Studies have demonstrated that CSCs depend on oxidative metabolism a show metabolic heterogeneity (Warrier et al. 2014). Since CSCs associated with drug resistance, tumor relapse as well as metastasis, there is a need for persistent strategies to eradicate aggressive cell population. The unique metabolic characters of CSCs establish convincing strategy for component of CSCs to eliminate cancer cells by targeting mitochondria, OXPHOS, lipid metabolism, glycolysis, redox hemostasis, alternative fuel (Onorati et al. 2018). CSCs finds an attractive area of research due to specific survival mechanism, high tumorigenicity, drug-resistant and epitomize key targets for development of novel therapeutics. They include self-renewal pathways, microenvironment and metabolic reprogramming of CSCs.

6.10 Targeting Autophagy in Colorectal Cancer

Functionally, autophagic mechanisms involve in delivery of cellular materials for lysosome-mediated degradation and turnover of cellular components to provide energy as well as precursors of macromolecules. In contrast, context-dependent functions and mediations for both stimulation and inhibition of autophagy have been recommended to exploit as cancer therapies (Mancias and Kimmelman 2011). Cysteine proteases, the homologs of Atg4, promotes mechanism of autophagy via cleavage of Atg8 homologs for conjugation as well as deconjugation of membrane

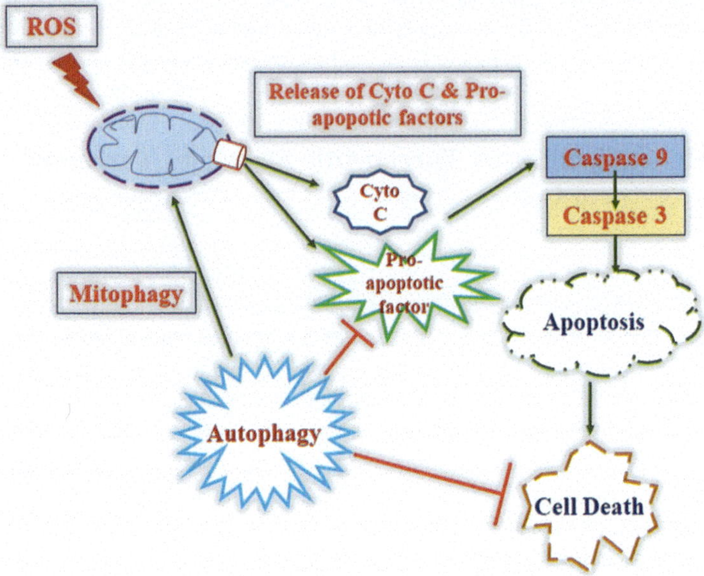

Fig. 6.1 Autophagy maintains basal turnover of cell components by delivering cellular materials to lysosomes for degradation. In cancer, autophagy has opposing, context-dependent roles. In colorectal cancer, autophagy can stimulate cell death by promoting mitophagy and inhibit cell death by interfering with pro-apoptotic proteins

lipids. ATG4B gene is highly expressed in CRC cells and critically involve autophagy. The autophagy inhibitor, S130 binds to ATG4B and induces apoptotic cell death (Singh et al. 2018). Apatinib, a novel inhibitor of receptor tyrosine kinase, has been recognized for treatment of advanced gastric carcinoma due to its efficacy and safety. Mechanistically, it induces both programmed cell death as well as autophagy in CRC by promoting ER stress. Furthermore, the combination of apatinib and chloroquine (CQ) showed prominent anti-tumor activity against CRC (Singh et al. 2018). In addition to provide metabolic precursors and energy, autophagy scavenges damaged proteins under stress conditions (Fig. 6.1). Therefore, autophagy serves as a pro-survival mediator under stress conditions, but promotes cell death depends on the cellular context. Therefore, autophagy medicated cancer cell death could be a novel area of research.

6.11 Conclusion

Metabolic reprogramming has a vital role in tumorigenesis. Thus, targeting bioenergetics is highly selective and effective approach in CRC therapy. Especially, inhibiting metabolic enzymes and pathways that are important for survival of CRC is highly promising. Inhibition of MYC which regulates genes associated with pyrimidine synthesis, blocks cell growth and implies its importance in CRC therapy. Targeting mitochondrial bioenergetics, application of miRNA technology, targeting novel molecular targets of lipid metabolism, would improve the efficacy of cancer treatment. Moreover, CSCs finds an attractive area for developing novel therapeutic strategies. In addition, autophagy is also a novel target of CRC because it mediates drug induced apoptosis.

6.12 Future Direction

Noteworthy advanced analytical methods are used to study of cellular metabolism and to explore the metabolic re-programs that are critical for tumorigenesis. Metabolite profiling greatly expands metabolites and facilitates identification of pathway associated with metabolites. These technical developments need further modification to expand cellular metabolism in real-time to drive future discovery as well as therapeutic development in cancer research.

Acknowledgement The authors are thankful to GITAM (Deemed to be University) for providing necessary support.

References

Aguirre-Portolés C, Fernández LP, De Molina AR (2017) Precision nutrition for targeting lipid metabolism in colorectal cancer. Nutrients. https://doi.org/10.3390/nu9101076

Bhuvanalakshmi G, Arfuso F, Millward M, Dharmarajan A, Warrier S (2015) Secreted frizzled-related protein 4 inhibits glioma stem-like cells by reversing epithelial to mesenchymal transition, inducing apoptosis and decreasing cancer stem cell properties. PLoS One. https://doi.org/10.1371/journal.pone.0127517

Caino MC, Ghosh JC, Chae YC, Vaira V, Rivadeneira DB, Faversani A, Rampini P, Kossenkov AV, Aird KM, Zhang R, Webster MR, Weeraratna AT, Bosari S, Languino LR, Altieri DC (2015) PI3K therapy reprograms mitochondrial trafficking to fuel tumor cell invasion. Proc Natl Acad Sci. https://doi.org/10.1073/pnas.1500722112

Dai Y, Grant S (2010) New insights into checkpoint kinase 1 in the DNA damage response signaling network. Clin Cancer Res 16:376–383. https://doi.org/10.1158/1078-0432.CCR-09-1029

Dang CV (2012) MYC on the path to cancer. Cell. https://doi.org/10.1016/j.cell.2012.03.003

Faber AC, Dufort FJ, Blair D, Wagner D, Roberts MF, Chiles TC (2006) Inhibition of phosphatidylinositol 3-kinase-mediated glucose metabolism coincides with resveratrol-induced cell cycle arrest in human diffuse large B-cell lymphomas. Biochem Pharmacol. https://doi.org/10.1016/j.bcp.2006.08.009

Fei X, Qi M, Wu B, Song Y, Wang Y, Li T (2012) MicroRNA-195-5p suppresses glucose uptake and proliferation of human bladder cancer T24 cells by regulating GLUT3 expression. FEBS Lett. https://doi.org/10.1016/j.febslet.2012.01.006

Gregersen LH, Jacobsen A, Frankel LB, Wen J, Krogh A, Lund AH (2012) MicroRNA-143 down-regulates hexokinase 2 in colon cancer cells. BMC Cancer. https://doi.org/10.1186/1471-2407-12-232

Han A, Bennett N, Ahmed B, Whelan J, Donohoe DR (2018) Butyrate decreases its own oxidation in colorectal cancer cells through inhibition of histone deacetylases. Oncotarget. https://doi.org/10.18632/oncotarget.25546

Hanahan D, Weinberg RA (2011) Hallmarks of cancer: the next generation. Cell. https://doi.org/10.1016/j.cell.2011.02.013

Hardie DG (2015) AMPK: positive and negative regulation, and its role in whole-body energy homeostasis. Curr Opin Cell Biol. https://doi.org/10.1016/j.ceb.2014.09.004

Hardie DG, Ross FA, Hawley SA (2012) AMPK: a nutrient and energy sensor that maintains energy homeostasis. Nat Rev Mol Cell Biol. https://doi.org/10.1038/nrm3311

Hsu PP, Sabatini DM (2008) Cancer cell metabolism: Warburg and beyond. Cell. https://doi.org/10.1016/j.cell.2008.08.021

Hua YL, Ma Q, Yuan ZW, Zhang XS, Yao WL, Ji P, Hu JJ, Wei YM (2019) A novel approach based on metabolomics coupled with network pharmacology to explain the effect mechanisms of Danggui Buxue tang in anaemia. Chin J Nat Med. https://doi.org/10.1016/S1875-5364(19)30031-7

Jager S, Handschin C, St.-Pierre J, Spiegelman BM (2007) AMP-activated protein kinase (AMPK) action in skeletal muscle via direct phosphorylation of PGC-1. Proc Natl Acad Sci. https://doi.org/10.1073/pnas.0705070104

Jiang S, Zhang LF, Zhang HW, Hu S, Lu MH, Liang S, Li B, Li Y, Li D, Wang ED, Liu MF (2012) A novel miR-155/miR-143 cascade controls glycolysis by regulating hexokinase 2 in breast cancer cells. EMBO J. https://doi.org/10.1038/emboj.2012.45

Jones RG, Thompson CB (2009) Tumor suppressors and cell metabolism: a recipe for cancer growth. Genes Dev. https://doi.org/10.1101/gad.1756509

Kitamura T, Kometani K, Hashida H, Matsunaga A, Miyoshi H, Hosogi H, Aoki M, Oshima M, Hattori M, Takabayashi A, Minato N, Taketo MM (2007) SMAD4-deficient intestinal tumors recruit CCR1+ myeloid cells that promote invasion. Nat Genet. https://doi.org/10.1038/ng1997

Kumari S, Kaladhar DSVGK, Solmon KS, Malla RR, Kishore G (2013) Anti-proliferative and metastatic protease inhibitory activities of protoberberines: an in silico and in vitro approaches. Process Biochem 48:1565–1571. https://doi.org/10.1016/j.procbio.2013.06.027

Kumari S, Badana AK, Mohan GM, Naik GS, Malla RR (2017) Synergistic effects of coralyne and paclitaxel on cell migration and proliferation of breast cancer cells lines. Biomed Pharmacother. https://doi.org/10.1016/j.biopha.2017.04.027

Kumari S, Badana AK, Mohan GM, Shailender G, Malla RR (2018) Reactive oxygen species: a key constituent in cancer survival. Biomark Insights. https://doi.org/10.1177/1177271918755391

Laplante M, Sabatini DM (2013) Regulation of mTORC1 and its impact on gene expression at a glance. J Cell Sci. https://doi.org/10.1242/jcs.125773

Lebleu VS, O'Connell JT, Gonzalez Herrera KN, Wikman H, Pantel K, Haigis MC, De Carvalho FM, Damascena A, Domingos Chinen LT, Rocha RM, Asara JM, Kalluri R (2014) PGC-1α mediates mitochondrial biogenesis and oxidative phosphorylation in cancer cells to promote metastasis. Nat Cell Biol. https://doi.org/10.1038/ncb3039

Lee JY, Yoon JK, Kim B, Kim S, Kim MA, Lim H, Bang D, Song YS (2015) Tumor evolution and intratumor heterogeneity of an epithelial ovarian cancer investigated using next-generation sequencing. BMC Cancer. https://doi.org/10.1186/s12885-015-1077-4

Li L, Zhang Y, Qiao J, Yang JJ, Liu ZR (2014) Pyruvate kinase M2 in blood circulation facilitates tumor growth by promoting angiogenesis. J Biol Chem. https://doi.org/10.1074/jbc.M114.576934

Lin S, Li Y, Zamyatnin AA, Werner J, Bazhin AV (2018) Reactive oxygen species and colorectal cancer. J Cell Physiol. https://doi.org/10.1002/jcp.26356

Liu YC, Li F, Handler J, Huang CRL, Xiang Y, Neretti N, Sedivy JM, Zeller KI, Dang CV (2008) Global regulation of nucleotide biosynthetic genes by c-myc. PLoS One. https://doi.org/10.1371/journal.pone.0002722

Liu H, Liu X, Zhang C, Zhu H, Xu Q, Bu Y, Lei Y (2017) Redox imbalance in the development of colorectal cancer. J Cancer. https://doi.org/10.7150/jca.18735

Locasale JW (2013) Serine, glycine and one-carbon units: cancer metabolism in full circle. Nat Rev Cancer. https://doi.org/10.1038/nrc3557

Malla RR, Pandrangi S, Kumari S, Gavara MM, Badana AK (2018) Exosomal tetraspanins as regulators of cancer progression and metastasis and novel diagnostic markers. Asia Pac J Clin Oncol. https://doi.org/10.1111/ajco.12869

Malla RR, Kumari S, Gavara MM, Badana AK, Gugalavath S, Kumar DKG, Rokkam P (2019a) A perspective on the diagnostics, prognostics, and therapeutics of microRNAs of triple-negative breast cancer. https://doi.org/10.1007/s12551-019-00503-8

Malla RR, Kumari S, Deepak KGK, Gavara MM, Guganavath S, Rokkam P (2019b) Terpenoids as potential targeted therapeutics of pancreatic cancer: current advances and future directions. In: Breaking Tolerance to Pancreatic Cancer Unresponsiveness to Chemotherapy. https://doi.org/10.1016/b978-0-12-817661-0.00007-x

Mancias JD, Kimmelman AC (2011) Targeting autophagy addiction in cancer. Oncotarget. https://doi.org/10.18632/oncotarget.384

Meyer N, Penn LZ (2008) Reflecting on 25 years with MYC. Nat Rev Cancer. https://doi.org/10.1038/nrc2231

Miura T, Chiba M, Kasai K, Nozaka H, Nakamura T, Shoji T, Kanda T, Ohtake Y, Sato T (2008) Apple procyanidins induce tumor cell apoptosis through mitochondrial pathway activation of caspase-3. Carcinogenesis 29:585–593. https://doi.org/10.1093/carcin/bgm198

Munkley J, Elliott DJ (2016) Hallmarks of glycosylation in cancer. Oncotarget. https://doi.org/10.18632/oncotarget.8155

Nct (2013) Prehabilitation to enhance postoperative functional capacity following radical cystectomy. https://Clinicaltrials.Gov/Show/Nct01836978

Nguyen LV, Vanner R, Dirks P, Eaves CJ (2012) Cancer stem cells: an evolving concept. Nat Rev Cancer. https://doi.org/10.1038/nrc3184

Onorati AV, Dyczynski M, Ojha R, Amaravadi RK (2018) Targeting autophagy in cancer. Cancer. https://doi.org/10.1002/cncr.31335

Ramos-Lopez O, Riezu Boj JI, Milagro FI, Martínez JA (2017) Epigenetics of undernutrition and obesity. Ann Nutr Metab 71(Suppl 2):1–1433

Singh SS, Vats S, Chia AYQ, Tan TZ, Deng S, Ong MS, Artuso F, Yap CT, Goh BC, Sethi G, Huang RYJ, Shen HM, Manjithaya R, Kumar AP (2018) Dual role of autophagy in hallmarks of cancer. Oncogene. https://doi.org/10.1038/s41388-017-0046-6

Soong R, Powell B, Elsaleh H, Gnanasampanthan G, Smith DR, Goh HS, Joseph D, Iacopetta B (2000) Prognostic significance of TP53 gene mutation in 995 cases of colorectal carcinomainfluence of tumour site, stage, adjuvant chemotherapy and type of mutation. Eur J Cancer. https://doi.org/10.1016/S0959-8049(00)00285-9

Terunuma A, Putluri N, Mishra P, Mathé EA, Dorsey TH, Yi M, Wallace TA, Issaq HJ, Zhou M, Keith Killian J, Stevenson HS, Karoly ED, Chan K, Samanta S, Prieto D, Hsu TYT, Kurley SJ, Putluri V, Sonavane R, Edelman DC, Wulff J, Starks AM, Yang Y, Kittles RA, Yfantis HG, Lee DH, Ioffe OB, Schiff R, Stephens RM, Meltzer PS, Veenstra TD, Westbrook TF, Sreekumar A, Ambs S (2014) MYC-driven accumulation of 2-hydroxyglutarate is associated with breast cancer prognosis. J Clin Invest. https://doi.org/10.1172/JCI71180

Vaira V, Webster MR, Rivadeneira DB, Ghosh JC, Kossenkov AV, Zhang R, Aird KM, Languino LR, Caino MC, Rampini P, Weeraratna AT, Faversani A, Bosari S, Chae YC, Altieri DC (2015) PI3K therapy reprograms mitochondrial trafficking to fuel tumor cell invasion. Proc Natl Acad Sci. https://doi.org/10.1073/pnas.1500722112

Vander Heiden MG, Locasale JW, Swanson KD, Sharfi H, Heffron GJ, Amador-Noguez D, Christofk HR, Wagner G, Rabinowitz JD, Asara JM, Cantley LC (2010) Evidence for an alternative glycolytic pathway in rapidly proliferating cells. Science 329(5998):1492–1499. https://doi.org/10.1126/science.1188015

Vazquez A, Kamphorst JJ, Markert EK, Schug ZT, Tardito S, Gottlieb E (2016) Cancer metabolism at a glance. J Cell Sci. https://doi.org/10.1242/jcs.181016

Visvader JE, Lindeman GJ (2012) Cancer stem cells: current status and evolving complexities. Cell Stem Cell. https://doi.org/10.1016/j.stem.2012.05.007

Walz S, Lorenzin F, Morton J, Wiese KE, Von Eyss B, Herold S, Rycak L, Dumay-Odelot H, Karim S, Bartkuhn M, Roels F, Wüstefeld T, Fischer M, Teichmann M, Zender L, Wei CL, Sansom O, Wolf E, Eilers M (2014) Activation and repression by oncogenic MYC shape tumour-specific gene expression profiles. Nature. https://doi.org/10.1038/nature13473

Warrier S, Bhuvanalakshmi G, Arfuso F, Rajan G, Millward M, Dharmarajan A (2014) Cancer stem-like cells from head and neck cancers are chemosensitized by the Wnt antagonist, sFRP4, by inducing apoptosis, decreasing stemness, drug resistance and epithelial to mesenchymal transition. Cancer Gene Ther. https://doi.org/10.1038/cgt.2014.42

Weinhouse S, Warburg O, Burk D, Schade AL (1956) On respiratory impairment in cancer cells. Science 124:267–272. https://doi.org/10.1126/science.124.3215.267

Chapter 7
Understanding Colorectal Cancer: The Basics

Mohan Krishna Ghanta, Santosh C. Gursale, and L.V.K.S. Bhaskar

Abstract Among all cancers, colorectal cancer (CRC) is the third most common cancer. Colorectal cancer is associated with various risk factors which may alter the physiological pathways and induce genetic mutations. The risk factors, genetic variations, gene related toxicities with treatment regimens of CRC, clinically relevant biomarker of CRC were discussed in this review. The aim of the present review is to provide a current update on the risk factors and their pathways, biomarkers and their association with treatment efficacy and toxicity in CRC, which may provide a view for the Precision Medicine in CRC.

Keywords Colorectal cancer · Epidemiology · Risk factors · Biomarkers · Toxicities

Abbreviations

18q LOH	Chromosome 18q loss of heterozygosity
5-FU	5-fluorouracil
ABCB1	ATP dependent translocase/drug transporter gene family
AKT	Protein kinase B
Anti-EGFR	Anti epidermal growth factor receptor
Anti-VEGF	Anti Vascular endothelial growth factor
APC	Adenomatous polyposis coli

M. K. Ghanta
Department of Pharmacology, Sri Ramachandra Medical College and Research Institute, SRIHER-(DU), Porur, Chennai, Tamil Nadu, India

S. C. Gursale
Department of Pharmacology, BKL Walawalkar Rural Medical College, Kasarwadi, Sawarde, Ratnagiri, Maharashtra, India

L.V.K.S. Bhaskar (✉)
Guru Ghasidas University, Bilaspur, Chhattisgarh, India

G. S. R. Raju, L.V.K.S. Bhaskar (eds.), *Theranostics Approaches to Gastric and Colon Cancer*, Diagnostics and Therapeutic Advances in GI Malignancies,
https://doi.org/10.1007/978-981-15-2017-4_7

BAT-25	Mononucleotide microsatellite
BAT-26	Mononucleotide microsatellite
Bcl-xl	B-cell lymphoma-extra large
BMI	Body mass index
BRAF	Gene encoding b-raf protein
CDA	Cytidine deaminase
CIMP	CpG island hypermethylation
CIN	Chromosomal instability
cmyc	Proto oncogene
CRC	Colorectal cancer
CRP	C-reactive protein
cyclin D1	cyclin D1
DNA	Deoxyribo nucleic acid
DPYD	Dihydropyrimidine dehydrogenase
EMT	Epithelial to mesenchymal transition
ENOSF1	Antisense RNA to thymidylate synthase
EPCAM	Epithelial cell adhesion molecule
ERK1/2	Extracellular signal-regulated kinase 2
ExoI	Exonuclease1
FAK	Focal adhesion kinase
FAP	Familial adenomatous polyposis
FRA1	Fos-related antigen 1
gp130	Glycoprotein130
GSK3B	Glycogen synthase kinase 3 beta
Gy	Gray
HAT	Histone acetyl transferase
HDAC	Histone deacetylase
HMGB1	High mobility group box 1 protein
hTERT	Telomerase reverse transcriptase
IL10	Interleukin10
IL-17F	Interleukin-17F
IL6	Interleukin6
IL6R	Interleukin6R
JAK	Janus kinase
K116	Lysine116
KRAS	K-Ras protein
MAPK	Mitogen activated protein kinase
mcl1	Antiapoptotic protein of Bcl2 family
miRNA	micro RNA
MLH1	MutL homolog 1
MONO-27	Mononucleotide microsatellite
MSH2	MutS protein homolog 2
MSH6	MutS Homolog 6
MSI	Microsatellite instability
MTHFR	Methylenetetrahydrofolate reductase

mVim	methylated VIMENTIN
NETRIN-1	Laminin-related secreted protein
NF-KB	Nuclear Factor kappa-light-chain-enhancer of activated B cells
Nlrp6	NOD-like receptor family pyrin6
NOS	Nitric oxide synthase
NR-21	Mononucleotide microsatellite
NR-24	Mononucleotide microsatellite
p53	Tumor protein 53
p65	Sub unit of NF-κB
PCNA	Proliferating cell nuclear antigen
PCR	Polymerase chain reaction
PMS2	PMS1 Homolog 2
Pol δ	DNA polymerase δ
PTEN	Phosphatase and tensin homolog
Ras	Ras proteim
RFC	Reduced folate carrier
ROS	Reactive oxygen species
SES	Socio-economic status
SMAD4	Mothers against decapentaplegic homolog 4
STAT3	Signal transducer and activator of transcription3
STK11	Serine/threonine kinase 11
Th17	T helper cell17
TNF α	Tumor necrosing factor α
TNF	Tumor necrosing factor
TNFβ	Tumor necrosing factorβ
TP53	Tumor protein 53
Treg	Regulatory T cells
TROP-2	Tumor-associated calcium signal transduction protein
TYMS	Thymidylate synthase
UC	Ulcerative colitis
UGT1A1	UDP-glucuronosyltransferase 1
USA	United States of America
VEGF	Vascular endothelial growth factor
VEGR2	Vascular endothelial growth factor receptor type 2
WHO	World Health Organisation
Wnt	Wingless (wg) and Int-1, traits

7.1 Introduction

Colorectal cancer (CRC) was described as a serious consequence of ulcerative colitis (UC) which is categorized under inflammatory bowel diseases. Co-occurrence of CRC and UC was first explained in 1925 by Crohn and Rosenberg (Crohn and

Rosenberg 1925). Later in subsequent years, other studies and case reports have also confirmed the relation between CRC and UC (Bargen 1994; Svartz and Ernberg 1949). One percent of diagnosed CRC cases have shown association with UC (Gyde et al. 1982). Before them, 100 years ago, Dr. Aldred Warthin suspected this disorder in a patient family and observed her family during 1895 to conclude the hereditary cancers. This study was named as "Family G". In 1913, their research publication revealed the familial susceptibility to some cancers which are inheritable. They also stated the specific localization of cancers on gender basis, males were more prone to gastrointestinal cancers and females were more prone to cancer of uterus (Warthin 1985). Later this project was updated till 1955 by Weller, and then this was revisited by Henry T. Lync. In this update, after thorough evaluation they suggested the localization of high risk cancers which were in colon and endometrium. They have also stated the association of Lynch syndrome (70–80%), sebaceous adenomas and sebaceous carcinomas with proximal colon cancers (Lynch and Krush 1971). Other cancer associated with lynch syndrome was endometrial cancer. Another study in 2000 reported the identified mutations in family G (Yan et al. 2000). The familial inheritance was explained by Knudson through two hit hypothesis, this introduced the role of tumor suppressor genes and second mutation in inheriting the colorectal cancer (Knudson 1974; Knudson and Strong 1972). Role of adenomatous polyposis coli gene was revised during 1990 which was discovered in 1930 as autosomal dominant familial polyposis, now known as familial adenomatous polyposis (Schlussel et al. 2014). Adenomatous polyposis coli gene was considered as first step in cancer pathogenesis (Fearon and Vogelstein 1990). The prognosis of the disease is poor, as stated in the study, the 5 year survival rate was 65% and 10 year survival rate was 58% (Miller et al. 2016). It is important to adopt effective programs in health care system focussing on prevention and effective treatment aimed to improve prognosis of the disease condition. To achieve this, better understanding of the disease is required which can aid to identify target population and early diagnosis. This chapter provides an update on the possible risk factors, biomarkers, mechanisms, therapeutic targets and available effective management of colorectal cancer. Literature search was done using databases of PubMed, PubMed Central, EMBASE and MEDLINE.

7.1.1 Epidemiology

The incidence of colorectal cancers were found to be increasing in the recent years and reached to a maximum threat level to human health. Among all cancers, incidence of colorectal cancer was third in position (Henrikson et al. 2015) and accounted for 7 lakh deaths worldwide (Arnold et al. 2017). In United States of America (USA), among cancer related death, colorectal cancer was second cause of mortality (Schoen et al. 2012). In New York the incidence rate was approximately 39/100,000 population during 1998. Later there was a decrease in incidence rate to approximately 30/100,000 population by 2007. This decrease in incidence rate was also seen in Osaka (Japan) from 31/100,000 to 27/100,000 population (approximately) during 1998 and 2008. But a study in 2015 revealed an incidence rate of 32.2/100,000 population in Japan (other area). In Kyadondo (Africa) the incidence rate was 10–11/100,000 population between 1998 and 2008. In Oxford the incidence rate was 30/100,000 population during 1998 and a decreasing trend was found in 2008 in comparison to 1998 census. An increasing trend in incidence rate was seen in Beijing (China) ranging from 9 to 17(approximately)/100,000 population during 1998–2008. A very low incidence rate (5/100,000 population) was reported in Mumbai (India) during 1998–2008. During this period the incidence rate was high in Eastern Europe, Western Asia, New York, Oxford and Osaka. A recent study has reported an increased incidence rate in North America, Europe New Zealand and Australia while lowest incidence rate was seen in South Central Asia and Africa (Fitzmaurice et al. 2017). Even recent estimates revealed overall worldwide colorectal cancer is third in incidence (6.1%) and second in mortality (9.2%) as per GLOBOCAN data 2018 (Bray et al. 2018). Recently in India, the incidence was 4.2/100,000 male population and 3.2/100,000 female population (Sharma and Singh 2017). The 5 year prevalence rate (2018) revealed that the prevalence of colorectal cancer in Asia, Europe, North America and Africa was 51.9, 188.7, 146.8, and 8.8 per 1 lakh population respectively. Among Asian countries, the 5 year prevalence of colorectal cancer in China, Japan, Korea, Iran and India was 87.7, 339.4, 240.3, 29.7 and 8.3 per 1 lakh population respectively (Wong et al. 2019). In United States of America, the incidence rate was 45.9 in males and 34.8 in females per 1 lakh population. The estimated incidence of new cases were 140,250 and deaths related to colorectal cancer were 50,630 (Siegel et al. 2018).

Gender difference in incidence rate of colorectal cancer varied among countries. Males were more affected than females (except in Africa), was reported in a study which analysed colorectal cancer incidence data of 1998–2008. A 10 year study has reported that there was a significant decrease of colorectal cancer incidence rates in male population of USA (3.0), Japan (0.6), Australia (0.5), New Zealand (1.3), France (0.6), Austria (1.8) and Czech Republic (0.8). A significant increase of colorectal cancer incidence rates in males was seen in Colombia, Costa Rica, Philippines, China, Latvia, United Kingdom (UK), Lithuania, Norway, Finland, Estonia, Sweden, Netherlands, Slovenia, Italy, Bulgaria, Poland, Russian Federation. In females, significant increase in incidence rates of colorectal cancer was reported in Brazil, Colombia, Costa Rica, Philippines, Latvia, UK, Estonia, Sweden, Denmark, Iceland, Netherlands, Spain, Slovenia, Italy, Bulgaria, Slovakia, Russian Federation and significant decrease in incidence rates was reported in USA, New Zealand, France, Austria, Czech Republic (Arnold et al. 2017).

Incidence rate of colorectal cancer in different age groups was elaborated by Vuik et al. They stated that age related incidence of colorectal cancer varied among age groups during 2016. Younger age group of 20–29 years had an incidence rate of 2.3/100,000 population. Age group of 30–39 had an incidence rate of 7.1/100,000 male population and 6.4/100,000 female population. Age group of 40–49 had shown an incidence rate of 19.2/100,000. Older age group have shown more incidence rate in comparison to younger age groups in this study and the incidence rate have increased with age (Vuik et al. 2019). A 20 year study in Taiwan revealed age related incidence of colorectal cancer larger in older age groups of greater than 75 years of age than the younger age group (Chang et al. 2012). A study retrieved data (1981–2013) from Florida cancer data system and grouped in to two groups of <50 years age group and > 50 years age group. It included 182,095 patients of which 7% were in <50 years age group and 93% were in >50 years age group (Moore et al. 2018). These studies reveal that colorectal cancer had more incidence in older age groups, but it is not uncommon in younger age groups.

7.2 Risk Factors of Colorectal Cancer

7.2.1 Inflammatory Bowel Disease (IBD)

Inflammatory bowel disease is chronic inflammatory disorder of intestinal tract and widely represented as Ulcerative colitis and Crohn's disease. This may result into low grade dysplasia and/or/to high grade dysplasia which converts to invasive adenocarcinoma (Hofseth et al. 2003; Xavier and Podolsky 2007). Various events in development of IBD to colorectal cancer may include derailed immune system; induction or activation of TNF, NF-κB, IL-6, netrin-1, stat3; reduction of IL10:Treg inhibition which cause antiapoptotic effects. On the other hand, oxidative stress due to increased nitric oxide synthase (NOS) and reactive oxygen species (ROS) were shown to cause tumor protein 53 (p53) mutation, MutL homolog 1 (MLH1) hypermethylation, and microsatellite instability/chromosomal instability which were identified in biopsies of colorectal cancer.

7.2.2 Familial Adenomatous Polyposis (FAP)

This is an autosomal dominant hereditary disorder with presence of large number of polyps in colon and rectum. Approximately 1% of CRC cases were due to FAP. Mutation in APC gene was considered responsible for CRC in FAP patients (Bojuwoye et al. 2018). This gene is responsible for developing CRC in 45–80% of all CRC cases.

7.2.3 Lynch Syndrome

Among all CRC cases, 1–3% are caused due to Lynch syndrome (Moller et al. 2017). The prominent role in DNA mismatch repair is held by various proteins with specific functions. The hMutSα, hMutSβ, hMutLα, hMutLβ, hMutLγ, ExoI, Pol δ, PCNA, RPA, HMGB1, RFC, DNA ligase I were the so far identified human proteins playing significant role in DNA mismatch repair. Strand discrimination and DNA helicase proteins are yet to be identified in human which were described in E.coli DNA mismatch repair. The microsatellite instability causing abnormalities and mutations in MutS protein homolog 2 (MSH2), MutS protein homolog 6 (MSH6), MutL homolog 1 (MLH1), or deletion in Epithelial cell adhesion molecule (EPCAM) gene results in lynch syndrome (Li 2008).

7.2.4 *Cholecystectomy*

Gall bladder diseases have become common and serious burden worldwide. In USA, about 15% of population are affected with gall bladder diseases (Acalovschi and Lammert 2012). A meta-analysis study including 524,649 study population, have shown an increased association and positive relationship (RR 1.22; 95% CI 1.08–1.38) between cholecystectomy and colorectal cancer (Zhang et al. 2017). The pathological events and biochemical variations due to cholecystectomy were considered responsible for development of colorectal cancer. In cholecystectomy patients, the secondary bile acids content were found high, the undigested fat content was found increased in stools, increased colonic microflora metabolic end products. These factors were found etiological for development of colorectal cancer, which are the consequences of cholecystectomy. Secondary bile acids were also shown to increase the risk of colorectal cancers in animal models (Schernhammer et al. 2003).

7.2.5 *Ureterocolic Anastomosis*

Colon cancer risk was higher up to 550 folds, as a late complication for ureterosigmoidostomy. The pathogenic mechanisms suggested for the development of colon cancer at the anastomosis site included production of N-nitrosocompounds by bacterial flora, chronic exposure to free radicals, chronic irritation, surgical or mechanical trauma, positivity for cytokeratin 20 and higher concentration of electrolytes. Some case reports have also suggested development of tumors at anastomosis site may also originate from urothelium (Makino et al. 2015).

7.2.6 *Diabetes Mellitus*

Diabetes mellitus (DM) was suggested as risk factor for the pathogenesis of CRC 34. Hyperglycemia has been reported to be strong stimulus favouring tumorigenesis. Hyperglycemia and advanced glycation end products (AGEPs) were revealed to cause oxidative stress and inflammation resulting in malignancy. The AGEPs were also shown to induce proliferation of CRC cells through carbohydrate response element-binding proteins (Chen et al. 2014; Mantovani et al. 2008; Rojas et al. 2010; Wu and Zhou 2009). The Wnt signalling pathway was found regulating the cell proliferation in CRC through miR-21 gene, and also other DM complications. Beta-catenin expression and cell proliferation was found higher in tumor surrounding unaffected cells of diabetic CRC patients in comparison to non diabetic CRC patients. Inflammation due to diabetes was major responsible factor for initiation and progression of CRC (Li et al. 2014). Chronic NF-κB/IL-6/STAT3 stimulation causes CRC (Liang et al. 2013). The microbiota dysregulation and deficiency of inflammasomes like Nlrp6, induces colon tumorigenesis through IL6 signalling activation (Chen et al. 2011; Elinav et al. 2011; Hu et al. 2013). Various genes responsible for development of CRC were also found to be affected by DM. Adenomatous polyposis coli (APC), serine/threonine kinase 11 (STK11), MSH2, MSH6, K-Ras protein (KRAS), and tumor protein53 (TP53) gene expressions were up-regulated in DM (Gonzalez et al. 2017).

7.2.7 *Pelvic Radiotherapy*

Colorectal cancer has been reported as a long-term complication following pelvic radiotherapy (Birgisson et al. 2005). Pathological macroscopic findings characterised radiation associated CRC as non-polypoid, diffusely infiltrated with well defined lesion margin (Tamai et al. 1999). Most of the cases manifested mucinous type of CRC (Castro et al. 1973). High dose irradiation (50–82 Gy) has been reported to cause severe vascular changes like subintimal fibrosis and hyaline sclerosis as late radiation injuries (Tamai et al. 1999; Tomori et al. 1999; Tsuji et al. 2003). Another study reported that incidence of radiation associated CRC was high with low dose radiation therapy (3.32%) in comparison to high dose radiation therapy (1.4%). Although p53 gene mutation was not specific mechanism for pathogenesis through pelvic radiation therapies, so far only p53 gene mutation has been reported in radiation associated CRC (Minami et al. 1998). As a precaution for radiation associated CRC, it was suggested to advise long-term follow up with colonoscopy for patients with chronic proctocolitis undergoing pelvic radiotherapy (Sasaki et al. 2017). Incidence of 5–11% radiation proctocolitis has been reported in patients treated for gynaecological cancers (Perez et al. 1984).

7.2.8 *Alcohol Consumption*

The alcohol and its metabolites in the body may cause CRC through various mechanisms like affecting cell signalling pathways, microbiota, causing epigenetic dysregulation and genetic abnormalities. The cell signalling pathways affected by metabolites of alcohol include IL6/STAT3, NF-kB, Ras, Wnt/B-Catenin/GSK3B, VEGF-VEGR2-AKT/FAK, ERK1/2, Notch. Microbial dysbiosis, bacterial overgrowth and increased inflammation due to alcohol and alcohol metabolites has been reported to cause CRC. Genetic abnormalities due to alcohol metabolites causing CRC may include mutation incidence, DNA adducts, impaired DNA synthesis and hindered DNA repair. Alcohol related epigenetic dysregulation mechanisms including global DNA hypomethylation, dysregulated miRNAs, altered levels of Histone acetyl transferase (HATs), Histone deacetylase (HDACs) and histone modifications has been reported to cause CRC (Rossi et al. 2018).

7.2.9 Smoking and Tobacco Consumption

Modest incidence rates of CRC (Relative Risk = 1.19, 95% CI = 1.05–1.35) were reported with smoking. Smoking or tobacco exposure causing genetic methylation and epigenetic modifications was found responsible for developing CRC. These include MSI-high (microsatellite instability, incidence: 15–25%), CIMP-positive (CpG island methylator phenotype, incidence: 20–30%) or *BRAF* (serine/threonine protein kinase gene) mutation positive CRCs (Limsui et al. 2010).

7.2.10 Food Habits

Most of the foods containing fruits and vegetables had inverse association with CRC (Robertson et al. 1998). High intake of fruits/vegetables of orange or yellow in color has been reported to be associated with CRC risk (OR = 1.61, 95% CI: 1.22–2.12). These kinds of foods include citrus fruit juices, orange, mandarin orange, kumquat, carrot, pumpkin, peach, persimmon, ginger. This risk was found in both sexes (OR = 2.41, 95% CI: 1.83–3.16 for male; OR = 2.28, 95% CI: 1.55–3.34 for female) (Lee et al. 2017). Fermented soy paste intake was also shown to be associated CRC risk. This contains high salts because of which it may cause colorectal carcinogenesis, also high intake of salts has been reported to cause CRC (Shin et al. 2015). A diet high in grains, low in fruits and vegetables was known as a high CRP – dietary pattern. This high CRP (C-reactive protein) – dietary pattern has been reported to be associated with increased CRC risk (OR (95% CI) = 3.58 (2.65–4.82). This risk was observed more in females than males. The pathological events in high CRP diet intake may include increased production of pro-inflammatory cytokines, decreased production of anti-inflammatory cytokines and genetic variations in IL-17F genes, which causes inflammation and cancer susceptibility. IL-17F rs763780 gene abnormality was also associated with the increased risk of Crohn's disease or UC (Cho et al. 2018). Diet with eggs, beef, pork or lamb has been reported to be associated with CRC risk. High dietary fat and high fat dairy products consumption was also shown to be associated with increased risk of CRC (Robertson et al. 1998).

7.2.11 Physical Activity and Obesity

Many studies have shown reduced risk of CRC with increased physical activity (Boyle et al. 2012; Wolin et al. 2009). This was supported by World Cancer Research Fund/American Institute for Cancer Research (Wiseman 2008). The observational epidemiological studies revealed 25–30% reduction in CRC risk with physical activity. Body mass index (BMI) $\geq$25 was associated with increased risk of CRC (Chao et al. 2004; Friedenreich et al. 2006; Wolin et al. 2007; Wolin et al. 2009). This was related to abnormal immune and endocrine systems which causes increase in pro-inflammatory adipokines (Tandon et al. 2015). Overweight (BMI, 25–30) association was about 9% and obesity (BMI $\geq$ 30) association was about 19% in comparison to normal BMI people (Moghaddam et al. 2007).

7.2.12 Socioeconomic Status

Low socio-economic status (SES) was associated with risk of CRC in comparison to high SES (Doubeni et al. 2012). Ecological analysis by Gorey et al., and Kee et al., found association between residence of lower SES and CRC. They reported men were more affected than women (Gorey et al. 1998; Kee et al. 1996). A study has related SES factor to incidence of right sided colon cancer (46.6%) (Doubeni et al. 2012).

7.3 Pathways Associated with Pathogenesis of Colorectal Cancer

The pathological pathways associated with development of CRC were reviewed to provide an overview of current knowledge on mechanism of CRC which may also be helpful in treatment. The risk factor associated mechanisms have been more targeted, so that the knowledge on the causal factor may be helpful for successful therapy as they differ from individual to individual (Fig. 7.1). Wang and Sun have elaborated the role of IL6 in development of CRC. IL6 upregulation has been reported to serve as autocrine tumorogenic factor. IL6 binds to IL6R, then this complex stimulates glycoprotein (gp130) which activates Janus kinase (JAK). This pathway results in phosphorylation of Signal transducer and activator of transcription3 (STAT3) and this activated STAT3 translocates to nucleus and causes transcription of genes like Bcl-xl, cyclic D1, Mcl1, c-myc, VEGF. These transcriptional events results in cancer cell proliferation (Wang and Sun 2014). Wang et al., have detailed the involvement of lysine residue acetyalation (K116) of FRA1 in development of CRC. They reported that chronic IL6 up-regulation increased FRA1 activity through STAT3, resulting in CRC. The HDAC6 has been reported to be key enzyme

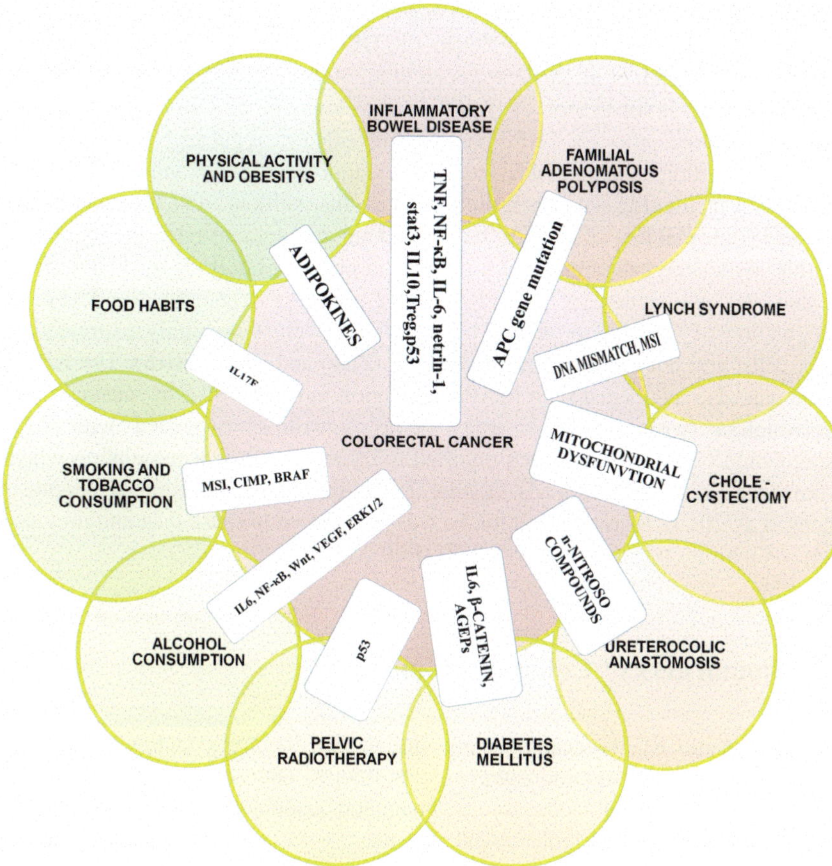

Fig. 7.1 Risk factors and associated pathways causing CRC

in this pathway and its mutation or dysregulation could also be aetiological to CRC (Wang et al. 2019). Chung et al., have demonstrated the synergistic activation of STAT3 by IL6 and TNFα. This activation leads to formation of NF-κB, STAT3 and STAT1 complex. This complex further binds to hTERT protein and increases telomerase activity. Increase in telomerase activity has been reported to develop CRC (Chung et al. 2017). De Simone et al., demonstrated the activation of STAT3 and NF-κB by IL6, TNFα and Th17 related cytokines and causing CRC. They believed that STAT3 synergising the actions of NF-κB through p65 acetylation, mediated by acetyltransferase p300 enzyme might result to development of CRC (De Simone et al. 2015). Yang et al., revealed the association between CRC and IL6/NF-κB pathway in dimethylhydrazine induced CRC mice model (Yang et al. 2014b). Yang et al., and Mager et al., have revealed the role of Th17 related pro-inflammatory cytokine IL17f in pathogenesis of CRC. NF-κB pathway was associated with this cytokine causing increased expression of antiapoptotic genes and tumorogenesis (Mager et al. 2016; Yang et al. 2014a). Kefeli et al., and Paradisi et al., have revealed that binding of Netrin-1 to DCC inhibited caspases and initiated tumorogenesis (Kefeli et al. 2017; Paradisi et al. 2009). Study by Buhrmann et al., has revealed the involvement of TNFβ in development of CRC. Activation of lymphotoxin β receptors by TNFβ has been revealed to regulate NF-κB pathway and FAK activation causing EMT. The EMT factors like vimetin and EMT specific transcription factor slug were increased through this pathway, and E-cadherin was decreased (Buhrmann et al. 2019). Zhao and Zhang has demonstrated the regulation of tumor-associated calcium signal transduction protein (TROP-2) by TNFα. They revealed the association of MAPK/ERK1/2 pathway triggered by TNFα in up-regulation of TROP-2 protein. The TROP-2 protein was highly expressed in cancer cells, and not seen in non-cancer cells (Zhao and Zhang 2018). The KRAS protein was also associated with MAPK activation through *BRAF* in causing CRC. The wingless (wg) and Int-1, traits (Wnt) signalling associated with APC dysregulation or absence has been reported to cause β catenin accumulation which on translocation to nucleus leads to gene transcription and inhibition of apoptosis (Cheng et al. 2019). Increased salt intake was also shown to cause β catenin induced apoptosis inhibition (Sumida et al. 2018) (Fig. 7.2).

7.4 Biomarkers of Colorectal Cancer

The diagnosis of CRC has been possible with various diagnostic methods like rectal examination, flexible sigmoidoscopy, colonoscopy, endoscopy capsule, barium enema, CT scan, virtual colonoscopy, magnetic resonance imaging, endorectal ultrasound, PET scan and biochemical diagnostic tests include guaiac test, immunohistochemistry, DNA analysis of fecal residues (Granados-Romero et al. 2017). Apart from pathological features, molecular analysis is more essential for accurate prognostic and predictive information of the disease. The analysis of biomarkers may provide information for early diagnosis, prognosis and therapeutic success in CRCs. Biomarkers that aid in early detection include methylated VIMENTIN (mVim). The early diagnosis has been inferred with most accurate method, colonoscopy in colorectal cancer screening. But the disadvantages of this method include poor patient compliance, complications and highly expensive. Identification of mVim marker through fecal occult blood test provides an inexpensive but low sensitive non-invasive colorectal cancer screening. This marker is identified with a PCR-based assay (Itzkowitz et al. 2008). The diagnostic markers include MSI, CIN, MMR genes (MLH1, MSH2, MSH6, PMS2), *BRAF*. These help in diagnosis and also in identifying the type of CRC. The MSI is assessed with 5 mononucleotide markers panel (BAT-25, BAT-26, NR-21, NR-24 and MONO-27) (Bacher et al. 2004). The CIN is assessed with DNA flow cytometry (Walther et al. 2008). The DNA MMR genes are suggestive for Lynch syndrome. These MMR genes were identified using DNA mutation analysis. The other common approach for diagnosis of Lynch syndrome includes first tier screening tests of immunohistochemistry for

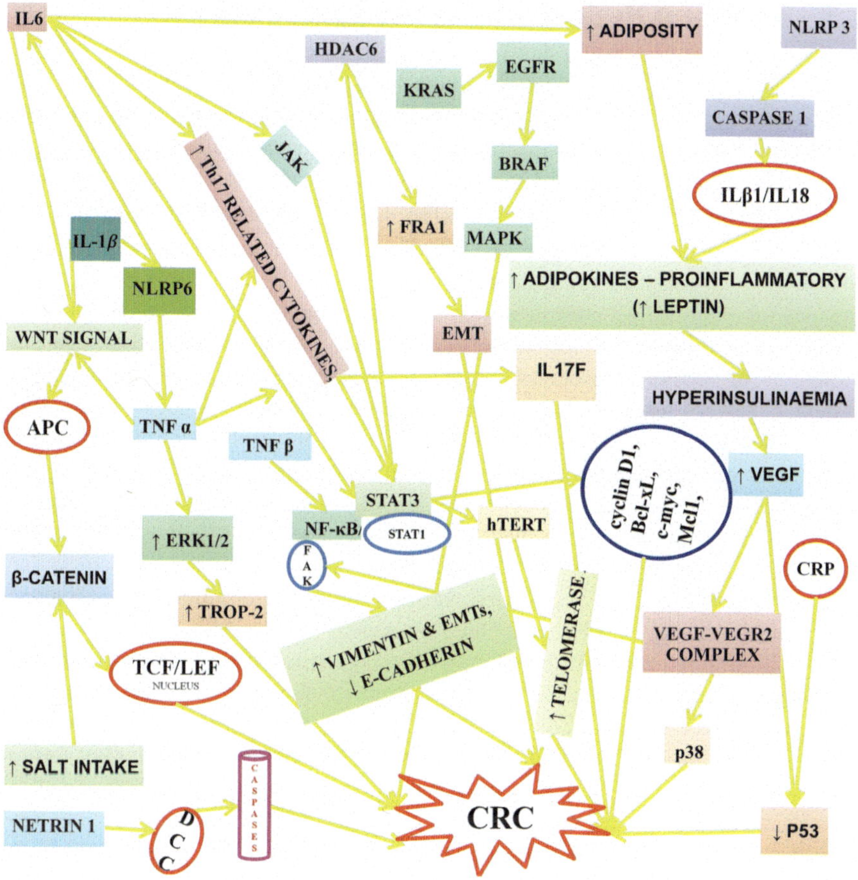

Fig. 7.2 Pathways, biomarkers associated with CRC

MMR genes evaluation and PCR for MSI identification. Then the tumors with MSI positive and loss of MLH1 protein will analysed for *BRAF* and MLH1-methylated status. Lynch syndrome MSI tumors do not exhibit *BRAF* mutation and MLH1-methylation. But sporadic MSI tumors were shown to be characterized with *BRAF* mutation and MLH1-methylation. And in those MSI tumors with loss of MSH2, MSH6 or PMS2 have high implication for germline mutation. This approach may exclude from highly expensive germline mutation testing (Baudhuin et al. 2005; Bettstetter et al. 2007; Hampel et al. 2005; Zhang 2008). The predictive markers include PTEN expression, SMAD4 expression evaluated through immunostaining, and *BRAF* (Bardelli and Siena 2010). Recently, it was reported that ETV6 gene was associated with CRC (Wang et al. 2016). These predictive markers may provide guidance for treatment selection.

7.5 Treatment Regimens and Associated Toxicities

Based on the histopathological findings, CRC was classified by WHO as, mucinous type, signet ring cell CRC, medullary CRC, micropapillary CRC, serrated CRC, cribriform comedo-type, adenosquamous type, spindle cell type, undifferentiated type. These histological findings and tumor staging of CRC represents as a factor for prognosis in CRC patients. The American Joint Committee on Cancer (AJCC) provided tumor-node metastasis (TNM). The T-stage represents depth of tumor invasion (T_x-primary tumor cannot be assessed, T_{is}-carcinoma in situ, T_1-tumor invades submucosa, T_2-tumor invades muscularis propia, T_3-tumor invades through the muscularis propria into the subserosa, T_4-tumor directly invades other organs or structures, or perforates visceral Peritoneum), N-stage represents risk for nodal and distant spread (N_x-regional lymph nodes cannot be assessed, N_0-no regional lymph node metastases, N_1-metastases in one to three regional lymph nodes, N_2-metastases in four or more regional lymph nodes). M-stage represents spread of tumor beyond colorectum and regional lymph nodes (M_x-presence or absence of distant metastases cannot be determined, M_0-no distant metastases detected, M_1-distant metastases detected) (Fleming et al. 2012).

The treatment of CRC was started with the use of 5-Fluorouracil (5-FU) alone. The standard dosing of 5-FU was based on patient's body surface area. Due to its short half life it required prolonged infusions and this resulted in patient discomfort and medical complications. Later it was combined with leucovorin which has synergistic effect, and this combination showed 11–23% increased response than 5-FU alone treatments (Ab Mutalib et al. 2017; Piedbois et al. 1992). Oral formulations of 5-FU prodrug (capecitabine) was introduced, and this reduced complications associated with prolonged central venous catheter infusions. This oral formulation was used as monotherapy for advance stage CRC (Hirsch and Zafar 2011). Drug combinations introduced for treating advanced staged CRC, like oxaliplatin/irinotecan have improved CRC treatment. Drug combinations 5-FU + leucovorin+irinotecan (FOLFIRI) and 5-FU + leucovorin+oxaliplatin (FOLFOX) have improved the treatment response. Another group of drugs introduced for treatment of CRC includes monoclonal antibodies like anti-VEGF antibody (bevacizumab), anti-EGFR antibodies (panitumumab, cetuximab), anti-VEGF agents (regorafenib, aflibercept) (Ab Mutalib et al. 2017). These treatment strategies (Table 7.1) aiding for the chance of cure or longer survival in CRC patients have shown negative impact in the form of toxicities. The most common toxicities included diarrhoea, neutropenia, thrombocytopenia, neurotoxicity, nausea, vomiting and infections. The 5-FU bolus regimens has been reported to produce more hematological toxicity like neutopenia and non-hematological toxicities like diarrhoea, mucositis in comparison to infusional 5-FU including regimens (Braun and Seymour 2011). Hand foot syndrome was more associated with infusional 5-FU regimen (de Gramont et al. 1997). Single agent therapy with irinotecan or in combination with 5-FU, has been reported to cause severe diarrhoea (Cunningham et al. 1998; Ledermann et al. 2001; Sargent et al. 2001). Treatments including

Table 7.1 CRC treatment regimens

S. No	Regimen	Description	Reference
1	LV5FU2	Leucovorin and 5-fluorouracil	(Piedbois et al. 1992)
2	de Gramont	Bolus dose of 5-FU 100 mg/m^2, followed by a 23 h 5-FU infusion on days 1 (400 mg/m^2) and 2 (600 mg/m^2) every 14 days, and subsequently a 46 h infusion	(de Gramont et al. 1997)
3	Lokich	Protracted venous infusion of 5-FU 300 mg/m^2/day	(Hale et al. 2002)
4	Raltitrexed	3 mg/m^2 i.v. q21d for an initial period of 12 weeks	
5	FOLFOX	Oxaliplatin with LV5FU2	(de Gramont et al. 2000)
6	FOLFOX-4	Oxaliplatin dose 85 mg/m^2	(Tournigand et al. 2006)
7	FOLFOX-7	Dose of oxaliplatin (130 mg/m^2) for six cycles	
8	Modified FOLFOX-7	Oxaliplatin dose 100 mg/m^2	(Chibaudel et al. 2009)
9	FOLFIRI	Irinotecan with an LV5FU2 infusion	(Tournigand et al. 2004)
10	IFL	Bolus of 5-FU 500 mg/m^2 and irinotecan 125 mg/m^2 given for 4 weeks out of 6	(Saltz et al. 2000)
11	Modified de Gramont	LV at a flat dose (350 mg d,l- or 175 mg l-LV, not adjusted for patient surface area), and on day 1 only.	(Cheeseman et al. 2002)
12	Mayo clinic	LV 20 mg/m^2 administered as a rapid i.v. infusion followed by a bolus injection of 5-FU 370–425 mg/m^2 administered daily on days 1–5 of a 28-day treatment cycle	(Cutsem et al. 2004)

oxaliplatin showed neurosensory toxicities like cold induced parasthesia and dose dependent chronic peripheral neuropathy (Grothey 2005). Toxicities related to prolonged chemotherapy were foot-hand syndrome, oxaliplatin related neuropathies, and fatigue (Braun and Seymour 2011). A case report has shown the association of ocular toxicity with 5-FU and folinic acid combination therapy in CRC patient (Baskin et al. 2015). The toxicities associated with chemotherapy for CRC, varied from individual to individual depending upon their genetic abnormalities or variabilities. So far identified genetic variations causing toxicities with CRC treatments include genes like DPYD (dihydropyrimidine dehydrogenase), TYMS (thymidylate synthase), ENOSF1 (antisense RNA to thymidylate synthase), MTHFR (methylenetetrahydrofolate reductase), ABCB1 (ATP-dependent translocase/drug transporter gene family), CDA (Cytidine deaminase), SMAD4 (Mothers against decapentaplegic homolog 4). These genes regulate the intensity or chances of toxicities in patients with CRC treatment (Ab Mutalib et al. 2017; Alhopuro et al. 2005). Dose related toxicity with irinotecan was associated with UGT1A1 (UDP-glucuronosyl transferase). Also consideration of tumor markers like 18q LOH, MSI, topoisomerase1 expression, may provide suitable drug for effective treatment in patient of CRC. Patients with MSI marker expression tumors were found ineffective to 5-FU or 5-FU based regimen treatment. The patients with 18q LOH marker expression showed toxicities with 5-FU based treatments. Patients with topoisomerase1 expression showed better responsiveness with irinotecan in comparison to patients without topoisomerase1 expression (Braun et al. 2008). Tumors expressing CIN marker shows poor prognosis to CRC treatment regimens. Lamivudine, dendritic cell vaccination are being investigated under phase II clinical trials for treatment of CRC (NCT03144804: A Phase 2 Study of Lamivudine in Patients With p53 Mutant Metastatic Colorectal Cancer, n.d.; NCT03730948: Pilot Study of Mature Dendritic Cell Vaccination for Resected Hypermutated Colorectal Cancer, n.d.). The combination of Dabrafenib + Trametinib + PDR001 is under phase II clinical trial investigation for treatment of CRC (Rodriguez et al. 2018).

7.6 Conclusion

In summary, it may be concluded that treatment of CRC and its efficacy differs from individual to individual due to genetic variability and other risk factors elucidated in this review. The pathological and definitive molecular diagnosis, complete history of patient for various risk factors before treatment initiation may provide a broad scope for precision medicine and therapeutic efficacy in treatment of CRC. Further studies may be required to evaluate the association of other risk factors with efficacy and toxicity of CRC treatments which may yield biomarkers helpful for personalized medicine in CRC.

References

Ab Mutalib NS, Md Yusof NF, Abdul SN, Jamal R (2017) Pharmacogenomics DNA biomarkers in colorectal Cancer: current update. Front Pharmacol 8:736

Acalovschi M, Lammert F (2012) The growing global burden of gallstone disease. World Gastroenterol News 17:6–9

Alhopuro P, Alazzouzi H, Sammalkorpi H, Davalos V, Salovaara R, Hemminki A, Jarvinen H, Mecklin JP, Schwartz S Jr, Aaltonen LA, Arango D (2005) SMAD4 levels and response to 5-fluorouracil in colorectal cancer. Clin Cancer Res 11:6311–6316

Arnold M, Sierra MS, Laversanne M, Soerjomataram I, Jemal A, Bray F (2017) Global patterns and trends in colorectal cancer incidence and mortality. Gut 66:683–691

Bacher JW, Flanagan LA, Smalley RL, Nassif NA, Burgart LJ, Halberg RB, Megid WM, Thibodeau SN (2004) Development of a fluorescent multiplex assay for detection of MSI-high tumors. Dis Markers 20:237–250

Bardelli A, Siena S (2010) Molecular mechanisms of resistance to cetuximab and panitumumab in colorectal cancer. J Clin Oncol Off J Am Soc Clin Oncol 28:1254–1261

Bargen JA (1994) Chronic ulcerative colitis associated with malignant disease. 1928. Dis Colon Rectum 37:727–730

Baskin Y, Amirfallah A, Unal OU, Calibasi G, Oztop I (2015) Dihydropyrimidine dehydrogenase 85T>C mutation is associated with ocular toxicity of 5-fluorouracil: a case report. Am J Ther 22:e36–e39

Baudhuin LM, Burgart LJ, Leontovich O, Thibodeau SN (2005) Use of microsatellite instability and immunohistochemistry testing for the identification of individuals at risk for Lynch syndrome. Familial Cancer 4:255–265

Bettstetter M, Dechant S, Ruemmele P, Grabowski M, Keller G, Holinski-Feder E, Hartmann A, Hofstaedter F, Dietmaier W (2007) Distinction of hereditary nonpolyposis colorectal cancer and sporadic microsatellite-unstable colorectal cancer through quantification of MLH1 methylation by real-time PCR. Clin Cancer Res 13:3221–3228

Birgisson H, Pahlman L, Gunnarsson U, Glimelius B (2005) Occurrence of second cancers in patients treated with radiotherapy for rectal cancer. J Clin Oncol Off J Am Soc Clin Oncol 23:6126–6131

Bojuwoye MO, Olokoba AB, Ogunlaja OA, Agodirin SO, Ibrahim OK, Okonkwo KC, Aliyu AM (2018) Familial adenomatous polyposis syndrome with colorectal cancer in two Nigerians: a report of two cases and review of literature. Pan Afr Med J 30:6

Boyle T, Keegel T, Bull F, Heyworth J, Fritschi L (2012) Physical activity and risks of proximal and distal colon cancers: a systematic review and meta-analysis. J Natl Cancer Inst 104:1548–1561

Braun MS, Seymour MT (2011) Balancing the efficacy and toxicity of chemotherapy in colorectal cancer. Ther Adv Med Oncol 3:43–52

Braun MS, Richman SD, Quirke P, Daly C, Adlard JW, Elliott F, Barrett JH, Selby P, Meade AM, Stephens RJ, Parmar MK, Seymour MT (2008) Predictive biomarkers of chemotherapy efficacy in colorectal cancer: results from the UK MRC FOCUS trial. J Clin Oncol Off J Am Soc Clin Oncol 26:2690–2698

Bray F, Ferlay J, Soerjomataram I, Siegel RL, Torre LA, Jemal A (2018) Global cancer statistics 2018: GLOBOCAN estimates of incidence and mortality worldwide for 36 cancers in 185 countries. CA Cancer J Clin 68:394–424

Buhrmann C, Yazdi M, Popper B, Kunnumakkara AB, Aggarwal BB, Shakibaei M (2019) Induction of the epithelial-to-Mesenchymal transition of human colorectal Cancer by human TNF-β (Lymphotoxin) and its reversal by resveratrol. Nutrients 11:704

Castro EB, Rosen PP, Quan SH (1973) Carcinoma of large intestine in patients irradiated for carcinoma of cervix and uterus. Cancer 31:45–52

Chang H-C, Horng J-T, Lin W-C, Lai H-W, Chang C-W, Chen T-A (2012) Evaluation of the appropriate age range of colorectal cancer screening based on the changing epidemiology in the past 20 years in Taiwan. ISRN Gastroenterol 2012:960867–960867

Chao A, Connell CJ, Jacobs EJ, McCullough ML, Patel AV, Calle EE, Cokkinides VE, Thun MJ (2004) Amount, type, and timing of recreational physical activity in relation to colon and rectal cancer in older adults: the Cancer prevention study II nutrition cohort. Cancer Epidemiol Biomarkers Prev 13:2187–2195

Cheeseman SL, Joel SP, Chester JD, Wilson G, Dent JT, Richards FJ, Seymour MT (2002) A 'modified de Gramont' regimen of fluorouracil, alone and with oxaliplatin, for advanced colorectal cancer. Br J Cancer 87:393–399

Chen GY, Liu M, Wang F, Bertin J, Nunez G (2011) A functional role for Nlrp6 in intestinal inflammation and tumorigenesis. J Immunol 186:7187–7194

Chen H, Wu L, Li Y, Meng J, Lin N, Yang D, Zhu Y, Li X, Li M, Xu Y, Wu Y, Tong X, Su Q (2014) Advanced glycation end products increase carbohydrate responsive element binding protein expression and promote cancer cell proliferation. Mol Cell Endocrinol 395:69–78

Cheng X, Xu X, Chen D, Zhao F, Wang W (2019) Therapeutic potential of targeting the Wnt/beta-catenin signaling pathway in colorectal cancer. Biomed Pharmacother 110:473–481

Chibaudel B, Maindrault-Goebel F, Lledo G, Mineur L, Andre T, Bennamoun M, Mabro M, Artru P, Carola E, Flesch M, Dupuis O, Colin P, Larsen AK, Afchain P, Tournigand C, Louvet C, de Gramont A (2009) Can chemotherapy be discontinued in unresectable metastatic colorectal cancer? The GERCOR OPTIMOX2 study. J Clin Oncol Off J Am Soc Clin Oncol 27:5727–5733

Cho YA, Lee J, Oh JH, Chang HJ, Sohn DK, Shin A, Kim J (2018) Inflammatory dietary pattern, IL-17F genetic variant, and the risk of colorectal Cancer. Nutrients 10:724

Chung SS, Wu Y, Okobi Q, Adekoya D, Atefi M, Clarke O, Dutta P, Vadgama JV (2017) Proinflammatory cytokines IL-6 and TNF-α increased telomerase activity through NF-κB/STAT1/STAT3 activation, and Withaferin a inhibited the signaling in colorectal Cancer cells. Mediat Inflamm 2017:5958429–5958429

Crohn BB, Rosenberg H (1925) The SIGMOIDOSCOPIC picture of chronic ulcerative colitis (non-specific). Am J Med Sci 170:220–227

Cunningham D, Pyrhonen S, James RD, Punt CJ, Hickish TF, Heikkila R, Johannesen TB, Starkhammar H, Topham CA, Awad L, Jacques C, Herait P (1998) Randomised trial of irinotecan plus supportive care versus supportive care alone after fluorouracil failure for patients with metastatic colorectal cancer. Lancet 352:1413–1418

Cutsem EV, Hoff PM, Harper P, Bukowski RM, Cunningham D, Dufour P, Graeven U, Lokich J, Madajewicz S, Maroun JA, Marshall JL, Mitchell EP, Perez-Manga G, Rougier P, Schmiegel W, Schoelmerich J, Sobrero A, Schilsky RL (2004) Oral capecitabine vs intravenous 5-fluorouracil and leucovorin: integrated efficacy data and novel analyses from two large, randomised, phase III trials. Br J Cancer 90:1190–1197

de Gramont A, Bosset JF, Milan C, Rougier P, Bouche O, Etienne PL, Morvan F, Louvet C, Guillot T, Francois E, Bedenne L (1997) Randomized trial comparing monthly low-dose leucovorin and fluorouracil bolus with bimonthly high-dose leucovorin and fluorouracil bolus plus continuous infusion for advanced colorectal cancer: a French intergroup study. J Clin Oncol Off J Am Soc Clin Oncol 15:808–815

de Gramont A, Figer A, Seymour M, Homerin M, Hmissi A, Cassidy J, Boni C, Cortes-Funes H, Cervantes A, Freyer G, Papamichael D, Le Bail N, Louvet C, Hendler D, de Braud F, Wilson C, Morvan F, Bonetti A (2000) Leucovorin and fluorouracil with or without oxaliplatin as first-line treatment in advanced colorectal cancer. J Clin Oncol Off J Am Soc Clin Oncol 18:2938–2947

De Simone V, Franze E, Ronchetti G, Colantoni A, Fantini MC, Di Fusco D, Sica GS, Sileri P, MacDonald TT, Pallone F, Monteleone G, Stolfi C (2015) Th17-type cytokines, IL-6 and TNF-alpha synergistically activate STAT3 and NF-kB to promote colorectal cancer cell growth. Oncogene 34:3493–3503

Doubeni CA, Laiyemo AO, Major JM, Schootman M, Lian M, Park Y, Graubard BI, Hollenbeck AR, Sinha R (2012) Socioeconomic status and the risk of colorectal cancer: an analysis of more than a half million adults in the National Institutes of Health-AARP diet and health study. Cancer 118:3636–3644

Elinav E, Strowig T, Kau AL, Henao-Mejia J, Thaiss CA, Booth CJ, Peaper DR, Bertin J, Eisenbarth SC, Gordon JI, Flavell RA (2011) NLRP6 inflammasome regulates colonic microbial ecology and risk for colitis. Cell 145:745–757

Fearon ER, Vogelstein B (1990) A genetic model for colorectal tumorigenesis. Cell 61:759–767

Fitzmaurice C, Allen C, Barber RM, Barregard L, Bhutta ZA, Brenner H, Dicker DJ, Chimed-Orchir O, Dandona R, Dandona L, Fleming T, Forouzanfar MH, Hancock J, Hay RJ, Hunter-Merrill R, Huynh C, Hosgood HD, Johnson CO, Jonas JB, Khubchandani J, Kumar GA, Kutz M, Lan Q, Larson HJ, Liang X, Lim SS, Lopez AD, MF MI, Marczak L, Marquez N, Mokdad AH, Pinho C, Pourmalek F, Salomon JA, Sanabria JR, Sandar L, Sartorius B, Schwartz SM, Shackelford KA, Shibuya K, Stanaway J, Steiner C, Sun J, Takahashi K, Vollset SE, Vos T, Wagner JA, Wang H, Westerman R, Zeeb H, Zoeckler L, Abd-Allah F, Ahmed MB, Alabed S, Alam NK, Aldhahri SF, Alem G, Alemayohu MA, Ali R, Al-Raddadi R, Amare A, Amoako Y, Artaman A, Asayesh H, Atnafu N, Awasthi A, Saleem HB, Barac A, Bedi N, Bensenor I, Berhane A, Bernabe E, Betsu B, Binagwaho A, Boneya D, Campos-Nonato I, Castaneda-Orjuela C, Catala-Lopez F, Chiang P, Chibueze C, Chitheer A, Choi JY, Cowie B, Damtew S, das Neves J, Dey S, Dharmaratne S, Dhillon P, Ding E, Driscoll T, Ekwueme D, Endries AY, Farvid M, Farzadfar F, Fernandes J, Fischer F, GH TT, Gebru A, Gopalani S, Hailu A, Horino M, Horita N, Husseini A, Huybrechts I, Inoue M, Islami F, Jakovljevic M, James S, Javanbakht M, Jee SH, Kasaeian A, Kedir MS, Khader YS, Khang YH, Kim D, Leigh J, Linn S, Lunevicius R, HMA ER, Malekzadeh R, Malta DC, Marcenes W, Markos D, Melaku YA, Meles KG, Mendoza W, Mengiste DT, Meretoja TJ, Miller TR, Mohammad KA,

Mohammadi A, Mohammed S, Moradi-Lakeh M, Nagel G, Nand D, Le Nguyen Q, Nolte S, Ogbo FA, Oladimeji KE, Oren E, Pa M, Park EK, Pereira DM, Plass D, Qorbani M, Radfar A, Rafay A, Rahman M, Rana SM, Soreide K, Satpathy M, Sawhney M, Sepanlou SG, Shaikh MA, She J, Shiue I, Shore HR, Shrime MG, So S, Soneji S, Stathopoulou V, Stroumpoulis K, Sufiyan MB, Sykes BL, Tabares-Seisdedos R, Tadese F, Tedla BA, Tessema GA, Thakur JS, Tran BX, Ukwaja KN, BSC U, Vlassov VV, Weiderpass E, Wubshet Terefe M, Yebyo HG, Yimam HH, Yonemoto N, Younis MZ, Yu C, Zaidi Z, MES Z, Zenebe ZM, CJL M, Naghavi M (2017) Global, regional, and National Cancer Incidence, mortality, years of life lost, years lived with disability, and disability-adjusted life-years for 32 Cancer groups, 1990 to 2015: a systematic analysis for the global burden of disease study. JAMA Oncol 3:524–548

Fleming M, Ravula S, Tatishchev SF, Wang HL (2012) Colorectal carcinoma: pathologic aspects. J Gastrointest Oncol 3:153–173

Friedenreich C, Norat T, Steindorf K, Boutron-Ruault M-C, Pischon T, Mazuir M, Clavel-Chapelon F, Linseisen J, Boeing H, Bergman M, Johnsen NF, Tjønneland A, Overvad K, Mendez M, Quirós JR, Martinez C, Dorronsoro M, Navarro C, Gurrea AB, Bingham S, Khaw K-T, Allen N, Key T, Trichopoulou A, Trichopoulos D, Orfanou N, Krogh V, Palli D, Tumino R, Panico S, Vineis P, Bueno-de-Mesquita HB, PHM P, Monninkhof E, Berglund G, Manjer J, Ferrari P, Slimani N, Kaaks R, Riboli E (2006) Physical activity and risk of Colon and Rectal cancers: the European prospective investigation into Cancer and nutrition. Cancer Epidemiol Biomarkers Prev 15:2398

Gonzalez N, Prieto I, Del Puerto-Nevado L, Portal-Nunez S, Ardura JA, Corton M, Fernandez-Fernandez B, Aguilera O, Gomez-Guerrero C, Mas S, Moreno JA, Ruiz-Ortega M, Sanz AB, Sanchez-Nino MD, Rojo F, Vivanco F, Esbrit P, Ayuso C, Alvarez-Llamas G, Egido J, Garcia-Foncillas J, Ortiz A (2017) 2017 update on the relationship between diabetes and colorectal cancer: epidemiology, potential molecular mechanisms and therapeutic implications. Oncotarget 8:18456–18485

Gorey KM, Holowaty EJ, Laukkanen E, Fehringer G, Richter NL (1998) Association between socioeconomic status and cancer incidence in Toronto, Ontario: possible confounding of cancer mortality by incidence and survival. Cancer Prev Control 2:236–241

Granados-Romero JJ, Valderrama-Treviño AI, Contreras-Flores EH, Barrera-Mera B, Herrera Enríquez M, Uriarte-Ruíz K, Ceballos-Villalba J, Estrada-Mata AG, Alvarado Rodríguez C, Arauz-Peña G (2017) Colorectal cancer: a review. Int J Res Med Sci 5:4667

Grothey A (2005) Clinical management of oxaliplatin-associated neurotoxicity. Clin Colorectal Cancer 5(Suppl 1):S38–S46

Gyde S, Prior P, Dew MJ, Saunders V, Waterhouse JA, Allan RN (1982) Mortality in ulcerative colitis. Gastroenterology 83:36–43

Hale JP, Cohen DR, Maughan TS, Stephens RJ (2002) Costs and consequences of different chemotherapy regimens in metastatic colorectal cancer. Br J Cancer 86:1684–1690

Hampel H, Frankel WL, Martin E, Arnold M, Khanduja K, Kuebler P, Nakagawa H, Sotamaa K, Prior TW, Westman J, Panescu J, Fix D, Lockman J, Comeras I, de la Chapelle A (2005) Screening for the Lynch syndrome (hereditary nonpolyposis colorectal cancer). N Engl J Med 352:1851–1860

Henrikson NB, Webber EM, Goddard KA, Scrol A, Piper M, Williams MS, Zallen DT, Calonge N, Ganiats TG, Janssens AC, Zauber A, Lansdorp-Vogelaar I, van Ballegooijen M, Whitlock EP (2015) Family history and the natural history of colorectal cancer: systematic review. Genet Med: J Am Coll Med Genet 17:702–712

Hirsch BR, Zafar SY (2011) Capecitabine in the management of colorectal cancer. Cancer Manag Res 3:79–89

Hofseth LJ, Saito S, Hussain SP, Espey MG, Miranda KM, Araki Y, Jhappan C, Higashimoto Y, He P, Linke SP, Quezado MM, Zurer I, Rotter V, Wink DA, Appella E, Harris CC (2003) Nitric oxide-induced cellular stress and p53 activation in chronic inflammation. Proc Natl Acad Sci U S A 100:143–148

Hu B, Elinav E, Huber S, Strowig T, Hao L, Hafemann A, Jin C, Wunderlich C, Wunderlich T, Eisenbarth SC, Flavell RA (2013) Microbiota-induced activation of epithelial IL-6 signaling links inflammasome-driven inflammation with transmissible cancer. Proc Natl Acad Sci U S A 110:9862–9867

Itzkowitz S, Brand R, Jandorf L, Durkee K, Millholland J, Rabeneck L, Schroy PC 3rd, Sontag S, Johnson D, Markowitz S, Paszat L, Berger BM (2008) A simplified, noninvasive stool DNA test for colorectal cancer detection. Am J Gastroenterol 103:2862–2870

Kee F, Wilson R, Currie S, Sloan J, Houston R, Rowlands B, Moorehead J (1996) Socioeconomic circumstances and the risk of bowel cancer in Northern Ireland. J Epidemiol Community Health 50:640–644

Kefeli U, Ucuncu Kefeli A, Cabuk D, Isik U, Sonkaya A, Acikgoz O, Ozden E, Uygun K (2017) Netrin-1 in cancer: potential biomarker and therapeutic target? Tumour Biol 39:1010428317698388

Knudson AG Jr (1974) Heredity and human cancer. Am J Pathol 77:77–84

Knudson AG Jr, Strong LC (1972) Mutation and cancer: a model for Wilms' tumor of the kidney. J Natl Cancer Inst 48:313–324

Ledermann JA, Leonard P, Seymour M (2001) Recommendation for caution with irinotecan, fluorouracil, and leucovorin for colorectal cancer. N Engl J Med 345:145–146

Lee J, Shin A, Oh JH, Kim J (2017) Colors of vegetables and fruits and the risks of colorectal cancer. World J Gastroenterol 23:2527–2538

Li GM (2008) Mechanisms and functions of DNA mismatch repair. Cell Res 18:85–98

Li JY, Yu T, Xia ZS, Chen GC, Yuan YH, Zhong W, Zhao LN, Chen QK (2014) Enhanced proliferation in colorectal epithelium of patients with type 2 diabetes correlates with beta-catenin accumulation. J Diabetes Complicat 28:689–697

Liang J, Nagahashi M, Kim EY, Harikumar KB, Yamada A, Huang WC, Hait NC, Allegood JC, Price MM, Avni D, Takabe K, Kordula T, Milstien S, Spiegel S (2013) Sphingosine-1-phosphate links persistent STAT3 activation, chronic intestinal inflammation, and development of colitis-associated cancer. Cancer Cell 23:107–120

Limsui D, Vierkant RA, Tillmans LS, Wang AH, Weisenberger DJ, Laird PW, Lynch CF, Anderson KE, French AJ, Haile RW, Harnack LJ, Potter JD, Slager SL, Smyrk TC, Thibodeau SN, Cerhan JR, Limburg PJ (2010) Cigarette smoking and colorectal cancer risk by molecularly defined subtypes. J Natl Cancer Inst 102:1012–1022

Lynch HT, Krush AJ (1971) Cancer family "G" revisited: 1895–1970. Cancer 27:1505–1511

Mager LF, Wasmer MH, Rau TT, Krebs P (2016) Cytokine-induced modulation of colorectal Cancer. Front Oncol 6:96

Makino K, Kume H, Morikawa T, Niimi A, Fujimura T, Nakagawa T, Fukuhara H, Homma Y (2015) Adenocarcinoma at anastomotic site of Ureterosigmoidostomy potentially of Urothelial origin spreading to the upper urinary tract. Case Rep Urol 2015:564082–564082

Mantovani A, Allavena P, Sica A, Balkwill F (2008) Cancer-related inflammation. Nature 454:436–444

Miller KD, Siegel RL, Lin CC, Mariotto AB, Kramer JL, Rowland JH, Stein KD, Alteri R, Jemal A (2016) Cancer treatment and survivorship statistics, 2016. CA Cancer J Clin 66:271–289

Minami K, Matsuzaki S, Hayashi N, Mokarim A, Ito M, Sekine I (1998) Immunohistochemical study of p53 overexpression in radiation-induced colon cancers. J Radiat Res 39:1–10

Moghaddam AA, Woodward M, Huxley R (2007) Obesity and risk of colorectal cancer: a meta-analysis of 31 studies with 70,000 events. Cancer Epidemiol Biomarkers Prev 16:2533–2547

Moller P, Seppala T, Bernstein I, Holinski-Feder E, Sala P, Evans DG, Lindblom A, Macrae F, Blanco I, Sijmons R, Jeffries J, Vasen H, Burn J, Nakken S, Hovig E, Rodland EA, Tharmaratnam K, de Vos Tot Nederveen Cappel WH, Hill J, Wijnen J, Green K, Lalloo F, Sunde L, Mints M, Bertario L, Pineda M, Navarro M, Morak M, Renkonen-Sinisalo L, Frayling IM, Plazzer JP, Pylvanainen K, Sampson JR, Capella G, Mecklin JP, Moslein G (2017) Cancer incidence and survival in Lynch syndrome patients receiving colonoscopic and gynaecological surveillance: first report from the prospective Lynch syndrome database. Gut 66:464–472

Moore K, Sussman D, Koru-Sengul T (2018) Age-specific risk factors for advanced stage colorectal Cancer, 1981–2013. Prev Chronic Dis 15:170274

NCT03144804: a phase 2 study of lamivudine in patients with p53 mutant metastatic colorectal cancer. https://clinicaltrials.gov/ct2/show/NCT03144804

NCT03730948: pilot study of mature dendritic cell vaccination for resected hypermutated colorectal cancer. https://clinicaltrials.gov/ct2/show/NCT03730948

Paradisi A, Maisse C, Coissieux M-M, Gadot N, Lépinasse F, Delloye-Bourgeois C, Delcros J-G, Svrcek M, Neufert C, Fléjou J-F, Scoazec J-Y, Mehlen P (2009) Netrin-1 up-regulation in inflammatory bowel diseases is required for colorectal cancer progression. Proc Natl Acad Sci U S A 106:17146–17151

Perez CA, Breaux S, Bedwinek JM, Madoc-Jones H, Camel HM, Purdy JA, Walz BJ (1984) Radiation therapy alone in the treatment of carcinoma of the uterine cervix. II. Analysis of complications. Cancer 54:235–246

Piedbois P, Buyse M, Rustum Y, Machover D, Erlichman C, Carlson RW, Valone F, Labianca R, Doroshow JH, Petrelli N (1992) Modulation of fluorouracil by leucovorin in patients with advanced colorectal cancer: evidence in terms of response rate. Advanced colorectal cancer meta-analysis project. J Clin Oncol Off J Am Soc Clin Oncol 10:896–903

Robertson I, Bound R, Segal L (1998) Colorectal cancer, diet and lifestyle factors: opportunities for prevention. Health Promot Int 13:141–150

Rodriguez J, Castañón E, Perez-Gracia JL, Rodriguez I, Viudez A, Alfaro C, Oñate C, Perez G, Rotellar F, Inogés S, López-Diaz de Cerio A, Resano L, Ponz-Sarvise M, Rodriguez-Ruiz ME, Chopitea A, Vera R, Melero I (2018) A randomized phase II clinical trial of dendritic cell vaccination following complete resection of colon cancer liver metastasis. J Immunother Cancer 6:96

Rojas A, Figueroa H, Morales E (2010) Fueling inflammation at tumor microenvironment: the role of multiligand/RAGE axis. Carcinogenesis 31:334–341

Rossi M, Jahanzaib Anwar M, Usman A, Keshavarzian A, Bishehsari F (2018) Colorectal Cancer and alcohol consumption-populations to molecules. Cancers 10:38

Saltz LB, Cox JV, Blanke C, Rosen LS, Fehrenbacher L, Moore MJ, Maroun JA, Ackland SP, Locker PK, Pirotta N, Elfring GL, Miller LL (2000) Irinotecan plus fluorouracil and leucovorin for metastatic colorectal cancer. Irinotecan study group. N Engl J Med 343:905–914

Sargent DJ, Niedzwiecki D, O'Connell MJ, Schilsky RL (2001) Recommendation for caution with irinotecan, fluorouracil, and leucovorin for colorectal cancer. N Engl J Med 345:144–145

Sasaki K, Ishihara S, Hata K, Kiyomatsu T, Nozawa H, Kawai K, Tanaka T, Nishikawa T, Otani K, Yasuda K, Kaneko M, Murono K, Abe H, Morikawa T, Watanabe T (2017) Radiation-associated colon cancer: a case report. Mol Clin Oncol 6:817–820

Schernhammer ES, Leitzmann MF, Michaud DS, Speizer FE, Giovannucci E, Colditz GA, Fuchs CS (2003) Cholecystectomy and the risk for developing colorectal cancer and distal colorectal adenomas. Br J Cancer 88:79–83

Schlussel AT, Gagliano RA Jr, Seto-Donlon S, Eggerding F, Donlon T, Berenberg J, Lynch HT (2014) The evolution of colorectal cancer genetics-part 1: from discovery to practice. J Gastrointest Oncol 5:326–335

Schoen RE, Pinsky PF, Weissfeld JL, Yokochi LA, Church T, Laiyemo AO, Bresalier R, Andriole GL, Buys SS, Crawford ED, Fouad MN, Isaacs C, Johnson CC, Reding DJ, O'Brien B, Carrick DM, Wright P, Riley TL, Purdue MP, Izmirlian G, Kramer BS, Miller AB, Gohagan JK, Prorok PC, Berg CD (2012) Colorectal-cancer incidence and mortality with screening flexible sigmoidoscopy. N Engl J Med 366:2345–2357

Sharma D, Singh G (2017) Clinico-pathological profile of colorectal cancer in first two decades of life: a retrospective analysis from tertiary health center. Indian J Cancer 54:397–400

Shin A, Lee J, Lee J, Park MS, Park JW, Park SC, Oh JH, Kim J (2015) Isoflavone and Soyfood intake and colorectal Cancer risk: a case-control study in Korea. PLoS One 10:e0143228–e0143228

Siegel RL, Miller KD, Jemal A (2018) Cancer statistics, 2018. CA Cancer J Clin 68:7–30

Sumida T, Lincoln MR, Ukeje CM, Rodriguez DM, Akazawa H, Noda T, Naito AT, Komuro I, Dominguez-Villar M, Hafler DA (2018) Activated β-catenin in Foxp3+ regulatory T cells links inflammatory environments to autoimmunity. Nat Immunol 19:1391–1402

Svartz N, Ernberg T (1949) Cancer coli in cases of colitis ulcerosa. Acta Med Scand 135:444–447

Tamai O, Nozato E, Miyazato H, Isa T, Hiroyasu S, Shiraishi M, Kusano T, Muto Y, Higashi M (1999) Radiation-associated rectal cancer: report of four cases. Dig Surg 16:238–243

Tandon K, Imam M, Ismail BES, Castro F (2015) Body mass index and colon cancer screening: the road ahead. World J Gastroenterol 21:1371–1376

Tomori H, Yasuda T, Shiraishi M, Isa T, Muto Y, Egawa H (1999) Radiation-associated ischemic coloproctitis: report of two cases. Surg Today 29:1088–1092

Tournigand C, Andre T, Achille E, Lledo G, Flesh M, Mery-Mignard D, Quinaux E, Couteau C, Buyse M, Ganem G, Landi B, Colin P, Louvet C, de Gramont A (2004) FOLFIRI followed by FOLFOX6 or the reverse sequence in advanced colorectal cancer: a randomized GERCOR study. J Clin Oncol Off J Am Soc Clin Oncol 22:229–237

Tournigand C, Cervantes A, Figer A, Lledo G, Flesch M, Buyse M, Mineur L, Carola E, Etienne PL, Rivera F, Chirivella I, Perez-Staub N, Louvet C, Andre T, Tabah-Fisch I, de Gramont A (2006) OPTIMOX1: a randomized study of FOLFOX4 or FOLFOX7 with oxaliplatin in a stop-and-go fashion in advanced colorectal cancer--a GERCOR study. J Clin Oncol Off J Am Soc Clin Oncol 24:394–400

Tsuji T, Sawai T, Nakagoe T, Hidaka S, Shibasaki S, Tanaka K, Nanashima A, Yamaguchi H, Yasutake T, Tagawa Y (2003) Genetic analysis of radiation-associated rectal cancer. J Gastroenterol 38:1185–1188

Vuik FER, Nieuwenburg SAV, Bardou M, Lansdorp-Vogelaar I, Dinis-Ribeiro M, Bento MJ, Zadnik V, Pellisé M, Esteban L, Kaminski MF, Suchanek S, Ngo O, Májek O, Leja M, Kuipers EJ, Spaander MCW (2019) Increasing incidence of colorectal cancer in young adults in Europe over the last 25 years. Gut gutjnl-2018:317592

Walther A, Houlston R, Tomlinson I (2008) Association between chromosomal instability and prognosis in colorectal cancer: a meta-analysis. Gut 57:941–950

Wang SW, Sun YM (2014) The IL-6/JAK/STAT3 pathway: potential therapeutic strategies in treating colorectal cancer (review). Int J Oncol 44:1032–1040

Wang M, Gu D, Du M, Xu Z, Zhang S, Zhu L, Lu J, Zhang R, Xing J, Miao X, Chu H, Hu Z, Yang L, Tang C, Pan L, Du H, Zhao J, Du J, Tong N, Sun J, Shen H, Xu J, Zhang Z, Chen J (2016) Common genetic variation in ETV6 is associated with colorectal cancer susceptibility. Nat Commun 7:11478

Wang T, Song P, Zhong T, Wang X, Xiang X, Liu Q, Chen H, Xia T, Liu H, Niu Y, Hu Y, Xu L, Shao Y, Zhu L, Qi H, Shen J, Hou T, Fodde R, Shao J (2019) The inflammatory cytokine IL-6 induces FRA1 deacetylation promoting colorectal cancer stem-like properties. Oncogene 38:4932–4947

Warthin AS (1985) Classicsin oncology: heredity with reference to carcinoma as shown by the study of the cases examined in the pathological laboratory of the university of Michigan, 1895-1913. CA Cancer J Clin 35:348–359

Wiseman M (2008) The second world cancer research fund/American Institute for Cancer Research expert report. food, nutrition, physical activity, and the prevention of cancer: a global perspective. Proc Nutr Soc 67:253–256

Wolin KY, Glynn RJ, Colditz GA, Lee IM, Kawachi I (2007) Long-term physical activity patterns and health-related quality of life in U.S. women. Am J Prev Med 32:490–499

Wolin KY, Yan Y, Colditz GA, Lee IM (2009) Physical activity and colon cancer prevention: a meta-analysis. Br J Cancer 100:611–616

Wong MC, Ding H, Wang J, Chan PS, Huang J (2019) Prevalence and risk factors of colorectal cancer in Asia. Intest Res 17:317

Wu Y, Zhou BP (2009) Inflammation: a driving force speeds cancer metastasis. Cell Cycle 8:3267–3273

Xavier RJ, Podolsky DK (2007) Unravelling the pathogenesis of inflammatory bowel disease. Nature 448:427–434

Yan H, Papadopoulos N, Marra G, Perrera C, Jiricny J, Boland CR, Lynch HT, Chadwick RB, de la Chapelle A, Berg K, Eshleman JR, Yuan W, Markowitz S, Laken SJ, Lengauer C, Kinzler KW, Vogelstein B (2000) Conversion of diploidy to haploidy. Nature 403:723–724

Yang B, Kang H, Fung A, Zhao H, Wang T, Ma D (2014a) The role of interleukin 17 in tumour proliferation, angiogenesis, and metastasis. Mediat Inflamm 2014:12

Yang H, Qi H, Ren J, Cui J, Li Z, Waldum HL, Cui G (2014b) Involvement of NF-κB/IL-6 pathway in the processing of colorectal carcinogenesis in colitis mice. Int J Inflamm 2014:7

Zhang L (2008) Immunohistochemistry versus microsatellite instability testing for screening colorectal cancer patients at risk for hereditary nonpolyposis colorectal cancer syndrome. Part II. The utility of microsatellite instability testing. J Mol Diagn: JMD 10:301–307

Zhang Y, Liu H, Li L, Ai M, Gong Z, He Y, Dong Y, Xu S, Wang J, Jin B, Liu J, Teng Z (2017) Cholecystectomy can increase the risk of colorectal cancer: a meta-analysis of 10 cohort studies. PLoS One 12:e0181852

Zhao P, Zhang Z (2018) TNF-α promotes colon cancer cell migration and invasion by upregulating TROP-2. Oncol Lett 15:3820–3827

Chapter 8
Molecular Signaling Pathways Involved in Gastric Cancer Chemoresistance

Henu Kumar Verma, Geppino Falco, and L.V.K.S. Bhaskar

Abstract Gastric cancer (GC) is the third most frequent cause of cancer-related deaths worldwide. The Molecular Mechanism of pathogenesis in GC is still unknown and unclear due to the Chemoresistance. Chemotherapy still remains only a single treatment for GC patients. Among those patients, most are becoming resisting to chemotherapeutic agents, nowadays chemoresistance causes recurrence and is a major challenge in the treatment of cancer because of the deregulations of numerous signaling pathways such as a tumor suppressor gene signaling, PI3K/Akt signaling, NF-κB signaling, Wnt/β-catenin signaling, mitogen-activated protein kinase (MAPK), Hedgehog signaling, Hippo signaling, Notch signaling pathways, and epidermal growth factor receptor (EGFR) have been found in GC. Epithelial-mesenchymal transition (EMT), as a major process during embryogenesis and tumor genesis, as well as is playing a vital role in chemoresistance of GC. In this chapter we summarize important molecular pathway aspects of multi-drug resistance (MDR). It is crucial for the identification of the new drug target, and combination therapy to clarify these complex molecular signaling mechanisms.

Keywords Gastric cancer · Chemoresistance

H. K. Verma
Institute of Experimental Endocrinology and Oncology, National Research Council, Naples, Italy

Section of Stem Cell and Development, Istituto di Ricerche Genetiche "Gaetano Salvatore" Biogem s.c. a.r.l., Ariano Irpino, Italy

G. Falco
Section of Stem Cell and Development, Istituto di Ricerche Genetiche "Gaetano Salvatore" Biogem s.c. a.r.l., Ariano Irpino, Italy

Department of Biology, University of Naples Federico II, Naples, Italy

L.V.K.S. Bhaskar (✉)
Guru Ghasidas University, Bilaspur, Chhattisgarh, India

G. S. R. Raju, L.V.K.S. Bhaskar (eds.), *Theranostics Approaches to Gastric and Colon Cancer*, Diagnostics and Therapeutic Advances in GI Malignancies, https://doi.org/10.1007/978-981-15-2017-4_8

Abbreviations

5-FU	Fluorouracil
Apa	Apatinib
ATM	Ataxia-telangiectasia mutated
BAD	Bcl-2-associated death promoter protein
CCL2	chemokine C-C motif ligand 2
CDDP	Cisplatin
DARPP-32	Dopamine and adenosine 3′, 5′-cyclic monophosphate-regulated phosphoprotein, Mr. 32000
EGFR	Epidermal growth factor receptor
EMT	Epithelial-mesenchymal transition
GC	Gastric cancer
HMMR	Hyaluronan-mediated motility receptor
KRAS	Kirsten-Ras
LOH	Loss of heterozygosity
LRIG1	Leucine-rich repeats and immunoglobulin-like domains 1
MAPK	Mitogen-activated protein kinase
MDR	Multidrug resistance
MSC	Mesenchymal stem cell
MST1	Mammalian ste20-like kinase 1
NF-κB	Nuclear factor-kappa B
PI3K/Akt	Phosphoinositide-3-kinase–protein kinase B
TGI	tumor growth inhibition
Tzb	trastuzumab
VEGF	Vascular endothelial growth factor

8.1 Introduction

Remarkable progress has been made on the development and progression of human gastric cancer (GC) over the past decade. Gastric cancer represents the third leading cause of cancer deaths worldwide (Sitarz et al. 2018). The incidence rates of GC vary in different regions, with a higher incidence in Eastern Asia, European and South American countries and a lower incidence in North America and some parts of Africa (Marques-Lespier et al. 2016; Torre et al. 2016). In the management of unresectable tumors, several chemotherapeutic strategies have been used to relieve symptoms, to decrease the risk of recurrence and distant metastasis (Hamamoto 2015; Liu et al. 2016; Shin et al. 2016). The 5-year overall survival (OS) rate varies from 20% to 35% in these patients (Chon et al. 2017; Kuo et al. 2014). chemoresistance is a major hindrance to effective and successful cancer treatment

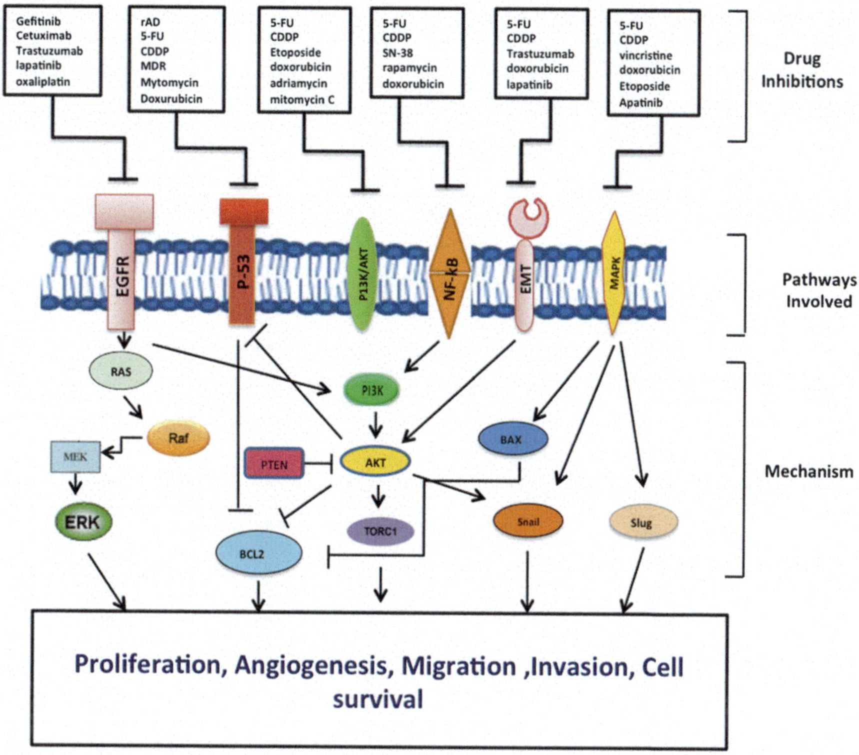

Fig. 8.1 Molecular mechanism of chemotherapeutic drugs involved in gastric cancer chemoresistance

in various cases. There are many reasons that lead to the failure of cancer chemotherapy (Fig. 8.1 and Table 8.1). Several lines of evidence report the involvement of tumor microenvironment (TME), Hedgehog (Hh), p53 oncogene, phosphatidylinositol-3 kinase (PI3K)/Akt, Notch signaling, mitogen-activated protein kinase (MAPK), Hippo signaling and WNT signaling pathways play role in GC chemoresistance (Gao et al. 2018; Martin et al. 2013). Hence, it is essential to understand its molecular mechanisms to identify a novel therapeutic target for cancer cell invasiveness and metastasis suppression. In this chapter, we summarized the major molecular signaling pathways that are involved in chemoresistance of GC.

8.2 Oncogenes p53

The p53 is one of the most well-known tumor suppressor genes involved in various important processes such as apoptosis, cell cycle regulation, and DNA repair. Hence p53 is also called as "guardian of the genome (Lane 1992). It has been observed that

Table 8.1 Summary of molecular signaling pathways and chemotherapeutic drugs associated with gastric cancer chemoresistance

Target pathways	Explanation	Chemo-drugs	Expression level	Reference
Tumor suppressor gene (P53)	Drug exposure increases mutant p53 levels by a translational mechanism	5-FU, Mytomycin, Cis-diclorodiamine platinum	Up-regulation	(Nabeya et al. 1995)
	Mutation of the p53 gene with increased overexpression of the Bcl-2 protein	CDDP	Up-regulation	(Ikeguchi et al. 1997)
		rAD	Up-regulation	(Chen et al. 2011)
	miR-27b up-regulation leads to increased miR-508-5p expression, mediated by CCNG1 and P53	MDR	Up-regulation	(Shang et al. 2016)
	Mutation of the p53 gene inhibits apoptosis	5-FU,CDDP	Up-regulation	(Matsuhashi et al. 2005)
	miR-19a/b suppressed drug-induced apoptosis by regulating Bcl-2 and Bax	Doxorubicin	Up-regulation	(Wang et al. 2013a)
EGFR	Up-regulated integrin beta4 expression inhibited apoptosis, and enhanced resistance	Gefitinib	Up-regulation	(Huafeng et al. 2018)
	Overexpression of EGFR and low levels of receptor activation, MET activation and mutations of KRAS and CDH1 was associated with resistance	Cetuximab	Up-regulation	(Heindl et al. 2012)
	Overexpression of DARPP-32, promote interaction between EGFR and ERBB3 and blocked gefitinib-induced apoptosis	Gefitinib	Up-regulation	(Zhu et al. 2011)
	HER2-positive cells prevent cell apoptosis via autophagic flux inhibition through mTOR pathways	Trastuzumab	Up-regulation	(Ye et al. 2018)
	Higher PD-L1 expression and dMMR and HER2 status indicate drug resistance	Trastuzumab	Up-regulation	(Wang et al. 2018)
	Up-regulation of HER2 and MET and Downregulation of FOXO1gene increased resistance	Apatinib	Up-regulation	(Park et al. 2018)
	Up-regulation of ATXN2L was responsible for intrinsic and acquired chemoresistance	Oxaliplatin	Up-regulation	(Lin et al. 2019)

(continued)

Table 8.1 (continued)

Target pathways	Explanation	Chemo-drugs	Expression level	Reference
P13K/AKT	Wortmannin promotes caspase-3, caspase-9 activation, and poly ADP-ribose polymerase cleavage and increased drug resistance	Etoposide, doxorubicin	Up-regulation	(Yu et al. 2008)
	AKT activation and LOH of PTEN enhanced chemoresistance	5-FU, adriamycin, mitomycin C and CDDP	Up-regulation	(Oki et al. 2005)
	pAkt- and p53-positive tumors, enhanced chemoresistance	5-FU	Down-regulation of AKT and up-regulation of p53	(Murakami et al. 2007)
	Down-regulation of PTEN and PIK3CA gene could activate the PI3-kinase/AKT signaling pathway of chemoresistance	CDDP	Up-regulation	(Byun et al. 2003)
	Suppression of the expression of p53 and promoting the expression of c-Myc enhanced chemoresistance	Etoposide	Up-regulation	(Liu et al. 2006)
	Up-regulation of Janus kinase 2 (JAK2)/signal transducer and activator of transcription 3 (STAT3) signaling pathway promotes drug resistance	CDDP	Up-regulation	(Zhang et al. 2013)
	P-AKT might be up-regulated of Bcl-2, and down-regulate of Bax protein and promote drug resistance	MDR	Up-regulation	(Han et al. 2007)
MAPK	Overexpression of the MDR1gene promotes resistance to P-gp-related drug and P-gp-unrelated drugs	Vincristine	Up-regulation	(Guo et al. 2008)
	Overexpression of the Bax and decrease Bcl-2 expression involved in drug resistance	Doxorubicin	Up-regulation	(Tan et al. 2014)
	Upregulation of miRNA-16 promotes drug resistance	5-FU, Etoposide	Up-regulation	(Wang et al. 2013b)
	High expression of DUSP1 may be responsible for drug resistance	Apatinib	Up-regulation	(Teng et al. 2018)
	Up-regulated miR-135b was regulating MST1 and increased drug resistance	CDDP	Up-regulation	(Zhou and Chen 2019)
	Down-regulation of miR-206 is associated with drug resistance	CDDP	Up-regulation	(Chen et al. 2019)

(continued)

Table 8.1 (continued)

Target pathways	Explanation	Chemo-drugs	Expression level	Reference
NF-kB	Up-regulation of PTEN expression via CBF-1 binding enhanced drug resistance	Doxorubicin	Up-regulation	(Zhou et al. 2013)
	The NF—kB pathway is activated in response to chemotherapy	5-FU, SN-38	Up-regulation	(Camp et al. 2004)
	ERas induces chemoresistance to CPT-11 via activation of the signaling pathway	Rapamycin	Up-regulation	(Kubota et al. 2011)
	Overexpression of APRIL decreased the efficacy of chemo drugs	CDDP	Up-regulation	(Zhi et al. 2015)
	Up-regulation of IL-8 increased ABCB1 expression which promotes chemoresistance	CDDP	Up-regulation	(Zhai et al. 2019)
EMT	HMMR activates the TGF-beta/Smad2 and promotes chemoresistance	5-FU,	Up-regulation	(Zhang et al. 2019)
	Testican-1 mediated drug-resistance in HER2 positive cells	Apatinib	Up-regulation	(Kim et al. 2014)
	Overexpression of the HER2/snail positive cell increased chemoresistance	CDDP	Up-regulation	(Huang et al. 2016)
	Up-regulation of eif5a2's expression enhanced the chemoresistance	CDDP	Up-regulation	(Sun et al. 2018a)
	Up-regulation of HER4, p-HER4, YAP1, and Vimentin were associated in chemoresistance	Trastuzumab	Up-regulation	(Shi et al. 2018)
	Up-regulation of DUSP4 promotes chemoresistance	Doxorubicin	Up-regulation	(Kang et al. 2017)
	Down-regulation of miR-30a promotes chemoresistance	CDDP	Up-regulation	(Wang et al. 2016)

p53 universally mutated in all categories of cancer including GC. The cyclin-dependent kinase inhibitor P21 is a major target of p53 activity and these are associated with cell cycle arrest and tumor growth inhibition (TGI). About 60% of GC tissues showed a reduction in P21 expression and it significantly correlates with tumor metastasis, invasiveness and poor prognosis (Gamboa-Dominguez et al. 2007).

Further, p53 gene mutations rate was found to be 0–77% in GC (Fenoglio-Preiser et al. 2003). Moreover, the function of p53 alterations causes by loss of heterozygosity (LOH) including a high incidence of p53 mutations and infrequently by DNA methylation. Several mutations may exist in a one tumor causing in the heterogeneity of the p53 position, high-expression of the p53 protein, and the low level of p53 function which are initial events in GC (Bellini et al. 2012). Although the multifaceted relationship between p53 and chemoresistance in GC has been studied for several years, the outcomes are inconsistent. Currently, an epidemiological study was conducted to explain the associations between p53 mutations and the response to chemotherapy. The results indicated that p53 might be a good prognostic biomarker for early response to chemotherapy in GC (Xu et al. 2014).

A study has examined in GC cell lines for p53 mutational status and the results indicate that wild-type p53 protein expression increase during treatment with 5-fluorouracil (5-FU), mitomycin C, and cis-dichlorodiammin-e platinum (CDDP), In contrast, these consequences show that the mutation of p53 is predictive of chemosensitivity in GC (Nabeya et al. 1995). p53 induced apoptosis has been confirmed in GC cells by down-regulating of Bcl-2 protein and the up-regulating the expression of mutated p53 genes in MKN-74 cells after CDDP treatment (Ikeguchi et al. 1997). Additionally, Chen et al. showed that treatment with rAd-p53 significantly increased the sensitivity of the GC to chemotherapy by enhancing Bax expression and inhibits apoptosis (Chen et al. 2011). A study reported that miR-27b plays a major role in tumor development to chemotherapy in vitro and in vivo. Interestingly, up-regulation of miR-27b leads to enhanced miR-508-5p expression, mediated by mutant P53 in GC-associated MDR (Shang et al. 2016). Matsuhashi et al. demonstrate that combined administration of 5-FU and CDDP, induce apoptosis in MKN45, but not in MKN28 cell line these data indicated that the mutated p53 may deliver confirmation of the idea that p53 expression is related to MDR in GC (Matsuhashi et al. 2005). Hamada et al. analyzed and detected 4 GC patients with p53 mutations out of 24 patients with other cancer by immune-histochemical staining and found that p53-inducible WAF1/CIP1 protein in wild type p53 expression but not in mutant-p53 these results suggest that mutations in p53 are associated with lower response or chemosensitivity in GC (Hamada et al. 1996). In addition, it shows that microRNAs are involved in the up-regulation of MDR1 and overexpression of miR-19a/b confers resistance to doxorubicin on GC cells and decreasing the expression of Bcl-2 and Bax gene (Wang et al. 2013a).

8.3 Growth Factor Receptor and Signaling

Epidermal growth factor receptor (EGFR) is a transmembrane protein. Up-regulation of EGFR has been reported in 9–30% of GC cases (Terashima et al. 2012). It is a kind of glycoprotein receptor with HER family of tyrosine kinase activity, When the EGFR extracellular domain binds to its ligands proteins such as transforming growth factor-α (TGF-α), it promotes dimer formation with other EGFR family members which leads to high expression of EGFR and activate PI3K/Akt/mTOR, JAK/stat3, SOS/Grb2/Ras, and src/FAK/ROS pathways, as well as involved in differentiation, proliferation adhesion and metastasis in GC (Lee et al. 2015; Roskoski 2014).

Further, overexpression of EGFR can trigger STAT3 and NFκB, which leads to chemoresistance and poor prognosis in GC. Recently, Huafeng et al. identified that the up-regulation of integrin β4 expression was promoted gefitinib resistance and proliferation by inhibiting apoptosis and showed a negative correlation between integrin β4 and EGFR in GC patients (Huafeng et al. 2018).

An in vitro study on cell lines showed High EGFR expression with MET activation and Kirsten-Ras (KRAS) and CDH1 gene mutations was positively associated with cetuximab resistance in GC (Heindl et al. 2012). To complement this result one study also showed that, activation of *KRAS* mutation promotes cetuximab resistance in GC cell line (Kneissl et al. 2012). Zhu et al. reported that gefitinib resistance was associated with up-regulation of Dopamine and adenosine 3′,5′-cyclic monophosphate-regulated phosphoprotein, Mr 32,000 (DARPP-32) through EGFR mediated phosphatidylinositol-3-kinase–AKT signaling pathways in GC cells lines (Zhu et al. 2011).

Furthermore, Ye et al. indicated that autophagy plays a vital role in the resistance of HER2-positive in NCI-N87 cell lines to trastuzumab (Tzb) and showed that Tzb drug prevents cell apoptosis by autophagic flux inhibition. Which activate the Akt/mTOR pathway in GC (Ye et al. 2018). Additionally, Wang et al. found that patients with Tzb resistance existing high HER2 somatic copy number alterations (SCNA) during development. The PIK3CA mutations were significantly advanced in patients with innate resistance, compared with standard, as well as NF1 mutations also contributed a role in Tzb resistance in GC (Wang et al. 2018). Moreover, a current study showed that FOXO1 gene works as connective linker among HER2 and MET signaling pathways and play a key role in the regulation of the Apatinib resistance in HER2-positive GC cells. These findings suggest a novel strategy for treatment to overcome apatinib resistance in GC patients (Park et al. 2018). Very recent a study demonstrates that EGFR can stimulate ATXN2L gene expression and promotes cell invasiveness which leads to oxaliplatin resistance. This data indicates poor prognosis for overall survival and recurrence in GC tissue (Lin et al. 2019). Another study observed that Leucine-rich repeats and immunoglobulin-like domains 1 (LRIG1) was up-regulated in chemosensitive GC MDR cell lines via EGFR-mediated PI3K/AKT and MAPK/ERK signaling pathways and decreased expression of LRIG1 (Zhou et al. 2018).

8.4 PI3K/AKT Signaling Pathway Activation

The PI3K/Akt is a serine/threonine-specific kinase protein work as a key regulator of cell growth, proliferation, migration, and survival; it has been observed that is frequently active in GC. Overexpression of PIK3CA is frequently detected with a poor outcome in GC (Tsujitani et al. 2012) the triggering of TKI activates PI3K, which initiates AKT. Activated Akt can phosphorylate various Bcl-2-associated death promoter protein (BAD) on ser136 for the detach from Bcl-X/Bcl-2 gene family and overcome apoptosis initiating signals of BAD. Mutations in PIK3CA lead to activation of the PI3K signaling activity reported in GC confirmed by microarray analysis (Li et al. 2005). It has been reported the aberrant activation of the Akt also able to stimulate NFkB up-regulation, which helps in the transcription of pro-survival genes and overexpression of Akt1 outcomes in drug resistance of GC cells to chemotherapeutic agents. Nam et al. found that high expression of p-AKT and AKT was 78% and 74% in GC, respectively (Nam et al. 2003). Yu et al. observations have proven that high expression and phosphorylation of Akt could be deactivated by etoposide, doxorubicin and wortmannin and could increase the resistance of GC cells to chemotherapy through PI3K/Akt signaling pathway (Yu et al. 2008).

Moreover, a study has revealed that activated AKT and LOH of PTEN plays a major role in broad-spectrum resistance to adriamycin, mitomycin C, cis-platinum and 5-FU chemo drugs, mediated by AKT/PI3K pathways in GC patients (Oki et al. 2005). Another report also compliments this result adding with fluorouracil resistance treatment (Murakami et al. 2007). and Down-regulation of PTEN can lead to CDDP resistance, in GC cells (Byun et al. 2003) Liu et al. demonstrated that Etoposide can stimulate activation of the PI3K/AKT signaling pathway, which reduced the chemo drugs sensitivity of SGC7901 and BGC823 GC cell lines (Liu et al. 2006) Further, in one study, demonstrate that the overexpression of AKT at the molecular and cellular level is associated with CDDP resistance through the JAK2/STAT3 pathway and decreased the chemosensitivity of GC cells in vitro and in vivo (Zhang et al. 2013). And the down-regulation of AKT1 significantly increased cell sensitivity towards AGS cells to adriamycin, cisplatin, 5-fluorouracil, and vincristine chemotherapeutic drugs (Han et al. 2006).

It has been noticed that NF-kB work as a chemotherapeutic inducer of AKT activation, degradation, and phosphorylation also involved in the chemoresistance of GC cells (Yu et al. 2010). Recently Song et al. found that loss of CD133 stem cell biomarker significantly increased the growth inhibition of chemo agents and knock-down of CD133 significantly reduced the PI3K activity in the GC (Song et al. 2018). However, the PI3K/AKT signaling pathway plays a crucial role in drug resistance, the molecular mechanism of PI3K/AKT activation in chemoresistance is not completely understood. According to a study, Survivin plays an important role in downstream of AKT. higher levels of survivin and p-AKT have been detected In CDDP-resistant GC (Sun et al. 2014) Additionally, the overexpression of p-AKT could be responsible for MDR in AGS GC cell lines by the up-regulation of BCL-2 expression (Han et al. 2007).

8.5 Mitogen-Activated Protein Kinase Signaling Pathway

The mitogen-activated protein kinase (MAPK) including p38 and JNK kinase signaling pathway responds to extracellular stimulation and broadly expressed in eukaryotic organisms (Johnson and Lapadat 2002). which plays a crucial role in several biological processes, such as cell proliferation, differentiation, and survival of tumor cells. Deregulation of the MAPK signaling is associated with the progression of GC (Yang and Huang 2015). Besides, these numerous studies have confirmed that the MAPK pathway is also involved in chemotherapy resistance in GC. According to Atmaca et al. phosphorylated MAPK was positive in 59.6% of Cases with metastatic GC and Overall survival was found 8.5 months also, it has been observed that the expression of p-MAPK in primary and metastatic tumors was similar. These results directed that p-MAPK expression may be a probable predictive marker in metastatic GC who is ongoing treatment with chemotherapy (Atmaca et al. 2011). Further, the overexpression of the p38-MAPK signaling pathway was found in vincristine-resistant SGC7901/VCR GC cell lines and to be responsible for the MDR (Guo et al. 2008). Tan et al. demonstrates that the up-regulation of p38 MAPK pathway was involved in doxorubicin resistance in GC cells (SGC7901, BGC823) and xenograft model besides inhibition of p38 MAPK increased GC cell sensitivity to doxorubicin through the induction of the BAX and decrease in BCL-2 expression (Tan et al. 2014). Wang et al. found that etoposide and 5-FU could be activated miRNA-16 expression in vitro and in vivo, and the overexpression of miRNA-16 is mediated by p38MAPK in GC (Wang et al. 2013b). Recently, Teng, et al. was shown, that the expression of DUSP1 gene was significantly higher in the early stages of GC and associated with apatinib (Apa) resistance in GC cells through activation of MAPK signaling pathways in vitro (Teng et al. 2018).

Another current study reported that mammalian ste20-like kinase 1 (MST1) play a vital role in the progression of GC and the reduced sensitivity to CDDP in MKN45 cell lines. Down-regulation of miR-135b resulting in the reverse of CDDP resistance and increases the cells death via activation of MST1-mediated MAPK signaling pathway (Zhou and Chen 2019). along with, Chen et al. found that The down-regulation of miR206 is significantly associated with CDDP resistance of GC cells via induction of MAPK3 pathway (Chen et al. 2019).

8.6 NF-κB Signaling Pathways

Nuclear factor kappa B (NF-κB) constitutes a family of transcription factors and regulated by polyubiquitination, proteasomal degradation, and phosphorylation, by IκB protein. Which form homo and heterodimers and responsible for up-regulation or suppression of many genes involved in inflammation, cell proliferation, cell survival and immunity (Neumann and Naumann 2007). The activation of NF-κB RelA homology domain driven cytokines to include IL-1, IL-6, IL-8, MCP-1, TNF, pro- and anti-apoptotic factors. Especially, deregulation of NF-κB signaling pathway promotes tumor genesis which Associated with poor prognosis of GC and chemoresistance (Kinoshita et al. 2010; Kwon et al. 2012; Maeda and Omata 2008). Zhou et al. revealed that AKT1 expression was induced by doxorubicin. The activation of AKT can increase the binding of NF-kB on Notch1 promoter. Further up-regulation of PTEN by Notch-activated transcription factor (CBF1) in vitro and in vivo results, suggested that an AKT1/NF-kB/PTEN play an important role in the development of chemoresistance in GC (Zhou et al. 2013). Camp et al. found that NF-kB is activated in NCI-N87 and AGS human GC cells in response to 5-FU and SN-38 chemotherapeutic drugs, and May outcome in inducible chemoresistance (Camp et al. 2004). A another study demonstrated that ERas-overexpressing clones were significantly more resistant to CPT-11 after treatment of rapamycin than the control in GCIY cells via activation of NF-κB/mTOR pathway (Kubota et al. 2011) A recent study reported that conditioned medium (CM) made by the metabolism of SGC-7901 GC cell lines increased drug resistance by activating the ataxia-telangiectasia mutated (ATM) and NF-κB pathways in GC cells (Zhuang et al. 2018) Zhi et al. found that Up-regulation of APRIL in AGS cells significantly decreased the efficacy of CDDP in vitro and in vivo and data showed that NF-κB pathway involved in APRIL-mediated chemo-resistance in GC patients (Zhi et al. 2015). Further, a study revealed that IL-8 was overexpressed in GC drug-resistant patients, and increased the IC50 of CDDP in AGS cells, located in cancer-associated fibroblasts (CAFs). Instantaneously, IL-8 therapeutic enriched the expression of PI3K, p-AKT, p-IKb, p-p65 and ABCB1, besides promotes chemoresistance through NF-κB activation and up-regulation of ABCB1 (Zhai et al. 2019). In Additional, Xu et al. reported that drug-resistant GC cells secrete more chemokine C-C motif ligand 2 (CCL2) than drug-sensitive cells and decreased the drug-induced cytotoxicity by inhibiting autophagy and increase SQSTM1 expression. Besides, these enhanced the expression of SQSTM1 in turn, activated CCL2 transcription via the NF-κB signal pathway, demonstrating as a positive feedback drug resistance (Xu et al. 2018). Hypoxia is another well-recognized common feature in tumor biology to be a key point of treatment resistance and poor prognosis in several cancer patients. Hypoxia leads to the expression of many genetic factors that are involved in tumor progression and metastasis in GC (Griffiths et al. 2005). HIF-1α can induce the vascular endothelial growth factor (VEGF) expression and inflammatory state through NF-kB signaling pathway which leads to the suppression of p53 and promotes 5-FU and CDDP chemoresistant in human GC cells (Rohwer et al. 2010). Nakamura et al. has been reported that HIF-1α leads to drug resistance against adjuvant chemotherapy using 5-FU in advance gastric tumor patients (Nakamura et al. 2009). overexpression of HIF-1α increases the expression of Bcl-2 and reduces the expression of Bax protein outlining hypoxia-induced drug resistance in GC (Liu et al. 2008).

8.7 EMT

Epithelial-mesenchymal transition (EMT) is a multistage reprogramming process play a vital role in the development of homogenous adhesion that is essential for embryonic expansion and fibrotic disease (Peng et al. 2014). During EMT progression, there is a loose cell polarity in epithelial cell junctions and increase invasive properties of the mesenchymal stem cell (MSCs). Consequently, the expression of epithelial marker such as E-cadherin showed down-regulation and activates β-catenin, which translocate into the nucleus and modulates the expression of VEGF, CD44, cyclin D1kinase, and c-Myc which leads to tumor initiation and progression. it has been observed that the expression of mesenchymal markers such as Snail, Slug, Vimentin, ZEB1, ZEB2 is up-regulated in tumor cells (Thiery et al. 2009).

However, the ability to self-renewal, an overexpression of Drug resistance genes, have shown that the EMT is a major molecular mechanism linked with metastasis and provide resistance to chemotherapy (Mitra et al. 2015). Recent studies have proved that hyaluronan-mediated motility receptor (HMMR) was up-regulated in 5-Fu resistant GC cell line. Further, biopsies sample observed that HMMR increased the cancer stem cell (CSCs) properties and resistance to chemotherapy via TGF-beta/Smad2-induced EMT in GC (Zhang et al. 2019). Kim et al. described the Testican-1 are responsible for EMT mediated signaling and confers acquired resistance to apatinib in HER2-positive gastric cancer in in-vitro (Kim et al. 2014).

Similarly, Huang et al. found that up-regulation of HER2/Snail double-positive patients had poor survival and significantly associated with CDDP-resistant in GC cells mediated by EMT (Huang et al. 2016). Eukaryotic translation initiation factor 5A2 (eIF5A2) is an essential tumor-promoting function in GC. One report showed that the Silencing of eIF5A2 factor enhanced the sensitivity of GC cells to cisplatin by mediating EMT (Sun et al. 2018b). A current study found that phosphorylated p-HER4, HER4, YAP1, and Vimentin were significantly higher and HER2 and E-cadherin were found down-regulated in response to the trastuzumab in vivo. These results revealed that the major role of the HER4-YAP1 in trastuzumab resistance of HER2-positive GC cells via induction of EMT (Shi et al. 2018). Kang et al. demonstrated that up-regulation of DUSP4 can enhance doxorubicin resistance by stimulating EMT in GC cells (Kang et al. 2017).

A study Report proposed that depletion of TAZ (transcriptional co-activator) caused partial Transition of EMT to MET in CDDP resistant GC cells, which are negatively regulated by the Hippo pathway (Ge et al. 2017). Wang et al. also showed that the chemoresistance to cisplatin-induced EMT in human GC cells (Wang et al. 2016). Additionally, has been observed that Doxorubicin is able to induce EMT in GC patients through inhibition of the β-catenin signaling pathway by indomethacin and inhibition of p300 in BGC-823 GC cell (Han et al. 2013; Han et al. 2014). Moreover, Fas belongs to a member of the TNF family, which stimulate tumor cell motility inducing EMT, and support metastasis formation in GC. down-regulation of Snail and Twist expression significantly decreased Fas-induced motility, as well as the use of oxaliplatin chemo drug prompt to induce EMT moderately resulting in chemo-resistant through Fas signaling pathway (Zheng et al. 2013).

8.8 Conclusion

The chemoresistance of tumor cells to chemotherapy occurs from a reduction in drug availability and induction of several oncogenic signaling pathways. Due to the cell specificity to chemoresistance. Major chemoresistance-related proteins are localized in the cell membrane; these proteins are complex and highly versatile in various events, including apoptosis, proliferation, autophagy, and EMT. Now it is essential for the cancer patients receiving targeted combine therapy to increase therapeutic efficacy and reduced tissue toxicity. This chapter may deliver a further understanding of molecular signaling network in Gastric cancer chemoresistance, which facilitates the establishment of novel therapeutic targets and potential chemo sensitive biomarkers to decrease the cancer recurrence and improve the patient lifespan.

References

Atmaca A, Pauligk C, Steinmetz K, Altmannsberger HM, Jager E, Al-Batran SE (2011) Prognostic impact of phosphorylated mitogen-activated protein kinase expression in patients with metastatic gastric cancer. Oncology 80:130–134

Bellini MF, Cadamuro AC, Succi M, Proenca MA, Silva AE (2012) Alterations of the TP53 gene in gastric and esophageal carcinogenesis. J Biomed Biotechnol 2012:891961

Byun DS, Cho K, Ryu BK, Lee MG, Park JI, Chae KS, Kim HJ, Chi SG (2003) Frequent monoallelic deletion of PTEN and its reciprocal associatioin with PIK3CA amplification in gastric carcinoma. Int J Cancer 104:318–327

Camp ER, Li J, Minnich DJ, Brank A, Moldawer LL, MacKay SL, Hochwald SN (2004) Inducible nuclear factor-kappaB activation contributes to chemotherapy resistance in gastric cancer. J Am Coll Surg 199:249–258

Chen GX, Zheng LH, Liu SY, He XH (2011) rAd-p53 enhances the sensitivity of human gastric cancer cells to chemotherapy. World J Gastroenterol 17:4289–4297

Chen Z, Gao YJ, Hou RZ, Ding DY, Song DF, Wang DY, Feng Y (2019) MicroRNA-206 facilitates gastric cancer cell apoptosis and suppresses cisplatin resistance by targeting MAPK2 signaling pathway. Eur Rev Med Pharmacol Sci 23:171–180

Chon SH, Berlth F, Plum PS, Herbold T, Alakus H, Kleinert R, Moenig SP, Bruns CJ, Hoelscher AH, Meyer HJ (2017) Gastric cancer treatment in the world: Germany. Trans Gastroenterol Hepatol 2:53

Fenoglio-Preiser CM, Wang J, Stemmermann GN, Noffsinger A (2003) TP53 and gastric carcinoma: a review. Hum Mutat 21:258–270

Gamboa-Dominguez A, Seidl S, Reyes-Gutierrez E, Hermannstadter C, Quintanilla-Martinez L, Busch R, Hofler H, Fend F, Luber B (2007) Prognostic significance of p21WAF1/CIP1, p27Kip1, p53 and E-cadherin expression in gastric cancer. J Clin Pathol 60:756–761

Gao JP, Xu W, Liu WT, Yan M, Zhu ZG (2018) Tumor heterogeneity of gastric cancer: from the perspective of tumor-initiating cell. World J Gastroenterol 24:2567–2581

Ge L, Li DS, Chen F, Feng JD, Li B, Wang TJ (2017) TAZ overexpression is associated with epithelial-mesenchymal transition in cisplatin-resistant gastric cancer cells. Int J Oncol 51:307–315

Griffiths EA, Pritchard SA, Welch IM, Price PM, West CM (2005) Is the hypoxia-inducible factor pathway important in gastric cancer? Eur J Cancer 41:2792–2805

Guo X, Ma N, Wang J, Song J, Bu X, Cheng Y, Sun K, Xiong H, Jiang G, Zhang B, Wu M, Wei L (2008) Increased p38-MAPK is responsible for chemotherapy resistance in human gastric cancer cells. BMC Cancer 8:375

Hamada M, Fujiwara T, Hizuta A, Gochi A, Naomoto Y, Takakura N, Takahashi K, Roth JA, Tanaka N, Orita K (1996) The p53 gene is a potent determinant of chemosensitivity and radiosensitivity in gastric and colorectal cancers. J Cancer Res Clin Oncol 122:360–365

Hamamoto Y (2015) Complications in advanced or recurrent gastric cancer patients with peritoneal metastasis during and after palliative systemic chemotherapy. Mol Clin Oncol 3:539–542

Han Z, Hong L, Wu K, Han S, Shen H, Liu C, Han Y, Liu Z, Han Y, Fan D (2006) Reversal of multidrug resistance of gastric cancer cells by downregulation of Akt1 with Akt1 siRNA. J Exp Clin Cancer Res 25:601–606

Han Z, Hong L, Han Y, Wu K, Han S, Shen H, Li C, Yao L, Qiao T, Fan D (2007) Phospho Akt mediates multidrug resistance of gastric cancer cells through regulation of P-gp, Bcl-2 and Bax. J Exp Clin Cancer Res 26:261–268

Han R, Xiong J, Xiao R, Altaf E, Wang J, Liu Y, Xu H, Ding Q, Zhang Q (2013) Activation of beta-catenin signaling is critical for doxorubicin-induced epithelial-mesenchymal transition in BGC-823 gastric cancer cell line. Tumour biology: the journal of the international society for. Oncodev Biol Med 34:277–284

Han RF, Ji X, Dong XG, Xiao RJ, Liu YP, Xiong J, Zhang QP (2014) An epigenetic mechanism underlying doxorubicin induced EMT in the human BGC-823 gastric cancer cell. Asian Pac J Cancer Prev: APJCP 15:4271–4274

Heindl S, Eggenstein E, Keller S, Kneissl J, Keller G, Mutze K, Rauser S, Gasteiger G, Drexler I, Hapfelmeier A, Hofler H, Luber B (2012) Relevance of MET activation and genetic alterations of KRAS and E-cadherin for cetuximab sensitivity of gastric cancer cell lines. J Cancer Res Clin Oncol 138:843–858

Huafeng J, Deqing Z, Yong D, Yulian Z, Ailing H (2018) A cross-talk between integrin beta4 and epidermal growth factor receptor induces gefitinib chemoresistance to gastric cancer. Cancer Cell Int 18:50

Huang D, Duan H, Huang H, Tong X, Han Y, Ru G, Qu L, Shou C, Zhao Z (2016) Cisplatin resistance in gastric cancer cells is associated with HER2 upregulation-induced epithelial-mesenchymal transition. Sci Rep 6:20502

Ikeguchi M, Tatebe S, Kaibara N, Ito H (1997) Changes in levels of expression of p53 and the product of the bcl-2 in lines of gastric cancer cells during cisplatin-induced apoptosis. Eur Surg Res 29:396–402

Johnson GL, Lapadat R (2002) Mitogen-activated protein kinase pathways mediated by ERK, JNK, and p38 protein kinases. Science 298:1911–1912

Kang X, Li M, Zhu H, Lu X, Miao J, Du S, Xia X, Guan W (2017) DUSP4 promotes doxorubicin resistance in gastric cancer through epithelial-mesenchymal transition. Oncotarget 8:94028–94039

Kim HP, Han SW, Song SH, Jeong EG, Lee MY, Hwang D, Im SA, Bang YJ, Kim TY (2014) Testican-1-mediated epithelial-mesenchymal transition signaling confers acquired resistance to lapatinib in HER2-positive gastric cancer. Oncogene 33:3334–3341

Kinoshita J, Fushida S, Harada S, Makino I, Nakamura K, Oyama K, Fujita H, Ninomiya I, Fujimura T, Kayahara M, Ohta T (2010) PSK enhances the efficacy of docetaxel in human gastric cancer cells through inhibition of nuclear factor-kappaB activation and survivin expression. Int J Oncol 36:593–600

Kneissl J, Keller S, Lorber T, Heindl S, Keller G, Drexler I, Hapfelmeier A, Hofler H, Luber B (2012) Association of amphiregulin with the cetuximab sensitivity of gastric cancer cell lines. Int J Oncol 41:733–744

Kubota E, Kataoka H, Tanaka M, Okamoto Y, Ebi M, Hirata Y, Murakami K, Mizoshita T, Shimura T, Mori Y, Tanida S, Kamiya T, Aoyama M, Asai K, Joh T (2011) ERas enhances resistance to CPT-11 in gastric cancer. Anticancer Res 31:3353–3360

Kuo CY, Chao Y, Li CP (2014) Update on treatment of gastric cancer. J Chin Med Assoc 77:345–353

Kwon HC, Kim SH, Oh SY, Lee S, Lee JH, Jang JS, Kim MC, Kim KH, Kim SJ, Kim SG, Kim HJ (2012) Clinicopathologic significance of expression of nuclear factor-kappaB RelA and its target gene products in gastric cancer patients. World J Gastroenterol 18:4744–4750

Lane DP (1992) Cancer. p53, guardian of the genome. Nature 358:15–16

Lee HC, Su MY, Lo HC, Wu CC, Hu JR, Lo DM, Chao TY, Tsai HJ, Dai MS (2015) Cancer metastasis and EGFR signaling is suppressed by amiodarone-induced versican V2. Oncotarget 6:42976–42987

Li VS, Wong CW, Chan TL, Chan AS, Zhao W, Chu KM, So S, Chen X, Yuen ST, Leung SY (2005) Mutations of PIK3CA in gastric adenocarcinoma. BMC Cancer 5:29

Lin L, Li X, Pan C, Lin W, Shao R, Liu Y, Zhang J, Luo Y, Qian K, Shi M, Bin J, Liao Y, Liao W (2019) ATXN2L upregulated by epidermal growth factor promotes gastric cancer cell invasiveness and oxaliplatin resistance. Cell Death Dis 10:173

Liu SQ, Yu JP, Yu HG, Lv P, Chen HL (2006) Activation of Akt and ERK signalling pathways induced by etoposide confer chemoresistance in gastric cancer cells. Dig Liver Dis 38:310–318

Liu L, Ning X, Sun L, Zhang H, Shi Y, Guo C, Han S, Liu J, Sun S, Han Z, Wu K, Fan D (2008) Hypoxia-inducible factor-1 alpha contributes to hypoxia-induced chemoresistance in gastric cancer. Cancer Sci 99:121–128

Liu D, Lu M, Li J, Yang Z, Feng Q, Zhou M, Zhang Z, Shen L (2016) The patterns and timing of recurrence after curative resection for gastric cancer in China. World J Surg Oncol 14:305

Maeda S, Omata M (2008) Inflammation and cancer: role of nuclear factor-kappaB activation. Cancer Sci 99:836–842

Marques-Lespier JM, Gonzalez-Pons M, Cruz-Correa M (2016) Current perspectives on gastric Cancer. Gastroenterol Clin N Am 45:413–428

Martin TAYL, Sanders AJ et al (2013) Cancer invasion and metastasis: molecular and cellular perspective. Landes Biosci

Matsuhashi N, Saio M, Matsuo A, Sugiyama Y, Saji S (2005) The evaluation of gastric cancer sensitivity to 5-FU/CDDP in terms of induction of apoptosis: time- and p53 expression-dependency of anti-cancer drugs. Oncol Rep 14:609–615

Mitra A, Mishra L, Li S (2015) EMT, CTCs and CSCs in tumor relapse and drug-resistance. Oncotarget 6:10697–10711

Murakami D, Tsujitani S, Osaki T, Saito H, Katano K, Tatebe S, Ikeguchi M (2007) Expression of phosphorylated Akt (pAkt) in gastric carcinoma predicts prognosis and efficacy of chemotherapy. Gastric Cancer 10:45–51

Nabeya Y, Loganzo F Jr, Maslak P, Lai L, de Oliveira AR, Schwartz GK, Blundell ML, Altorki NK, Kelsen DP, Albino AP (1995) The mutational status of p53 protein in gastric and esophageal adenocarcinoma cell lines predicts sensitivity to chemotherapeutic agents. Int J Cancer 64:37–46

Nakamura J, Kitajima Y, Kai K, Mitsuno M, Ide T, Hashiguchi K, Hiraki M, Miyazaki K (2009) Hypoxia-inducible factor-1alpha expression predicts the response to 5-fluorouracil-based adjuvant chemotherapy in advanced gastric cancer. Oncol Rep 22:693–699

Nam SY, Lee HS, Jung GA, Choi J, Cho SJ, Kim MK, Kim WH, Lee BL (2003) Akt/PKB activation in gastric carcinomas correlates with clinicopathologic variables and prognosis. APMIS 111:1105–1113

Neumann M, Naumann M (2007) Beyond IkappaBs: alternative regulation of NF-kappaB activity. FASEB J 21:2642–2654

Oki E, Baba H, Tokunaga E, Nakamura T, Ueda N, Futatsugi M, Mashino K, Yamamoto M, Ikebe M, Kakeji Y, Maehara Y (2005) Akt phosphorylation associates with LOH of PTEN and leads to chemoresistance for gastric cancer. Int J Cancer 117:376–380

Park J, Choi Y, Ko YS, Kim Y, Pyo JS, Jang BG, Kim MA, Lee JS, Chang MS, Park JW, Lee BL (2018) FOXO1 suppression is a determinant of acquired Lapatinib-resistance in HER2-positive gastric Cancer cells through MET Upregulation. Cancer Res Treat 50:239–254

Peng Z, Wang CX, Fang EH, Wang GB, Tong Q (2014) Role of epithelial-mesenchymal transition in gastric cancer initiation and progression. World J Gastroenterol 20:5403–5410

Rohwer N, Dame C, Haugstetter A, Wiedenmann B, Detjen K, Schmitt CA, Cramer T (2010) Hypoxia-inducible factor 1alpha determines gastric cancer chemosensitivity via modulation of p53 and NF-kappaB. PLoS One 5:e12038

Roskoski R Jr (2014) The ErbB/HER family of protein-tyrosine kinases and cancer. Pharmacol Res 79:34–74

Shang Y, Feng B, Zhou L, Ren G, Zhang Z, Fan X, Sun Y, Luo G, Liang J, Wu K, Nie Y, Fan D (2016) The miR27b-CCNG1-P53-miR-508-5p axis regulates multidrug resistance of gastric cancer. Oncotarget 7:538–549

Shi J, Li F, Yao X, Mou T, Xu Z, Han Z, Chen S, Li W, Yu J, Qi X, Liu H, Li G (2018) The HER4-YAP1 axis promotes trastuzumab resistance in HER2-positive gastric cancer by inducing epithelial and mesenchymal transition. Oncogene 37:3022–3038

Shin CH, Lee WY, Hong SW, Chang YG (2016) Characteristics of gastric cancer recurrence five or more years after curative gastrectomy. Chin J Cancer Res 28:503–510

Sitarz R, Skierucha M, Mielko J, Offerhaus GJA, Maciejewski R, Polkowski WP (2018) Gastric cancer: epidemiology, prevention, classification, and treatment. Cancer Manag Res 10:239–248

Song S, Pei G, Du Y, Wu J, Ni X, Wang S, Jiang B, Luo M (2018) Yu J: interaction between CD133 and PI3K-p85 promotes chemoresistance in gastric cancer cells. Am J Transl Res 10:304–314

Sun XP, Dong X, Lin L, Jiang X, Wei Z, Zhai B, Sun B, Zhang Q, Wang X, Jiang H, Krissansen GW, Qiao H, Sun X (2014) Up-regulation of survivin by AKT and hypoxia-inducible factor 1alpha contributes to cisplatin resistance in gastric cancer. FEBS J 281:115–128

Sun J, Xu Z, Lv H, Wang Y, Wang L, Ni Y, Wang X, Hu C, Chen S, Teng F, Chen W (2018a) Cheng X: eIF5A2 regulates the resistance of gastric cancer cells to cisplatin via induction of EMT. Am J Transl Res 10:4269–4279

Sun J, Xu Z, Lv H, Wang Y, Wang L, Ni Y, Wang X, Hu C, Chen S, Teng F, Chen W (2018b) Cheng X: eIF5A2 regulates the resistance of gastric cancer cells to cisplatin via induction of EMT. Am J Transl Res 10:4269–4279

Tan W, Yu HG, Luo HS (2014) Inhibition of the p38 MAPK pathway sensitizes human gastric cells to doxorubicin treatment in vitro and in vivo. Mol Med Rep 10:3275–3281

Teng F, Xu Z, Chen J, Zheng G, Zheng G, Lv H, Wang Y, Wang L, Cheng X (2018) DUSP1 induces apatinib resistance by activating the MAPK pathway in gastric cancer. Oncol Rep 40:1203–1222

Terashima M, Kitada K, Ochiai A, Ichikawa W, Kurahashi I, Sakuramoto S, Katai H, Sano T, Imamura H, Sasako M, Group A-G (2012) Impact of expression of human epidermal growth factor receptors EGFR and ERBB2 on survival in stage II/III gastric cancer. Clin Cancer Res 18:5992–6000

Thiery JP, Acloque H, Huang RY, Nieto MA (2009) Epithelial-mesenchymal transitions in development and disease. Cell 139:871–890

Torre LA, Siegel RL, Ward EM, Jemal A (2016) Global Cancer incidence and mortality rates and trends--an update. Cancer epidemiology, biomarkers & prevention: a publication of the American Association for Cancer Research, cosponsored by the American Society of Preventive. Oncology 25:16–27

Tsujitani S, Saito H, Wakatsuki T, Ikeguchi M, Shirabe K, Morita M, Kakeji Y, Yano T, Maehara Y (2012) Relationship between expression of apoptosis-related proteins and the efficacy of postoperative chemotherapy in patients with T3 gastric cancer. Surg Today 42:225–232

Wang F, Li T, Zhang B, Li H, Wu Q, Yang L, Nie Y, Wu K, Shi Y, Fan D (2013a) MicroRNA-19a/b regulates multidrug resistance in human gastric cancer cells by targeting PTEN. Biochem Biophys Res Commun 434:688–694

Wang F, Song X, Li X, Xin J, Wang S, Yang W, Wang J, Wu K, Chen X, Liang J, Tian J, Cao F (2013b) Noninvasive visualization of microRNA-16 in the chemoresistance of gastric cancer using a dual reporter gene imaging system. PLoS One 8:e61792

Wang LL, Zhang XH, Zhang X, Chu JK (2016) MiR-30a increases cisplatin sensitivity of gastric cancer cells through suppressing epithelial-to-mesenchymal transition (EMT). Eur Rev Med Pharmacol Sci 20:1733–1739

Wang L, Zhang Q, Ni S, Tan C, Cai X, Huang D, Sheng W (2018) Programmed death-ligand 1 expression in gastric cancer: correlation with mismatch repair deficiency and HER2-negative status. Cancer Med 7:2612–2620

Xu HY, Xu WL, Wang LQ, Chen MB, Shen HL (2014) Relationship between p53 status and response to chemotherapy in patients with gastric cancer: a meta-analysis. PLoS One 9:e95371

Xu W, Wei Q, Han M, Zhou B, Wang H, Zhang J, Wang Q, Sun J, Feng L, Wang S, Ye Y, Wang X, Zhou J, Jin H (2018) CCL2-SQSTM1 positive feedback loop suppresses autophagy to promote chemoresistance in gastric cancer. Int J Biol Sci 14:1054–1066

Yang M, Huang CZ (2015) Mitogen-activated protein kinase signaling pathway and invasion and metastasis of gastric cancer. World J Gastroenterol 21:11673–11679

Ye H, Chai X, Wang X, Zheng Q, Zheng D, Wu F, Zheng C, Chen P (2018) Autophagy flux inhibition augments gastric cancer resistance to the anti-human epidermal growth factor receptor 2 antibody trastuzumab. Oncol Lett 15:4143–4150

Yu HG, Ai YW, Yu LL, Zhou XD, Liu J, Li JH, Xu XM, Liu S, Chen J, Liu F, Qi YL, Deng Q, Cao J, Liu SQ, Luo HS, Yu JP (2008) Phosphoinositide 3-kinase/Akt pathway plays an important role in chemoresistance of gastric cancer cells against etoposide and doxorubicin induced cell death. Int J Cancer 122:433–443

Yu LL, Dai N, Yu HG, Sun LM, Si JM (2010) Akt associates with nuclear factor kappaB and plays an important role in chemoresistance of gastric cancer cells. Oncol Rep 24:113–119

Zhai J, Shen J, Xie G, Wu J, He M, Gao L, Zhang Y, Yao X, Shen L (2019) Cancer-associated fibroblasts-derived IL-8 mediates resistance to cisplatin in human gastric cancer. Cancer Lett 454:37–43

Zhang LL, Zhang J, Shen L, Xu XM, Yu HG (2013) Overexpression of AKT decreases the chemosensitivity of gastric cancer cells to cisplatin in vitro and in vivo. Mol Med Rep 7:1387–1390

Zhang H, Ren L, Ding Y, Li F, Chen X, Ouyang Y, Zhang Y, Zhang D (2019) Hyaluronan-mediated motility receptor confers resistance to chemotherapy via TGFbeta/Smad2-induced epithelial-mesenchymal transition in gastric cancer. FASEB J 33:6365–6377

Zheng HX, Cai YD, Wang YD, Cui XB, Xie TT, Li WJ, Peng L, Zhang Y, Wang ZQ, Wang J, Jiang B (2013) Fas signaling promotes motility and metastasis through epithelial-mesenchymal transition in gastrointestinal cancer. Oncogene 32:1183–1192

Zhi X, Tao J, Xiang G, Cao H, Liu Z, Yang K, Lv C, Ni S (2015) APRIL induces cisplatin resistance in gastric cancer cells via activation of the NF-kappaB pathway. Cell Physiol Biochem 35:571–585

Zhou J, Chen Q (2019) Poor expression of microRNA-135b results in the inhibition of cisplatin resistance and proliferation and induces the apoptosis of gastric cancer cells through MST1-mediated MAPK signaling pathway. FASEB J 33:3420–3436

Zhou W, Fu XQ, Zhang LL, Zhang J, Huang X, Lu XH, Shen L, Liu BN, Liu J, Luo HS, Yu JP, Yu HG (2013) The AKT1/NF-kappaB/Notch1/PTEN axis has an important role in chemoresistance of gastric cancer cells. Cell Death Dis 4:e847

Zhou L, Li X, Zhou F, Jin Z, Chen D, Wang P, Zhang S, Zhuge Y, Shang Y, Zou X (2018) Downregulation of leucine-rich repeats and immunoglobulin-like domains 1 by microRNA-20a modulates gastric cancer multidrug resistance. Cancer Sci 109:1044–1054

Zhu S, Belkhiri A, El-Rifai W (2011) DARPP-32 increases interactions between epidermal growth factor receptor and ERBB3 to promote tumor resistance to gefitinib. Gastroenterology 141:1738–1748. e1731–1732

Zhuang X, Li X, Zhang J, Hu Y, Hu B, Shi Y, Sun Y, Hong G (2018) Conditioned medium mimicking the tumor microenvironment augments chemotherapeutic resistance via ataxiatelangiectasia mutated and nuclear factorkappaB pathways in gastric cancer cells. Oncol Rep 40:2334–2342

Chapter 9
Meta-Analysis Reveals no Significant Association of EPHX1 Tyr113His and His139Arg Polymorphisms with the Colorectal Cancer Risk

L.V.K.S. Bhaskar, Akriti Gupta, and Smaranika Pattnaik

Abstract The Tyr113His and His139Arg polymorphisms in microsomal epoxide gene (EPHX1) have been reported to be associated with colorectal cancer (CRC) risk, but the results are inconclusive. Considering the functional importance of these polymorphisms and heterogeneity in genetic association studies, we performed a meta-analysis to investigate the association between the EPHX1 Tyr113His and His139Arg polymorphisms and CRC susceptibility. A comprehensive literature search of PubMed, Embase, and Google Scholar databases were conducted before May 10, 2019. Twenty eligible studies were finally included in this meta-analysis. The pooled odds ratio (OR) with 95% confidence intervals (CIs) were calculated. In the overall analysis, both Tyr113His and His139Arg polymorphisms were not associated with CRC in allelic and dominant genetic models. On subgroup analysis, no significant associations were observed in Asians and Caucasians in any of the genetic models for these polymorphisms. Our results were confirmed by sensitivity analysis and no publication bias was found. Taken together, our data indicate that EPHX1 Tyr113His and His139Arg polymorphisms are not associated with the susceptibility to colorectal cancer. Further well-designed studies with large sample size are warranted to establish the role of EPHX1 polymorphisms in CRC, especially for Tyr113His and His139Arg.

Keywords Colorectal cancer · Meta-analysis · Microsomal epoxide hydrolase · EPHX1 · Polymorphism · Susceptibility

---

L.V.K.S. Bhaskar
Guru Ghasidas University, Bilaspur, Chhattisgarh, India

A. Gupta
Sickle Cell Institute Chhattisgarh, Raipur, India

S. Pattnaik (✉)
Department of Biotechnology and Bioinformatics, Sambalpur University, Sambalpur, Odisha, India

G. S. R. Raju, L.V.K.S. Bhaskar (eds.), *Theranostics Approaches to Gastric and Colon Cancer*, Diagnostics and Therapeutic Advances in GI Malignancies,
https://doi.org/10.1007/978-981-15-2017-4_9

Abbreviations

Arg	Arginine
CI	Confidence intervals
CRC	Colorectal cancer
FEM	Fixed effects model
HCAs	heterocyclic amines
His	Histidin
HPFS	Health Professionals Follow-up Study
HWE	Hardy-Weinberg equilibrium
mEH	microsomal epoxide gene
NHS	Nurses' Health Study
OR	Odds ratio
PAHs	polycyclic aromatic hydrocarbons
PC	Pancreatic cancer
PLCO trial	Prostate, Lung, Colorectal, and Ovarian Cancer Screening Trial
REM	Random effects model
SNPs	Single nucleotide polymorphisms
Tyr	Tyrosine

9.1 Introduction

Colorectal cancer (CRC) is most common malignancy worldwide. The incidence of CRS varies over tenfold in different geographical regions. Developed countries such as Australia, Europe, and North America have higher incidence rates compared to the developing countries like Africa and South-Central Asia (Fitzmaurice et al. 2017). Further, due to early detection of CRC, polypectomy and introduction of effective primary and adjuvant treatments, the death rates from CRC was declined in western countries (Siegel et al. 2019). However in countries lacks strong healthcare infrastructure and limited resources a continuous increment in mortality rates was documented (Center et al. 2009). Although majority of CRCs are sporadic, a considerable inherited susceptibility has been observed in the CRC patients. Hence the likelihood of CRC development is the net results of environmental and genetic factors (Chan and Giovannucci 2010). Modern western lifestyles and clinical environmental factors are often associated with the increased risk of CRC. Several lines of evidences have demonstrated the long-term consumption of processed foods and foods cooked at high temperatures are implicated in CRC risk (Joshi et al. 2015). Cooking meats at high temperatures produce some compound such as polycyclic aromatic hydrocarbons (PAHs) and heterocyclic amines (HCAs) which has carcinogenic and mutagenic properties (Adeyeye 2018).

Microsomal epoxide hydrolase (mEH) (EPHX1; EC 3.3.2.3) is a phase II biotransformation enzyme that detoxifies epoxides, including PAHs and carcinogens

(Okat 2018). The mEH provide protection against the toxicities of reactive epoxide intermediates by converting them as less reactive and less toxic intermediates (Oesch et al. 2004). The EPHX1 gene encoding mEH is positioned at chromosome 1q42.1 and possesses two functional polymorphisms (Hartsfield et al. 1998). The coding region of the EPHX1 gene has two genetic variants (Tyr113His and His139Arg) that alter enzyme activity. In vitro expression studies revealed that the Tyr113His polymorphism decreased mEH enzymatic activity by 40%, while His139Arg polymorphism increased mEH activity by 25% (Hassett et al. 1994). Both polymorphisms exhibit differences in alleles and genotypes among different ethnic populations (Bhaskar et al. 2013; Lakkakula et al. 2013). As mEH involved in detoxification of epoxides together with carcinogens such as PAHs and HCAs present in cigarette smoke also in cooked meats, the functional polymorphisms of EPHX modulate the rate of PAHs metabolism and subsequently modulate CRC risk. A number of studies have analysed the association between EPHX1 gene polymorphisms and the risk of various cancers, but the results are inconclusive. As the results from the previous studies investigating the correlation between colorectal cancer and EPHX1 polymorphism were not similarly conclusive (Harrison et al. 1999; Ikeda et al. 2008; Kiss et al. 2007; Mitrou et al. 2007), we performed a meta-analysis of all available data to investigate the role of *EPHX1* Tyr113His and His139Arg polymorphisms with respect to the colorectal cancer risk.

9.2 Materials and Methods

9.2.1 Data Extraction

Studies related to association between EPHX1 polymorphisms and colorectal cancer risk were collected by searching PubMed, Embase, and Google Scholar. To harvest more comprehensive information published till May 2019, search terms such as "EPHX1 or mEH", "polymorphism or mutation" and "colorectal cancer or carcinoma," were used without any language restrictions. To facilitate the proper elucidation of results, potentially relevant studies were selected based the following criteria: (i) evaluation of the *EPHX1* Tyr113His or His139Arg and risk of CRC, (ii) case-control study, and (iv) availability of genotypes. The studies matching with the above mentioned basic criteria were included in this meta-analysis. After assessing the methodological quality of individual papers, first author's name, publication year, country of origin, and genotype frequencies were collected independently by two authors.

9.2.2 Statistical Analysis

To find the departure of Hardy–Weinberg equilibrium (HWE) in the control groups of EPHX polymorphisms, Chi-Square goodness of fit test was performed. To measure the strength of the association between the EPHX1 polymorphisms and CRC risk of cancer, odds ratios (ORs) and their corresponding 95% confidence intervals (CIs) were calculated for each study. The pooled OR with 95% CI was calculated in allelic and dominant genetic models. Between study heterogeneity in these genetic models were calculated using Chi-square test and I2 test. Random effects model (REM) or Fixed effects model (FEM) was selected respectively in the presence or absence of the heterogeneity. Subgroup analysis was also conducted according to ethnicity. To assess the influence of the individual studies to the pooled results, sensitivity analysis was conducted by omitting one study at a time. To test the publication bias, Begg's and Egger's tests were used. MetaGenyo, a web tool was used to calculate results of the meta-analysis in this study (Martorell-Marugan et al. 2017).

9.3 Results

9.3.1 Characteristics of Studies

A total 20 publications dealing with EPHX1 polymorphisms and CRC risk were included in the meta-analysis. The workflow of study identification is illustrated in Fig. 9.1. The characteristics of each study were summarized in Table 9.1. For Tyr113His, 20 publications and for His139Arg, 14 publications from several countries involving Caucasian and Asian subjects were investigated. All genotype distribution in controls was in accordance with HWE with the exception of 4 studies for Tyr113His (Kiss et al. 2007; Sachse et al. 2002; Sahin et al. 2012; Tranah et al. 2005).

9.3.2 Meta-Analysis of EPHX1 Tyr113His Polymorphism with CRC Risk

In this meta-analysis, a total of 20 studies (Cleary et al. 2010; Cotterchio et al. 2008; Fernandes et al. 2016; Harrison et al. 1999; Hlavata et al. 2010; Huang et al. 2005; Ikeda et al. 2008; Kiss et al. 2007; Kury et al. 2008; Landi et al. 2005; Mitrou et al. 2007; Nisa et al. 2013; Northwood et al. 2010; Pande et al. 2008; Sachse et al. 2002; Sahin et al. 2012; Skjelbred et al. 2007; Tranah et al. 2005; van der Logt et al. 2006; Wang et al. 2012) involving 9770 CRC patients and 11,634 controls were included to investigate the associations between EPHX1 Tyr113His and the risk of CRC

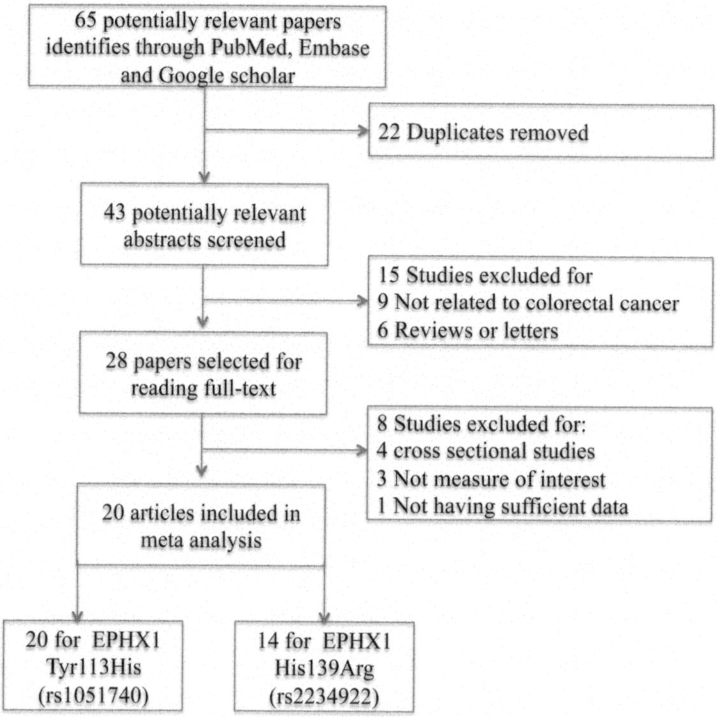

Fig. 9.1 Flow chart of the study selection process identifying studies comparing EPHX1 Tyr113His and His139Arg polymorphisms with the colorectal cancer

(Table 9.1). Pooled data showed that EPHX1 Tyr113His polymorphism was not significantly associated with an increased risk of CRC in both allelic and dominant genetic models (allelic C versus T: OR = 0.95, 95% CI: 0.89–1.02 and dominant **CC + CT** versus TT: OR = 0.94, 95% CI: 0.85–1.03) (Table 9.2; Fig. 9.2). In addition, a further subgroup analysis by ethnicity in Caucasians, Asians and mixed populations indicated that no association between EPHX1 Tyr113His polymorphism and CRC was observed for both allelic and dominant genetic models.

9.3.3 *Meta-Analysis Between EPHX1 His139Arg Polymorphism and CRC Risk*

A meta-analysis of the association between EPHX1 His139Arg polymorphism and CRC risk included 14 independent studies (Fernandes et al. 2016; Harrison et al. 1999; Hlavata et al. 2010; Huang et al. 2005; Kiss et al. 2007; Landi et al. 2005; Nisa et al. 2013; Northwood et al. 2010; Pande et al. 2008; Sachse et al. 2002; Sahin et al. 2012; Skjelbred et al. 2007; Tranah et al. 2005; van der Logt et al. 2006) with a total

Table 9.1 Characteristics of the studies included in the meta-analysis and their genotype distributions of the EPHX1 Tyr113His and His139Arg gene polymorphisms

S. No.	Author's Name	Country	Ethnicity	Method	Case			Control			HWE p-value	References
Tyr113His (rs1051740)					CC	CT	TT	CC	CT	TT		
1	Fernandes et al. 2016	Brazil	Caucasian	PCR-RFLP	13	88	126	28	158	214	0.874	(Fernandes et al. 2016)
2	Nisa et al. 2013	Japan	Asian	TaqMan	115	342	228	143	396	239	0.347	(Nisa et al. 2013)
3	Wang et al. 2012	USA	Mixed	TaqMan	28	108	167	29	141	188	0.723	(Wang et al. 2012)
4	Sahin et al. 2012	Turkey	Caucasian	PCR-RFLP	12	29	27	8	64	44	0.017	(Sahin et al. 2012)
5	Cleary et al. 2010	Canada	Caucasian	TaqMan	561	502	100	625	549	118	0.871	(Cleary et al. 2010)
6	Hlavata et al. 2010	Czech Republic	European	TaqMan	50	224	221	52	212	231	0.746	(Hlavata et al. 2010)
7	Northwood et al. 2010	Scotland	Caucasian	TaqMan	24	118	166	26	112	158	0.341	(Northwood et al. 2010)
8	Kury et al. 2008	France	Caucasian	TaqMan	89	409	525	88	469	564	0.486	(Kury et al. 2008)
9	Ikeda et al. 2008	Japan	Asian	TaqMan	22	80	120	24	56	26	0.557	(Ikeda et al. 2008)
10	Cotterchio et al. 2008	Ontario	Caucasian	TaqMan	74	354	404	113	526	610	0.980	(Cotterchio et al. 2008)
11	Pande et al. 2008	Texas	Mixed	Pyrosequencing	8	46	66	19	108	130	0.595	(Pande et al. 2008)
12	Skjelbred et al. 2007	Norway	Caucasian	PCR-RFLP	9	41	52	33	134	132	0.908	(Skjelbred et al. 2007)
13	Mitrou et al. 2007	UK	Caucasian	PCR-RFLP	40	295	509	31	285	555	0.450	(Mitrou et al. 2007)
14	Kiss et al. 2007	Hungary	Caucasian	PCR-RFLP	53	227	220	31	221	248	0.046	(Kiss et al. 2007)
15	van der Logt et al. 2006	Netherlands	Caucasian	PCR-RFLP	39	141	185	32	165	194	0.709	(van der Logt et al. 2006)
16	Huang et al. 2005	USA	Caucasian	TaqMan	56	299	357	80	291	358	0.076	(Huang et al. 2005)
17	Landi et al. 2005	Spain	Caucasian	APEX	29	129	168	33	153	177	0.994	(Landi et al. 2005)
18	Tranah et al. 2005	USA	Mixed	TaqMan	160	303	422	208	425	585	<0.001	(Tranah et al. 2005)
19	Sachse et al. 2002	UK	Caucasian	PCR-RFLP	74	187	228	129	193	270	<0.001	(Sachse et al. 2002)

20	Harrison et al. 1999	UK	Caucasian	PCR-RFLP	42	38	21	91	99	13	0.040	(Harrison et al. 1999)
His139Arg (rs2234922)												
					AA	AG	GG	AA	AG	GG		
1	Fernandes et al. 2016	Brazil	Caucasian	PCR-RFLP	153	67	7	235	145	20	0.696	(Fernandes et al. 2016)
2	Nisa et al. 2013	Japan	Asian	TaqMan	485	182	18	525	224	29	0.405	(Nisa et al. 2013)
3	Sahin et al. 2012	Turkey	Caucasian	PCR-RFLP	52	15	1	72	44	0	0.012	(Sahin et al. 2012)
4	Hlavata et al. 2010	Czech	European	TaqMan	297	173	25	290	183	22	0.306	(Hlavata et al. 2010)
5	Northwood et al. 2010	Scotland	Caucasian	TaqMan	185	109	14	173	106	17	0.886	(Northwood et al. 2010)
6	Pande et al. 2008	Texas	Mixed	Pyrosequencing	77	37	6	155	92	10	0.420	(Pande et al. 2008)
7	Skjelbred et al. 2007	Norway	Caucasian	PCR-RFLP	56	27	18	190	90	19	0.068	(Skjelbred et al. 2007)
8	Kiss et al. 2007	Hungary	Caucasian	PCR-RFLP	337	157	6	329	161	10	0.054	(Kiss et al. 2007)
9	van der Logt et al. 2006	Netherland	Caucasian	PCR-RFLP	241	106	24	269	128	17	0.719	(van der Logt et al. 2006)
10	Huang et al. 2005	USA	Caucasian	TaqMan	31	228	443	24	221	479	0.809	(Huang et al. 2005)
11	Landi et al. 2005	Spain	Caucasian	APEX	250	98	13	225	85	11	0.403	(Landi et al. 2005)
12	Tranah et al. 2005	USA	Mixed	TaqMan	584	275	42	789	406	40	0.160	(Tranah et al. 2005)
13	Sachse et al. 2002	UK	Caucasian	PCR-RFLP	329	142	19	378	200	15	0.055	(Sachse et al. 2002)
14	Harrison et al. 1999	UK	Caucasian	PCR-RFLP	147	53	3	76	21	4	0.121	(Harrison et al. 1999)

Table 9.2 Meta-analysis of the relationships of the EPHX1 Tyr113His and His139Arg polymorphisms with the colorectal cancer

EPHX1 Tyr113His	Overall (REM)	Asian	Caucasian	Mixed
Number of studies	20	2	15	3
Allele model (C vs. T)				
I2	0.62	0.95	0.34	<0.001
PHeterogeneity	<0.001	<0.001	0.093	0.659
OR (95% CI)	0.95 (0.89–1.02)	0.62(0.28–1.38)	0.99 (0.93–1.05)	1.00 (0.90–1.12)
Association p value	0.192	0.242	0.755	0.984
Egger's p value	0.078	NA	0.390	0.076
Dominant model (CC + CT vs. TT)				
I2	0.59	0.94	0.35	0.00
PHeterogeneity	<0.001	<0.001	0.091	0.633
OR (95% CI)	0.94 (0.85–1.03)	0.51 (0.16–1.59	0.99 (0.91–1.08)	0.97 (0.84–1.12)
Association p value	0.169	0.245	0.821	0.658
Egger's p value	0.003	NA	0.024	0.070
EPHX1 His139Arg	Overall (FEM)	Asian	Caucasian	Mixed
Number of studies	14	1	11	2
Allele contrast (G vs. A)				
I2	0.32	NA	0.42	0.00
PHeterogeneity	0.115	NA	0.067	0.637
OR (95% CI)	0.95 (0.89–1.01)	0.85 (0.70–1.04)	0.96 (0.86–1.07)	1.00 (0.87–1.15)
Association p value	0.127	0.113	0.440	0.980
Egger's p value	0.998	NA	0.832	NA
Dominant model (GG + GA vs. AA)				
I2	0.00	NA	0.13	0.00
PHeterogeneity	0.505	NA	0.317	0.616
OR (95% CI)	0.92 (0.85–0.99)	0.86 (0.69–1.07)	0.92 (0.83–1.02)	0.94 (0.80–1.12)
Association p value	0.035	0.170	0.106	0.498
Egger's p value	0.617	NA	0.703	NA

of 5532 CRC cases and 6529 controls (Table 9.1). Pooled data revealed that EPHX1 His139Arg polymorphism was correlated with the risk of CRC in the both allelic and dominant model (allelic G versus A: OR = 0.95, 95% CI: 0.89–1.01; dominant GG + GA versus AA: OR = 0.92, 95% CI: 0.85–0.99). In addition, a further subgroup analysis by ethnicity in Caucasians, Asians and mixed populations indicated that no association between EPHX1 His139Arg polymorphism and CRC was observed for both allelic and dominant genetic models.

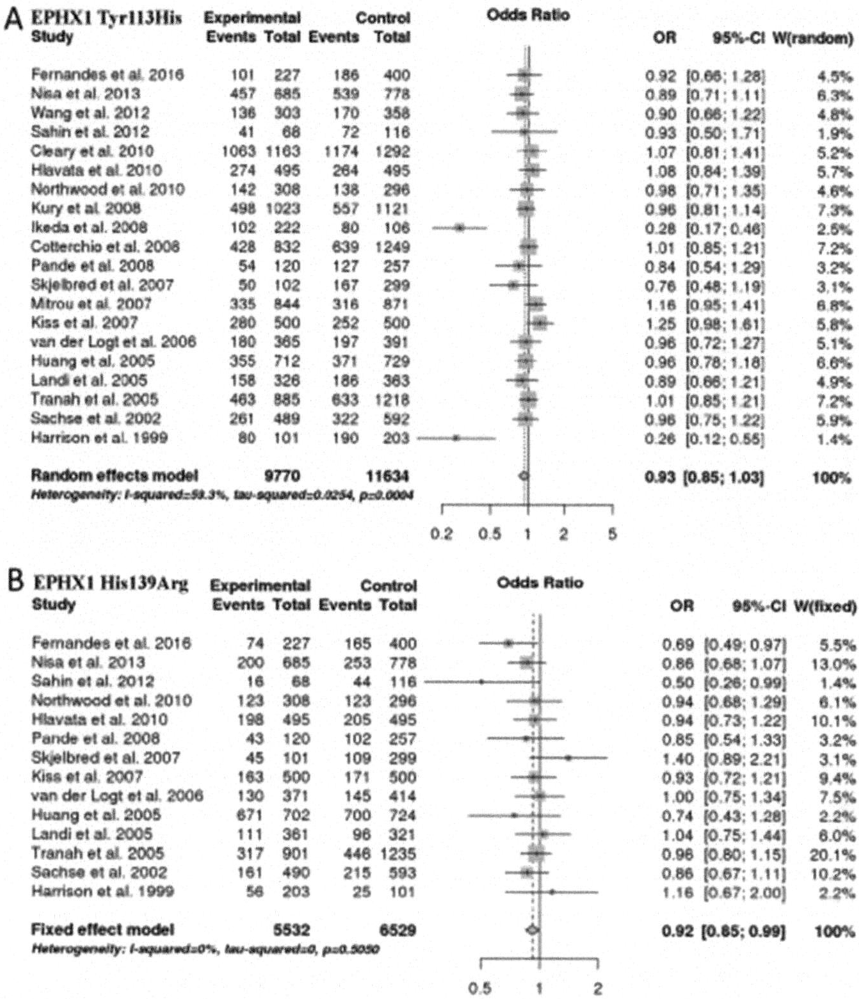

Fig. 9.2 Forest plot of the studies assessing the association between colorectal cancer and EPHX1 gene polymorphisms. (**a**) EPHX1 Tyr113His and colorectal cancer; (**b**) His139Arg and colorectal cancer

9.3.4 Heterogeneity and Sensitivity Analysis

The heterogeneities that observed within the *EPHX1* Tyr113His or His139Arg studies and within each subgroup of studies are shown in Table 9.2. The heterogeneity test showed a moderate heterogeneity between *EPHX1* Tyr113His studies (Dominant model $P_{heterogeneity} = 0.003$, I-squared = 59%). However, no heterogeneity was detected in *EPHX1* His139Arg studies (Dominant model $P_{heterogeneity} = 0.505$, I-squared = 0%). Hence in our meta-analysis, to calculate

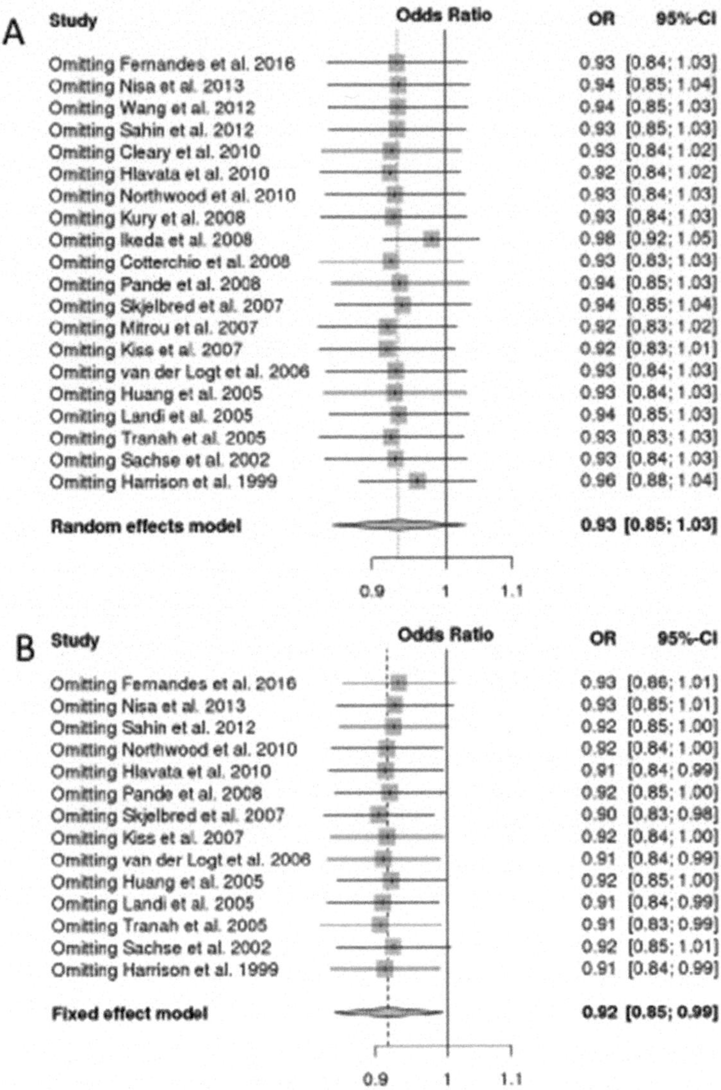

Fig. 9.3 Sensitivity analysis diagram by omitting a single study. (**a**) Sensitivity analysis of the OR coefficients for the association between EPHX1 Tyr113His and colorectal cancer; (**b**) Sensitivity analysis of the OR coefficients for the association between His139Arg and colorectal cancer

the summary ORs we applied REM and FEM respectively for Tyr113His and His139Arg polymorphisms. Further, to evaluate the sensitivity of this meta-analysis, we conducted pooled analyses by omitting one study each time (leave-one-out method). The results of the leave-one-out method for Tyr113His and His139Arg polymorphisms was shown in Fig. 9.3a and b. For both polymorphisms, there is no change in statistical significance of the results when any single study was omitted indicating the stability and reliability of the results.

9.3.5 Publication Bias

The results of publication bias for *EPHX1* Tyr113His and His139Arg studies were shown in Table 9.2 and Fig. 9.4a and b. The shape of Begg's funnel plot did not reveal any obvious asymmetry for both *EPHX1* Tyr113His and His139Arg studies (Fig. 9.4a and b). Further, Egger's linear regression tests did not reveal publication bias for both *EPHX1* Tyr113His and His139Arg studies in all genetic models tested (Table 9.2).

9.4 Discussion

Indeed cancer initiation and progression has been linked to activation of the immune system and oxidative stress. Throughout the life, the colon is exposed to microbiota and free radicals that respectively cause immune responses and oxidative damage. Although the etiology of CRC is not fully known, several lines of evidences indicated that the red meat intake of one of the major risk factors for the CRC (Bernstein et al. 2015; Demeyer et al. 2016; Domingo and Nadal 2017; Zhao et al. 2017a). In contrast to this some studies showed an inverse association of fiber intake with risk of CRC (Lee et al. 2017; Song et al. 2015). The meat cooking processes produce several carcinogens such as PAHs, HCAs and dioxin-like compounds. Dietary consumption is one of the highest sources of these environmental carcinogens (Zhang et al. 2013). To become carcinogenic, these PAHs and aromatic amines have to undergo phase I and phase II biotransformation reactions (Turesky 2004).

Microsomal epoxide hydrolase is one of the biotransformation enzymes that involved either detoxification or bio-activation of a wide range of substrates. Hence it is hypothesized that the polymorphisms of EPHX1 are crucial for the susceptibility of colon cancer. The Tyr113His variation is linked with low enzyme activity and His139Arg variation increases enzyme activity. Two nested case-control studies from the cohorts Nurses' Health Study and Health Professionals' Follow-up Study, did not reveal significant association EPHX1 gene polymorphisms and CRC risk (Tranah et al. 2004). However, individuals with ≥25 *pack-year smoking* history showed increased CRC risk (Tranah et al. 2004). The *Prostate, Lung, Colorectal and* Ovarian (PLCO) Cancer Screening Trial demonstrated that the EPHX1 polymorphisms increased risk of CRC in non-Hispanic current and recent smokers (Huang et al. 2005). Subsequent studies showed that the EPHX1 polymorphisms are not associated with CRC risk (Landi et al. 2005; Mitrou et al. 2007; van der Logt et al. 2006). Meta analysis of interaction of smoking with the EPHX1 Tyr113His polymorphisms showed that the carriers of low metabolizer allele had slightly lower risk

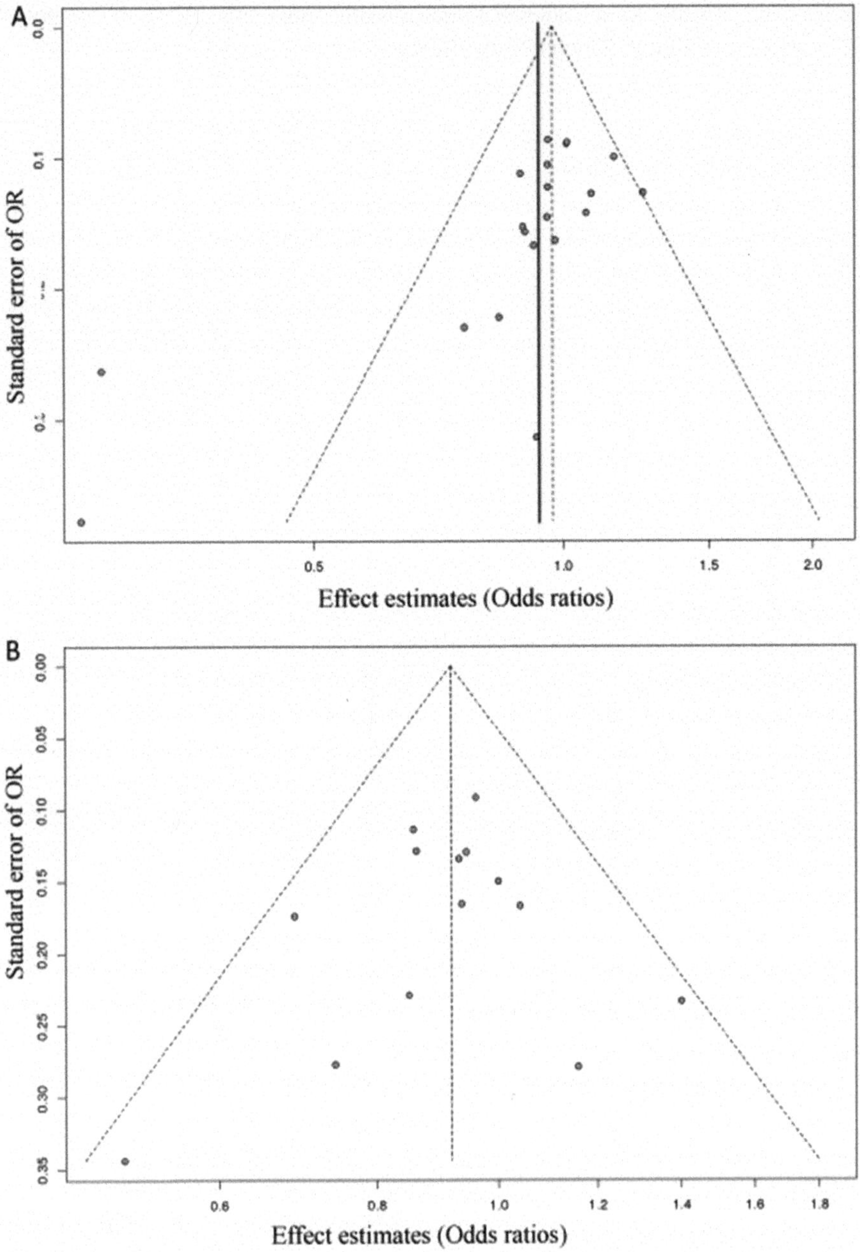

Fig. 9.4 Funnel plots analyses to assess the publication bias between colorectal cancer and EPHX1 gene polymorphisms. (**a**) Funnel plot based on EPHX1 Tyr113His and colorectal cancer; (**b**) Funnel plot based on His139Arg and colorectal cancer. The dotted vertical line indicates the triangular region within which 95% of studies are expected to lie in the absence of bias and heterogeneity. The solid vertical line corresponds to no intervention effect

of CRC compared to its high metabolizer (Raimondi et al. 2009). Analysis of interactions among the charred meat consumption, smoking, EPHX1 polymorphisms and CRC, did not reveal significant association between EPHX1 genotype and colorectal polyps (Burnett-Hartman et al. 2011). However, meta-analysis of observational studies demonstrated that the increased intake of red and processed meat is associated with significantly increased risk of CRC (Xu et al. 2013). Further, meta-analysis also indicated that the red and processed meat intake was associated with an increased CRC incidence but not for CRC recurrence (Zhao et al. 2017b).

This meta-analysis included 20 independent case-control studies of CRC to investigate its correlation with EPHX1 variants. The results of our meta-analysis showed that EPHX1 gene Tyr113His and His139Arg polymorphisms were not associated with the risk of CRC in allelic and dominant genetic comparison models. Further in subgroup analysis by ethnicity, these polymorphisms were not associated with the risk of CRC in both Asian and Caucasians. There was no evidence of publication bias.

9.5 Conclusion

In summary, the results of this meta-analysis demonstrated that no evidence supporting the relationship between EPHX1 polymorphism and CRC risk was detected. As the pathogenesis of CRC is complex and involving interactions of gene with gene and gene with dietary factors some limitations should be taken into consideration when interpreting results. A major limitation of current meta-analysis is non-availability of the data on meat consumption, alcohol intake and smoking, which limited the evaluation of the potential interactions between these risk factors and EPHX1 polymorphisms. Well-designed studies with large sample size are warranted to establish the role of EPHX1 polymorphisms in CRC, especially for Tyr113His and His139Arg.

Conflict of Interest There are no conflicts of interests.

References

Adeyeye SAO (2018) Heterocyclic amines and polycyclic aromatic hydrocarbons in cooked meat products: a review. Polycycl Aromat Compd:1–11

Bernstein AM, Song M, Zhang X, Pan A, Wang M, Fuchs CS, Le N, Chan AT, Willett WC, Ogino S, Giovannucci EL, Wu K (2015) Processed and unprocessed red meat and risk of colorectal Cancer: analysis by tumor location and modification by time. PLoS One 10:e0135959

Bhaskar LV, Thangaraj K, Patel M, Shah AM, Gopal K, Saikrishna L, Tamang R, Singh L, Rao VR (2013) EPHX1 gene polymorphisms in alcohol dependence and their distribution among the Indian populations. Am J Drug Alcohol Abuse 39:16–22

Burnett-Hartman AN, Newcomb PA, Mandelson MT, Adams SV, Wernli KJ, Shadman M, Wurscher MA, Makar KW (2011) Colorectal polyp type and the association with charred meat consumption, smoking, and microsomal epoxide hydrolase polymorphisms. Nutr Cancer 63:583–592

Center MM, Jemal A, Smith RA, Ward E (2009) Worldwide variations in colorectal cancer. CA Cancer J Clin 59:366–378

Chan AT, Giovannucci EL (2010) Primary prevention of colorectal cancer. Gastroenterology 138:2029–2043. e2010

Cleary SP, Cotterchio M, Shi E, Gallinger S, Harper P (2010) Cigarette smoking, genetic variants in carcinogen-metabolizing enzymes, and colorectal cancer risk. Am J Epidemiol 172:1000–1014

Cotterchio M, Boucher BA, Manno M, Gallinger S, Okey AB, Harper PA (2008) Red meat intake, doneness, polymorphisms in genes that encode carcinogen-metabolizing enzymes, and colorectal cancer risk. Cancer epidemiology, biomarkers & prevention: a publication of the American Association for Cancer Research, cosponsored by the American Society of Preventive. Oncology 17:3098–3107

Demeyer D, Mertens B, De Smet S, Ulens M (2016) Mechanisms linking colorectal Cancer to the consumption of (processed) red meat: a review. Crit Rev Food Sci Nutr 56:2747–2766

Domingo JL, Nadal M (2017) Carcinogenicity of consumption of red meat and processed meat: a review of scientific news since the IARC decision. Food Chem Toxicol 105:256–261

Fernandes GM, Russo A, Proenca MA, Gazola NF, Rodrigues GH, Biselli-Chicote PM, Silva AE, Netinho JG, Pavarino EC, Goloni-Bertollo EM (2016) CYP1A1, CYP2E1 and EPHX1 polymorphisms in sporadic colorectal neoplasms. World J Gastroenterol 22:9974–9983

Fitzmaurice C, Allen C, Barber RM, Barregard L, Bhutta ZA, Brenner H, Dicker DJ, Chimed-Orchir O, Dandona R, Dandona L, Fleming T, Forouzanfar MH, Hancock J, Hay RJ, Hunter-Merrill R, Huynh C, Hosgood HD, Johnson CO, Jonas JB, Khubchandani J, Kumar GA, Kutz M, Lan Q, Larson HJ, Liang X, Lim SS, Lopez AD, MacIntyre MF, Marczak L, Marquez N, Mokdad AH, Pinho C, Pourmalek F, Salomon JA, Sanabria JR, Sandar L, Sartorius B, Schwartz SM, Shackelford KA, Shibuya K, Stanaway J, Steiner C, Sun J, Takahashi K, Vollset SE, Vos T, Wagner JA, Wang H, Westerman R, Zeeb H, Zoeckler L, Abd-Allah F, Ahmed MB, Alabed S, Alam NK, Aldhahri SF, Alem G, Alemayohu MA, Ali R, Al-Raddadi R, Amare A, Amoako Y, Artaman A, Asayesh H, Atnafu N, Awasthi A, Saleem HB, Barac A, Bedi N, Bensenor I, Berhane A, Bernabe E, Betsu B, Binagwaho A, Boneya D, Campos-Nonato I, Castaneda-Orjuela C, Catala-Lopez F, Chiang P, Chibueze C, Chitheer A, Choi JY, Cowie B, Damtew S, das Neves J, Dey S, Dharmaratne S, Dhillon P, Ding E, Driscoll T, Ekwueme D, Endries AY, Farvid M, Farzadfar F, Fernandes J, Fischer F, TT GH, Gebru A, Gopalani S, Hailu A, Horino M, Horita N, Husseini A, Huybrechts I, Inoue M, Islami F, Jakovljevic M, James S, Javanbakht M, Jee SH, Kasaeian A, Kedir MS, Khader YS, Khang YH, Kim D, Leigh J, Linn S, Lunevicius R, El Razek HMA, Malekzadeh R, Malta DC, Marcenes W, Markos D, Melaku YA, Meles KG, Mendoza W, Mengiste DT, Meretoja TJ, Miller TR, Mohammad KA, Mohammadi A, Mohammed S, Moradi-Lakeh M, Nagel G, Nand D, Le Nguyen Q, Nolte S, Ogbo FA, Oladimeji KE, Oren E, Pa M, Park EK, Pereira DM, Plass D, Qorbani M, Radfar A, Rafay A, Rahman M, Rana SM, Soreide K, Satpathy M, Sawhney M, Sepanlou SG, Shaikh MA, She J, Shiue I, Shore HR, Shrime MG, So S, Soneji S, Stathopoulou V, Stroumpoulis K, Sufiyan MB, Sykes BL, Tabares-Seisdedos R, Tadese F, Tedla BA, Tessema GA, Thakur JS, Tran BX, Ukwaja KN, Uzochukwu BSC, Vlassov VV, Weiderpass E, Wubshet Terefe M, Yebyo HG, Yimam HH, Yonemoto N, Younis MZ, Yu C, Zaidi Z, Zaki MES, Zenebe ZM, Murray CJL, Naghavi M (2017) Global, regional, and National Cancer Incidence, mortality, years of life lost, years lived with disability, and disability-adjusted life-years for 32 Cancer groups, 1990 to 2015: a systematic analysis for the global burden of disease study. JAMA Oncol 3:524–548

Harrison DJ, Hubbard AL, MacMillan J, Wyllie AH, Smith CA (1999) Microsomal epoxide hydrolase gene polymorphism and susceptibility to colon cancer. Br J Cancer 79:168–171

Hartsfield JK Jr, Sutcliffe MJ, Everett ET, Hassett C, Omiecinski CJ, Saari JA (1998) Assignment1 of microsomal epoxide hydrolase (EPHX1) to human chromosome 1q42.1 by in situ hybridization. Cytogenet Cell Genet 83:44–45

Hassett C, Aicher L, Sidhu JS, Omiecinski CJ (1994) Human microsomal epoxide hydrolase: genetic polymorphism and functional expression in vitro of amino acid variants. Hum Mol Genet 3:421–428

Hlavata I, Vrana D, Smerhovsky Z, Pardini B, Naccarati A, Vodicka P, Novotny J, Mohelnikova-Duchonova B, Soucek P (2010) Association between exposure-relevant polymorphisms in CYP1B1, EPHX1, NQO1, GSTM1, GSTP1 and GSTT1 and risk of colorectal cancer in a Czech population. Oncol Rep 24:1347–1353

Huang WY, Chatterjee N, Chanock S, Dean M, Yeager M, Schoen RE, Hou LF, Berndt SI, Yadavalli S, Johnson CC, Hayes RB (2005) Microsomal epoxide hydrolase polymorphisms and risk for advanced colorectal adenoma. Cancer epidemiology, biomarkers & prevention: a publication of the American Association for Cancer Research, cosponsored by the American Society of Preventive. Oncology 14:152–157

Ikeda S, Sasazuki S, Natsukawa S, Shaura K, Koizumi Y, Kasuga Y, Ohnami S, Sakamoto H, Yoshida T, Iwasaki M, Tsugane S (2008) Screening of 214 single nucleotide polymorphisms in 44 candidate cancer susceptibility genes: a case-control study on gastric and colorectal cancers in the Japanese population. Am J Gastroenterol 103:1476–1487

Joshi AD, Kim A, Lewinger JP, Ulrich CM, Potter JD, Cotterchio M, Le Marchand L, Stern MC (2015) Meat intake, cooking methods, dietary carcinogens, and colorectal cancer risk: findings from the colorectal Cancer family registry. Cancer Med 4:936–952

Kiss I, Orsos Z, Gombos K, Bogner B, Csejtei A, Tibold A, Varga Z, Pazsit E, Magda I, Zolyomi A, Ember I (2007) Association between allelic polymorphisms of metabolizing enzymes (CYP 1A1, CYP 1A2, CYP 2E1, mEH) and occurrence of colorectal cancer in Hungary. Anticancer Res 27:2931–2937

Kury S, Buecher B, Robiou-du-Pont S, Scoul C, Colman H, Le Neel T, Le Houerou C, Faroux R, Ollivry J, Lafraise B, Chupin LD, Sebille V, Bezieau S (2008) Low-penetrance alleles predisposing to sporadic colorectal cancers: a French case-controlled genetic association study. BMC Cancer 8:326

Lakkakula S, Maram R, Munirajan AK, Pathapati RM, Visweswara SB, Lakkakula BVKS (2013) EPHX1 gene polymorphisms among south Indian populations. Mol Cell Toxicol 9:219–225

Landi S, Gemignani F, Moreno V, Gioia-Patricola L, Chabrier A, Guino E, Navarro M, de Oca J, Capella G, Canzian F (2005) A comprehensive analysis of phase I and phase II metabolism gene polymorphisms and risk of colorectal cancer. Pharmacogenet Genomics 15:535–546

Lee J, Shin A, Oh JH, Kim J (2017) Colors of vegetables and fruits and the risks of colorectal cancer. World J Gastroenterol 23:2527–2538

Martorell-Marugan J, Toro-Dominguez D, Alarcon-Riquelme ME, Carmona-Saez P (2017) MetaGenyo: a web tool for meta-analysis of genetic association studies. BMC Bioinformatics 18:563

Mitrou PN, Watson MA, Loktionov AS, Cardwell C, Gunter MJ, Atkin WS, Macklin CP, Cecil T, Bishop DT, Primrose J, Bingham SA (2007) Role of NQO1C609T and EPHX1 gene polymorphisms in the association of smoking and alcohol with sporadic distal colorectal adenomas: results from the UKFSS study. Carcinogenesis 28:875–882

Nisa H, Budhathoki S, Morita M, Toyomura K, Nagano J, Ohnaka K, Kono S, Ueki T, Tanaka M, Kakeji Y, Maehara Y, Okamura T, Ikejiri K, Futami K, Maekawa T, Yasunami Y, Takenaka K, Ichimiya H, Terasaka R (2013) Microsomal epoxide hydrolase polymorphisms, cigarette smoking, and risk of colorectal cancer: the Fukuoka colorectal cancer study. Mol Carcinog 52:619–626

Northwood EL, Elliott F, Forman D, Barrett JH, Wilkie MJ, Carey FA, Steele RJ, Wolf R, Bishop T, Smith G (2010) Polymorphisms in xenobiotic metabolizing enzymes and diet influence colorectal adenoma risk. Pharmacogenet Genomics 20:315–326

Oesch F, Hengstler JG, Arand M (2004) Detoxication strategy of epoxide hydrolase-the basis for a novel threshold for definable genotoxic carcinogens. Nonlinearity Biol Toxicol Med 2:21–26

Okat Z (2018) The connection between microsomal epoxide hydrolase enzyme and cancer risk. Int Phys Med Rehab J 3:402–404

Pande M, Amos CI, Osterwisch DR, Chen J, Lynch PM, Broaddus R, Frazier ML (2008) Genetic variation in genes for the xenobiotic-metabolizing enzymes CYP1A1, EPHX1, GSTM1, GSTT1, and GSTP1 and susceptibility to colorectal cancer in Lynch syndrome. Cancer epidemiology, biomarkers & prevention: a publication of the American Association for Cancer Research, cosponsored by the American Society of Preventive. Oncology 17:2393–2401

Raimondi S, Botteri E, Iodice S, Lowenfels AB, Maisonneuve P (2009) Gene-smoking interaction on colorectal adenoma and cancer risk: review and meta-analysis. Mutat Res 670:6–14

Sachse C, Smith G, Wilkie MJ, Barrett JH, Waxman R, Sullivan F, Forman D, Bishop DT, Wolf CR (2002) A pharmacogenetic study to investigate the role of dietary carcinogens in the etiology of colorectal cancer. Carcinogenesis 23:1839–1849

Sahin O, Arikan S, Oltulu YM, Coskunpinar E, Eren A, Cacina C, Guler E, Yaylim I (2012) Investigation of a possible relationship between EPHX1 gene polymorphisms and colorectal cancer in Turkish society. Genet Test Mol Biomarkers 16:423–428

Siegel RL, Miller KD, Jemal A (2019) Cancer statistics, 2019. CA Cancer J Clin 69:7–34

Skjelbred CF, Saebo M, Hjartaker A, Grotmol T, Hansteen IL, Tveit KM, Hoff G, Kure EH (2007) Meat, vegetables and genetic polymorphisms and the risk of colorectal carcinomas and adenomas. BMC Cancer 7:228

Song Y, Liu M, Yang FG, Cui LH, Lu XY, Chen C (2015) Dietary fibre and the risk of colorectal cancer: a case- control study. Asian Pac J Cancer Prev: APJCP 16:3747–3752

Tranah GJ, Giovannucci E, Ma J, Fuchs C, Hankinson SE, Hunter DJ (2004) Epoxide hydrolase polymorphisms, cigarette smoking and risk of colorectal adenoma in the Nurses' health study and the health professionals follow-up study. Carcinogenesis 25:1211–1218

Tranah GJ, Chan AT, Giovannucci E, Ma J, Fuchs C, Hunter DJ (2005) Epoxide hydrolase and CYP2C9 polymorphisms, cigarette smoking, and risk of colorectal carcinoma in the Nurses' health study and the Physicians' health study. Mol Carcinog 44:21–30

Turesky RJ (2004) The role of genetic polymorphisms in metabolism of carcinogenic heterocyclic aromatic amines. Curr Drug Metab 5:169–180

van der Logt EM, Bergevoet SM, Roelofs HM, Te Morsche RH, Dijk Y, Wobbes T, Nagengast FM, Peters WH (2006) Role of epoxide hydrolase, NAD(P)H:quinone oxidoreductase, cytochrome P450 2E1 or alcohol dehydrogenase genotypes in susceptibility to colorectal cancer. Mutat Res 593:39–49

Wang J, Joshi AD, Corral R, Siegmund KD, Marchand LL, Martinez ME, Haile RW, Ahnen DJ, Sandler RS, Lance P, Stern MC (2012) Carcinogen metabolism genes, red meat and poultry intake, and colorectal cancer risk. Int J Cancer 130:1898–1907

Xu X, Yu E, Gao X, Song N, Liu L, Wei X, Zhang W, Fu C (2013) Red and processed meat intake and risk of colorectal adenomas: a meta-analysis of observational studies. Int J Cancer 132:437–448

Zhang X, Lin S, Funk WE, Hou L (2013) Environmental and occupational exposure to chemicals and telomere length in human studies. Occup Environ Med 70:743–749

Zhao Z, Feng Q, Yin Z, Shuang J, Bai B, Yu P, Guo M, Zhao Q (2017a) Red and processed meat consumption and colorectal cancer risk: a systematic review and meta-analysis. Oncotarget 8:83306–83314

Zhao Z, Yin Z, Hang Z, Zhang C, Zhao Q (2017b) Association between red and processed meat intake and colorectal adenoma incidence and recurrence: a systematic review and meta-analysis. Oncotarget 9:32373–32382

Chapter 10
Meta-Analysis of Genetic Variants in Alcohol Metabolizing Enzymes and their Association with Colorectal Cancer Risk

L.V.K.S. Bhaskar, Shubhangi Sharma, Neha Merchant, and Smaranika Pattnaik

Abstract Several prospective cohort and case-control studies have provided strong evidence describing an association between alcohol consumption and risk of colorectal cancer (CRC). Studies on the association of ADH1B Arg47His and ALDH2 Glu487Lys variants with CRC risk were inconclusive. In this chapter we aimed to elucidate the genetic associations of the alcohol metabolic enzymes with CRC risk. An organized literature search was performed in PubMed, Embase and Google scholar bibliographic databases to retrieve various case-control studies associated with alcohol metabolism genes and CRC. A meta-analysis was performed based on the random-effect model to calculate the overall association of ADH1B Arg47His and ALDH2 Glu487Lys variants with CRC risk. No association was found between ADH1B and ALDH2 gene polymorphisms and susceptibility of CRC. Stratified analysis by ethnicity also did not reveal a significant association between these variants and CRC risk. Cochrane Q-test and I^2 statistics revealed heterogeneity in association with these variants ($P < 0.001$). Publication bias tests did not reveal evidence for bias. Thus, ADH1B Arg47His and ALDH2 Glu487Lys genetic variants are not considered as risk factors for CRC.

Keywords Colorectal cancer · Meta-analysis · Alcohol metabolism · ADH1B · ALDH2 · Polymorphism · Susceptibility

---

L.V.K.S. Bhaskar
Guru Ghasidas University, Bilaspur, Chhattisgarh, India

S. Sharma
Department of Biotechnology, Pt. J. N. M. Medical College, Raipur, India

N. Merchant
Department of Hematology and Medical Oncology, Winship Cancer Institute, Emory University, Atlanta, GA, USA

S. Pattnaik (✉)
Department of Biotechnology and Bioinformatics, Sambalpur University, Sambalpur, Odisha, India

G. S. R. Raju, L.V.K.S. Bhaskar (eds.), *Theranostics Approaches to Gastric and Colon Cancer*, Diagnostics and Therapeutic Advances in GI Malignancies,
https://doi.org/10.1007/978-981-15-2017-4_10

Abbreviations

ADH	alcohol dehydrogenase
ALDH	aldehyde dehydrogenase
Arg	Arginine
CI	Confidence intervals
CRC	Colorectal cancer
FEM	Fixed effects model
Glu	Glutamic acid
His	Histidine
HWE	Hardy-Weinberg equilibrium
Lys	Lysine
NAD	Nicotinamide adenine dinucleotide
OR	Odds ratio
PRISMA	Preferred reporting items for systematic reviews and meta-analyses
REM	Random effects model
Tyr	Tyrosine

10.1 Introduction

Colorectal cancer (CRC) is a highly lethal disease that constitutes a significant fraction of cancer morbidity and mortality worldwide (Favoriti et al. 2016). In many countries, the incidence of CRC is strongly associated with the Western lifestyle and dietary factors (Tuan and Chen 2016). Although the physiology of CRC remains unclear, numerous studies have shown that the risk of developing CRC is influenced by both environmental and genetic factors (Ramzi et al. 2014). Several prospective cohort and case-control studies has provided evidence revealing an association between alcohol consumption and risk of CRC (Park et al. 2019; Rossi et al. 2018; Wang et al. 2015). Alcohol intake is pervasive and the mechanisms of the carcinogenic effect of alcohol have not been entirely understood. The assumed mechanisms of alcohol-induced toxicity and carcinogenicity include (a) alcohol favors uptake of environmental carcinogens by changing permeability of mucosal cells (Seitz and Stickel 2007), (b) CYP2E1 induction by chronic alcohol intake, resulting in the generation of reactive oxygen and nitrogen species (Linhart et al. 2014) and (c) acetaldehyde induced DNA damage in the digestive tract (Na and Lee 2017). Upon consumption, alcohol is rapidly absorbed in the stomach (20%) and intestines (80%) through passive diffusion (Haass-Koffler et al. 2017). Majority of the absorbed alcohol is removed primarily by metabolism (95–98%), whereas the remaining (2–5%) amount is excreted in its original form (Holford 1987).

Alcohol metabolism occurs, principally by alcohol dehydrogenase (ADH) enzyme, in the liver, and converted into acetaldehyde. Acetaldehyde is further metabolized in a second reaction, which uses aldehyde dehydrogenase (ALDH) enzymes as catalyst, and oxidizes into acetate. Both ADH and ALDH reactions reduce NAD+ to NADH, which in turn alters the cellular redox state and triggers a number of adverse effects, related to alcohol consumption (Liu et al. 2017). Multiple ADH and ALDH enzymes encoded by various genes regulate these reactions. Functional polymorphisms in the genes encoding ADH and ALDH, contributes to the differences in functional activity of enzyme and the net blood acetaldehyde levels after consumption. The ADH1B Exon3 variant (Arg47His; rs1229984), is functionally validated in terms of genotype-phenotype associations and the Vmax values of this functional variant (47His) is higher than their counterparts (47Arg) (Li et al. 2011). On the other hand, ALDH2 exon 12 variant (Glu487Lys; rs671) encode active (487Glu) and inactive (487Lys) alleles (Farres et al. 1994). Interestingly both ADH1B *47His and ALDH2 *487Lys alleles are prevalent in East Asians but are absent in Europeans and Africans (Bhaskar et al. 2007; Peng et al. 2010; Rao et al. 2007).

The association between alcohol consumption and risk of CRC may perhaps be modulated by genetic factors (ADH1B Arg47His and ALDH2 Glu487Lys) involved in the conversion of ethanol to acetaldehyde, and its immediate alteration into acetate (Rossi et al. 2018). Till date numerous studies have investigated the links between ADH1B and ALDH2 gene variants with the risk of CRC. A meta-analysis by Guo et al. (2013a), showed that the ALDH2 487Lys allele carriers are associated with decreased risk of CRC, but no significant associations were observed between ADH1B polymorphism and risk of CRC. On the contrary, another meta-analysis by Chen et al. (2014b) showed significant association between CRC risk and ADH1B Arg47His polymorphism. Considering the conflicting findings of the prior studies on ADH1B and ALDH2 gene variants with CRC, we performed the current comprehensive meta-analysis to assess the relation of ADH1B Arg47His and ALDH2 Glu487Lys variants with the risk of CRC.

10.2 Materials and Methods

10.2.1 Search Strategy and Selection Criteria

An organized literature search was conducted using PubMed, Embase and Google scholar bibliographic databases to retrieve papers related to alcohol metabolism genes and CRC. The search terms used for the study included: ADH1B, ADH2, Arg47His, rs1229984 or ALDH2, Glu487Lys, rs671 in combination with CRC. Two investigators individually extracted and assessed the quality of selected studies. In order to obtain more relevant articles the references of the original papers and reviews articles were extracted manually. Studies were eligible for inclusion if they met the following criteria: (1) studies on the association between ADH1B Arg47His or ALDH2 Glu487Lys polymorphisms and CRC risk; (2) containing genotype data in both CRC and controls to allow appropriate estimation of the odds ratio (OR);(3) distributions of genotypes of controls follow Hardy-Weinberg equilibrium (HWE) proportions. Studies were omitted from the primary analysis in case of the following conditions (1) case-only studies; (2) insufficient genotype or allele data; (3) a duplicated publication. Name of first author, published year, country, ethnicity, genotyping method, alleles and genotypes distribution in both CRC and controls were collected from eligible studies. The meta-analysis was conducted based on guidelines laid down in preferred reporting items for systematic reviews and meta-analyses (PRISMA) statement (Moher et al. 2009).

10.2.2 Statistical Analysis

The association between CRC and alcohol metabolizing gene polymorphisms (ADH1B Arg47His and ALDH2 Glu487Lys) were evaluated via calculating ORs and 95%CIs. The pooled OR and 95% CI were estimated in allelic, dominant and recessive models. Between-study heterogeneity was assessed with Cochrane's Q test and the Higgins and Thompson I^2 statistics. Subgroup analysis was performed after classifying studies based on ethnicity (Caucasians and Asians). To assess the robustness of findings in terms of the primary endpoint, sensitivity analysis was conducted using "leave-one-out" meta-analysis. Publication bias was evaluated using Begg's funnel plots and Egger's regression test for funnel plot asymmetry. All statistical analyses were done using a web tool MetaGenyo (Martorell-Marugan et al. 2017). P-value <0.05 was considered as statistically significant.

10.3 Results

10.3.1 Characteristics of Included Studies

Our literature search in PubMed, Embase and Google scholar, identified 112 articles. After exclusion of 28 duplicate articles, 94 articles were selected for reading titles and abstracts. After reading abstract, 56 irrelevant papers were excluded and finally 38 full papers were selected for further evaluation. After reading full texts, 18 articles were excluded that were not linked with the measure of interest or those that lacked sufficient genotyping data. Finally about 20 papers that met the inclusion criteria were selected for our meta-analysis (Fig. 10.1). Characteristics of studies included in the meta-analysis are summarized in Table 10.1.

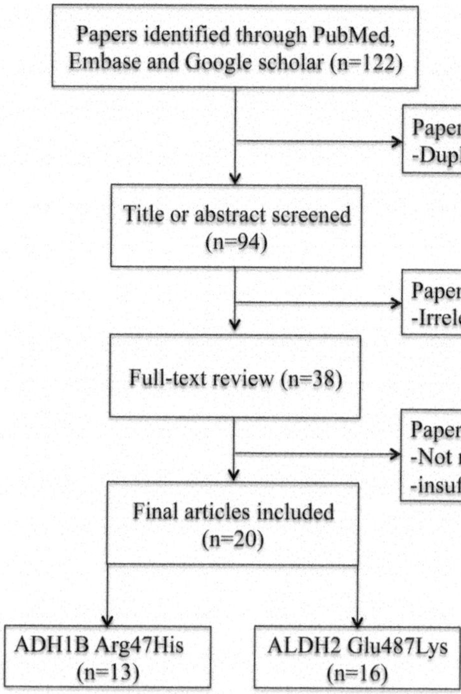

Fig. 10.1 Flow chart of the literature search used in this meta-analysis

10.3.2 Heterogeneity Analysis

With respect to ADH1B Arg47His polymorphism, statistically significant between-study heterogeneity was observed in the pooled analyses (allele model: $P_{Heterogeneity} < 0.001$, $I^2 = 92.4\%$; dominant model: $P_{Heterogeneity} < 0.001$, $I^2 = 92.9\%$; recessive model: $P_{Heterogeneity} < 0.001$, $I^2 = 78.3\%$). Further, ALDH2 Glu487Lys polymorphism also showed significant between-study heterogeneity in all three genetic models (allele model: $P_{Heterogeneity} < 0.001$, $I^2 = 65.1\%$; dominant model: $P_{Heterogeneity} = 0.001$, $I^2 = 59.8\%$; recessive model: $P_{Heterogeneity} = 0.032$, $I^2 = 44.6\%$). Hence, random effects model was used to synthesize the data.

10.3.3 Overall and Subgroup Analyses for ADH1B Arg47His Polymorphism

To evaluate potential associations between ADH1B Arg47His polymorphism and the risk of CRC, 13 studies were enrolled for overall analyses (Chiang et al. 2012;

Table 10.1 Characteristics of the studies for the association of ADH1B Arg47His and ALDH2 Glu487Lys genetic polymorphisms with colorectal cancer risk

S.No	Author's Name	Country	Ethnicity	Genotyping	Case			Control			HWE p value	Reference
					AA	AG	GG	AA	AG	GG		
ADH1B rs1229984 (Arg47His)												
1	Zhong et al. 2016	China	Asian	PCR-RFLP	64	125	85	34	172	152	0.140	(Zhong et al. 2016)
2	Svensson et al. 2016	Japan	Asian	TaqMan	194	139	23	400	257	51	0.276	(Svensson et al. 2016)
3	Shiotani et al. 2015	Japan	Asian	PCR-RFLP	137	74	10	100	65	14	0.455	(Shiotani et al. 2015)
4	Crous-Bou et al. 2013	Spain	Caucasian	BeadXpress	79	324	457	54	360	513	0.378	(Crous-Bou et al. 2013)
5	Ferrari et al. 2012	France	Caucasian	TaqMan	5	97	1129	6	76	180	0.539	(Ferrari et al. 2012)
6	Chiang et al. 2012	China	Asian	PCR-RFLP	62	34	7	297	205	43	0.363	(Chiang et al. 2012)
7	Yin et al. 2011	Japan	Asian	PCR-RFLP	268	161	25	588	393	71	0.628	(Yin et al. 2011)
8	Yang et al. 2009	China	Asian	SNPlex	205	181	39	370	319	62	0.559	(Yang et al. 2009)
9	Gao et al. 2008	China	Asian	DHPLC	102	73	15	93	109	20	0.132	(Gao et al. 2008)
10	Yin et al. 2007	Japan	Asian	PCR-RFLP	345	294	40	425	289	37	0.172	(Yin et al. 2007)
11	Jiang et al. 2007	China	Asian	DHPLC	107	18	85	204	194	40	0.526	(Jiang et al. 2007)
12	Matsuo et al. 2006	Japan	Asian	PCR-RFLP	136	102	19	437	259	36	0.765	(Matsuo et al. 2006)
13	Landi et al. 2005	France	Caucasian	Amplifluor®	2	54	292	3	48	263	0.626	(Landi et al. 2005)
ALDH2 rs671 (Glu487Lys)												
1	Zhong et al. 2016	China	Asian	PCR-RFLP	13	67	194	9	79	270	0.273	(Zhong et al. 2016)
2	Svensson et al. 2016	Japan	Asian	TaqMan	12	96	248	22	222	464	0.462	(Svensson et al. 2016)
3	Shiotani et al. 2015	Japan	Asian	PCR-RFLP	10	70	141	15	59	105	0.115	(Shiotani et al. 2015)
4	Chiang et al. 2012	China	Asian	PCR-RFLP	7	53	43	23	218	304	0.036	(Chiang et al. 2012)
5	Miyasaka et al. 2010	Japan	Asian	PCR-RFLP	2	22	24	15	125	112	0.009	(Miyasaka et al. 2010)
6	Yang et al. 2009	China	Asian	PCR-RFLP	33	119	274	35	261	489	0.982	(Yang et al. 2009)
7	Gao et al. 2008	China	Asian	DHPLC	5	54	131	9	90	123	0.132	(Gao et al. 2008)
8	Yin et al. 2007	Japan	Asian	PCR-RFLP	28	257	400	53	309	416	0.668	(Yin et al. 2007)
9	Jiang et al. 2007	China	Asian	DHPLC	6	61	143	23	149	266	0.720	(Jiang et al. 2007)
10	Matsuo et al. 2006	Japan	Asian	PCR-RFLP	24	104	129	71	314	383	0.567	(Matsuo et al. 2006)

11	Otani et al. 2005	Japan	Asian	TaqMan	9	36	61	15	72	137	0.198	(Otani et al. 2005)
12	Kuriki et al. 2005	Japan	Asian	PCR-CTPP	3	24	45	8	44	64	0.907	(Kuriki et al. 2005)
13	Hirose et al. 2005	Japan	Asian	PCR-RFLP	16	137	299	55	390	605	0.442	(Hirose et al. 2005)
14	Matsuo et al. 2002	Japan	Asian	PCR-CTPP	6	54	81	19	96	126	0.905	(Matsuo et al. 2002)
15	Murata et al. 1999	Japan	Asian	PCR-CTPP	15	89	166	3	41	77	0.364	(Murata et al. 1999)
16	Yokoyama et al. 1998	Japan	Asian	PCR-RFLP	0	10	36	0	44	443	0.296	(Yokoyama et al. 1998)

Table 10.2 Summary about meta-analysis results of ADH1B Arg47His and ALDH2 Glu487Lys polymorphisms and risk of colorectal cancer risk in different genetic models

	ADH1B Arg47His			ALDH2 Glu487Lys
		By ethnicity		
	Overall	Asian	Caucasian	Overall (All Asian)
Number of studies	13	10	3	16
Allele contrast (A vs. G)				
I^2	0.924	0.853	0.978	0.651
$P_{Heterogeneity}$	<0.001	<0.001	<0.001	<0.001
OR (95% CI)	0.92 (0.73–1.16)	1.02 (0.85–1.23)	0.63 (0.22–1.81)	0.94 (0.83–1.06)
Test of association p-val	0.488	0.814	0.392	0.307
Egger's test p-val	0.575	0.418	0.584	0.216
Dominant model (AA + AG vs. GG)				
I^2	0.929	0.899	0.975	0.598
$P_{Heterogeneity}$	<0.001	<0.001	<0.001	0.001
OR (95% CI)	0.81 (0.52–1.25)	0.90 (0.54–1.48)	0.60 (0.20–1.79)	0.92 (0.80–1.056)
Test of association p-val	0.346	0.666	0.360	0.226
Egger's test p-val	0.861	0.877	0.641	0.128
Recessive model (AA vs. AG + GG)				
I^2	0.783	0.777	0.848	0.446
$P_{Heterogeneity}$	<0.001	<0.001	0.001	0.032
OR (95% CI)	1.12 (0.91–1.38)	1.14 (0.93–1.39)	0.59 (0.12–2.88)	0.93 (0.71–1.22)
Test of association p-val	0.275	0.199	0.518	0.598
Egger's test p-val	0.613	0.014	0.380	0.743

Crous-Bou et al. 2013; Ferrari et al. 2012; Gao et al. 2008; Jiang et al. 2007; Landi et al. 2005; Matsuo et al. 2006; Shiotani et al. 2015; Svensson et al. 2016; Yang et al. 2009; Yin et al. 2007, 2011; Zhong et al. 2016). In overall analyses, no significant associations with the risk of CRC were detected for ADH1B Arg47His polymorphism (allele model: OR = 0.92, 95% CI 0.73–1.16, p = 0.488; dominant model: OR = 0.81, 95% CI 0.52–1.25, p = 0.346; recessive model: OR = 1.12, 95% CI 0.91–1.38, p = 0.275) (Table 10.2 and Fig. 10.2a). Further subgroup analyses according to ethnicity of study populations (Asian 10 and Caucasian 3 studies) the ADH1B Arg47His polymorphism was not associated with the risk of CRC in both Asians and Caucasians (Table 10.2).

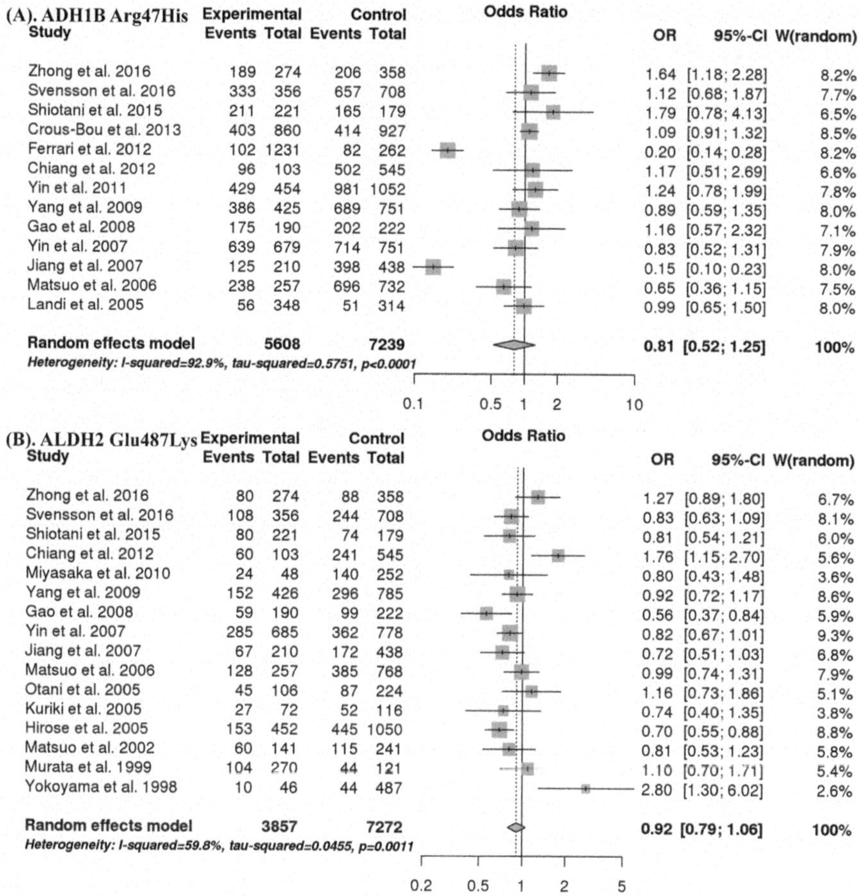

Fig. 10.2 Forest plot of the association between alcohol metabolizing gene polymorphisms and CRC in dominant model. (**a**) ADH1B Arg47His polymorphism; (**b**) ALDH2 Glu487Lys polymorphism

10.3.4 Overall and Subgroup Analyses for ALDH2 Glu487Lys Polymorphism

To evaluate potential associations between ALDH2 Glu487Lys polymorphism and the risk of CRC, 16 studies were enrolled for overall analyses (Chiang et al. 2012; Gao et al. 2008; Hirose et al. 2005; Jiang et al. 2007; Kuriki et al. 2005; Matsuo et al. 2002, 2006; Miyasaka et al. 2010; Murata et al. 1999; Otani et al. 2005; Shiotani et al. 2015; Svensson et al. 2016; Yang et al. 2009; Yin et al. 2007; Yokoyama et al. 1998; Zhong et al. 2016). In overall analyses, no significant association with the risk of CRC was detected for ALDH2 Glu487Lys polymorphism (allele model: OR = 0.94, 95% CI 0.83–1.06, p = 0.307; dominant model: OR = 0.92, 95% CI

0.80–1.056, p = 0.226; recessive model: OR = 0.93, 95% CI 0.71–1.22, p = 0.598) (Table 10.2 and Fig. 10.2b).

10.3.5 *Sensitivity Analysis and Publication Bias*

Sensitivity analyses showed no drastic changes in OR after removing either of the study, which indicates that our findings were statistically stable (Fig. 10.3a and b). To assess potential publication biases funnel plots were used. The shape of the funnel plots did not reveal any evidence of obvious asymmetry, which suggests that our findings in the meta-analysis were less likely to be affected by publication bias (Fig. 10.4a and b). Further, Egger's test results suggested the absence of publication bias for the meta-analysis evaluating the association between the ADH1B His47 vs. Arg47 allele (p = 0.448), the ADH1B Arg47His + His47His vs. Arg47Arg genotype (p = 0.861), the ADH1B His47His vs. Arg47Arg + Arg47His genotype (p = 0.613), the ALDH2 Lys487 vs. Glu487 allele (p = 0.216), the ALDH2 Glu487Lys + Lys487Lys vs. Glu487Glu genotype (p = 0.128), or the ALDH2 Lys487Lys vs. Glu487Glu + Glu487Lys genotype (p = 0.743) and CRC (Table 10.2).

10.4 Discussion

A positive association between alcohol and CRC was demonstrated in many studies (Park et al. 2019). As the ADH1B and ALDH2 are involved in ethanol degradation, it becomes obvious that the kinetics of these enzymes have a major role on the body's exposure to alcohol or its metabolites. Our meta-analysis examined the association of being homozygous carrier of the alleles of these two variants (ADH1B His48His and ALDH2 Lys487Lys). However, in this meta-analysis, both ADH1B and ALDH2 polymorphisms did not confer the risk for CRC.

Elevated amount of alcohol in the colon has been reports after alcohol absorption. Further, higher acetaldehyde concentration in the mucosa of large intestine was also observed (Pronko et al. 2002). Besides, inhibition of ALDH2 activity induced the colonic tumorigenesis in rats (Seitz et al. 1990)., many bacteria in the large intestine possess marked alcohol dehydrogenase activity leading to accumulation of acetaldehyde in the colon (Salaspuro 1997). Evidences reveal that alcohol intake is associated with increased risk of tumorigenesis of the colon and rectum as compared to no alcohol intake (Choi et al. 2017; Fedirko et al. 2011; Naing et al. 2017). The CRC risk imposed by both frequent and high amount of alcohol intake is positively associated with the risk of distal colon cancer in men and rectal cancer in women according to an investigation conducted in Korea (Shin et al. 2011). In Japanese men, alcohol intake is associated with significantly increased risk of distal colon and rectum cancer, but not the proximal colon cancer (Akhter et al. 2007). Further, there

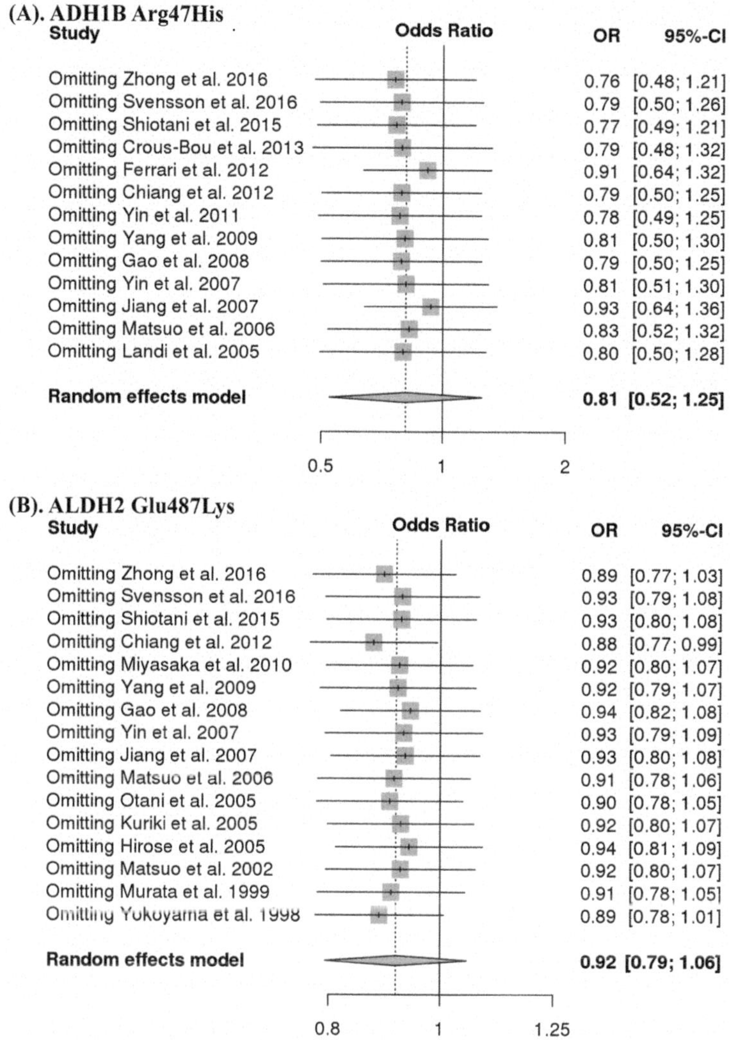

Fig. 10.3 Plot of sensitivity analysis by excluding one study each time. (**a**) ADH1B Arg47His polymorphism; (**b**) ALDH2 Glu487Lys polymorphism

is a dose-response relationship between alcohol intake and CRC as observed in Japanese patients (Mizoue et al. 2008). Additionally, estimation of the population attributable faction in Chinese CRC patients indicated that alcohol intake is accountable for higher incidence and mortality of CRC (Gu et al. 2018). However, there is no consensus on the evidence of associations between ADH1B and ALDH2 polymorphisms and the risk of CRC in many populations. A meta-analysis involving 6 studies with 1864 CRC cases and 3502 controls did not reveal significant association between ADH1B Arg47His polymorphism and CRC risk (Guo et al. 2013b).

(A). ADH1B Arg47His

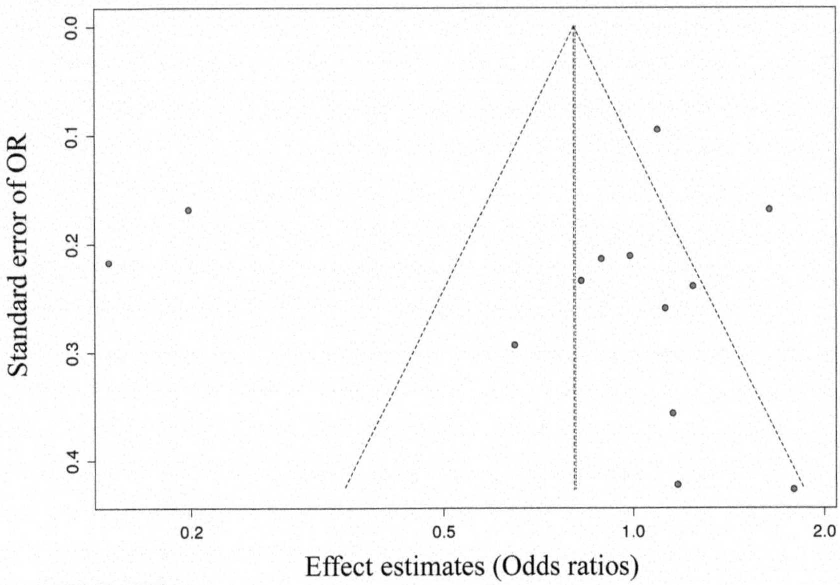

(B). ALDH2 Glu487Lys

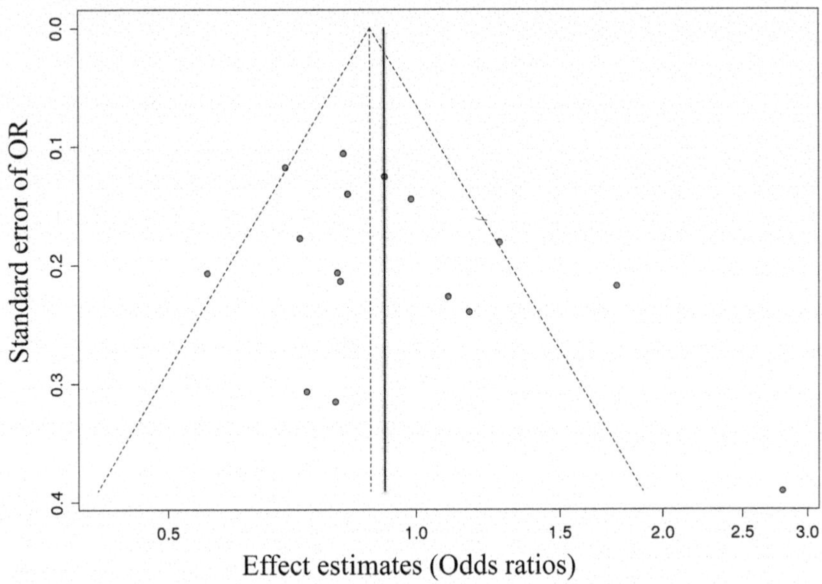

Fig. 10.4 Funnel plot analysis to detect publication bias for alcohol metabolizing gene polymorphisms and CRC in dominant model. (**a**) ADH1B Arg47His polymorphism; (**b**) ALDH2 Glu487Lys polymorphism

In contrast, another meta-analysis showed increased risk of CRC associated with ADH1B 47His allele (Chen et al. 2014a). With regards to ALDH2 gene, a meta-analysis of 11 case-control studies representing a total of 2909 CRC cases and 4903 controls, revealed that the ALDH2 487lys allele reduced CRC (Zhao et al. 2014). On contrary, a recent meta-analysis showed no association between ALDH2 487Lys allele and CRC (Zuo et al. 2019).

In conclusion, this meta-analysis demonstrated that the ADH1B and ALDH2 polymorphisms have no role in CRC. Further, we identified many studies that has analyzed the association of CRC and ADH1B and ALDH2 predominantly in Asian samples as compared to other ethnic groups. As ADH1B Arg47His and ALDH2 Glu487Lys are very rare in other ethnic groups, the results of the present meta analysis can only be assumed for Asian populations. However, large and well-designed prospective studies are needed to validate these findings.

Conflict of Interest There are no conflicts of interests.

References

Akhter M, Kuriyama S, Nakaya N, Shimazu T, Ohmori K, Nishino Y, Tsubono Y, Fukao A, Tsuji I (2007) Alcohol consumption is associated with an increased risk of distal colon and rectal cancer in Japanese men: the Miyagi cohort study. Eur J Cancer 43:383–390

Bhaskar LV, Thangaraj K, Osier M, Reddy AG, Rao AP, Singh L, Rao VR (2007) Single nucleotide polymorphisms of the ALDH2 gene in six Indian populations. Ann Hum Biol 34:607–619

Chen C, Wang L, Liao Q, Xu L, Huang Y, Zhang C, Ye H, Xu X, Ye M, Duan S (2014a) Association between six genetic polymorphisms and colorectal cancer: a meta-analysis. Genet Test Mol Biomarkers 18:187–195

Chen C, Wang L, Liao Q, Xu L, Huang Y, Zhang C, Ye H, Xu X, Ye M, Duan S (2014b) Association between six genetic polymorphisms and colorectal cancer: a meta-analysis. Genet Test Mol Biomarkers 18:187–195

Chiang CP, Jao SW, Lee SP, Chen PC, Chung CC, Lee SL, Nieh S, Yin SJ (2012) Expression pattern, ethanol-metabolizing activities, and cellular localization of alcohol and aldehyde dehydrogenases in human large bowel: association of the functional polymorphisms of ADH and ALDH genes with hemorrhoids and colorectal cancer. Alcohol (Fayetteville, NY) 46:37–49

Choi YJ, Lee DH, Han K-D, Kim HS, Yoon H, Shin CM, Park YS, Kim N (2017) The relationship between drinking alcohol and esophageal, gastric or colorectal cancer: a nationwide population-based cohort study of South Korea. PLoS One 12:e0185778–e0185778

Crous-Bou M, Rennert G, Cuadras D, Salazar R, Cordero D, Saltz Rennert H, Lejbkowicz F, Kopelovich L, Monroe Lipkin S, Bernard Gruber S, Moreno V (2013) Polymorphisms in alcohol metabolism genes ADH1B and ALDH2, alcohol consumption and colorectal cancer. PLoS One 8:e80158

Farres J, Wang X, Takahashi K, Cunningham SJ, Wang TT, Weiner H (1994) Effects of changing glutamate 487 to lysine in rat and human liver mitochondrial aldehyde dehydrogenase. A model to study human (oriental type) class 2 aldehyde dehydrogenase. J Biol Chem 269:13854–13860

Favoriti P, Carbone G, Greco M, Pirozzi F, Pirozzi RE, Corcione F (2016) Worldwide burden of colorectal cancer: a review. Updat Surg 68:7–11

Fedirko V, Tramacere I, Bagnardi V, Rota M, Scotti L, Islami F, Negri E, Straif K, Romieu I, La Vecchia C, Boffetta P, Jenab M (2011) Alcohol drinking and colorectal cancer risk: an overall and dose-response meta-analysis of published studies. Ann Oncol 22:1958–1972

Ferrari P, McKay JD, Jenab M, Brennan P, Canzian F, Vogel U, Tjonneland A, Overvad K, Tolstrup JS, Boutron-Ruault MC, Clavel-Chapelon F, Morois S, Kaaks R, Boeing H, Bergmann M, Trichopoulou A, Katsoulis M, Trichopoulos D, Krogh V, Panico S, Sacerdote C, Palli D, Tumino R, Peeters PH, van Gils CH, Bueno-de-Mesquita B, Vrieling A, Lund E, Hjartaker A, Agudo A, Suarez LR, Arriola L, Chirlaque MD, Ardanaz E, Sanchez MJ, Manjer J, Lindkvist B, Hallmans G, Palmqvist R, Allen N, Key T, Khaw KT, Slimani N, Rinaldi S, Romieu I, Boffetta P, Romaguera D, Norat T, Riboli E (2012) Alcohol dehydrogenase and aldehyde dehydrogenase gene polymorphisms, alcohol intake and the risk of colorectal cancer in the European prospective investigation into cancer and nutrition study. Eur J Clin Nutr 66:1303–1308

Gao CM, Takezaki T, Wu JZ, Zhang XM, Cao HX, Ding JH, Liu YT, Li SP, Cao J, Matsuo K, Hamajima N, Tajima K (2008) Polymorphisms of alcohol dehydrogenase 2 and aldehyde dehydrogenase 2 and colorectal cancer risk in Chinese males. World J Gastroenterol 14:5078–5083

Gu MJ, Huang QC, Bao CZ, Li YJ, Li XQ, Ye D, Ye ZH, Chen K, Wang JB (2018) Attributable causes of colorectal cancer in China. BMC Cancer 18:38

Guo X-F, Wang J, Yu S-J, Song J, Ji M-Y, Zhang J-X, Cao Z, Wang J, Dong W-G (2013a) Meta-analysis of the ADH1B and ALDH2 polymorphisms and the risk of colorectal cancer in east Asians. Intern Med 52:2693–2699

Guo XF, Wang J, Yu SJ, Song J, Ji MY, Zhang JX, Cao Z, Wang J, Dong WG (2013b) Meta-analysis of the ADH1B and ALDH2 polymorphisms and the risk of colorectal cancer in east Asians. Intern Med (Tokyo, Japan) 52:2693–2699

Haass-Koffler CL, Akhlaghi F, Swift RM, Leggio L (2017) Altering ethanol pharmacokinetics to treat alcohol use disorder: can you teach an old dog new tricks? J Psychopharmacol 31:812–818

Hirose M, Kono S, Tabata S, Ogawa S, Yamaguchi K, Mineshita M, Hagiwara T, Yin G, Lee KY, Tsuji A, Ikeda N (2005) Genetic polymorphisms of methylenetetrahydrofolate reductase and aldehyde dehydrogenase 2, alcohol use and risk of colorectal adenomas: self-defense forces health study. Cancer Sci 96:513–518

Holford NH (1987) Clinical pharmacokinetics of ethanol. Clin Pharmacokinet 13:273–292

Jiang A, Gao C, Wu J, Takezaki T, Cao J, Tajima K, Shen H (2007) Association between ADH2/ALDH2 genetic polymorphism and habit of alcohol drinking and the susceptibility of rectal cancer. Chin J Cancer Prev Treat 14:1445–1449

Kuriki K, Hamajima N, Chiba H, Kanemitsu Y, Hirai T, Kato T, Saito T, Matsuo K, Koike K, Tokudome S, Tajima K (2005) Relation of the CD36 gene A52C polymorphism to the risk of colorectal cancer among Japanese, with reference to with the aldehyde dehydrogenase 2 gene Glu487Lys polymorphism and drinking habit. Asian Pac J Cancer Prev: APJCP 6:62–68

Landi S, Gemignani F, Moreno V, Gioia-Patricola L, Chabrier A, Guino E, Navarro M, de Oca J, Capella G, Canzian F (2005) A comprehensive analysis of phase I and phase II metabolism gene polymorphisms and risk of colorectal cancer. Pharmacogenet Genomics 15:535–546

Li D, Zhao H, Gelernter J (2011) Strong association of the alcohol dehydrogenase 1B gene (ADH1B) with alcohol dependence and alcohol-induced medical diseases. Biol Psychiatry 70:504–512

Linhart K, Bartsch H, Seitz HK (2014) The role of reactive oxygen species (ROS) and cytochrome P-450 2E1 in the generation of carcinogenic etheno-DNA adducts. Redox Biol 3:56–62

Liu H, Liu X, Zhang C, Zhu H, Xu Q, Bu Y, Lei Y (2017) Redox imbalance in the development of colorectal cancer. J Cancer 8:1586–1597

Martorell-Marugan J, Toro-Dominguez D, Alarcon-Riquelme ME, Carmona-Saez P (2017) MetaGenyo: a web tool for meta-analysis of genetic association studies. BMC Bioinformatics 18:563

Matsuo K, Hamajima N, Hirai T, Kato T, Koike K, Inoue M, Takezaki T, Tajima K (2002) Aldehyde dehydrogenase 2 (ALDH2) genotype affects rectal cancer susceptibility due to alcohol consumption. J Epidemiol 12:70–76

Matsuo K, Wakai K, Hirose K, Ito H, Saito T, Suzuki T, Kato T, Hirai T, Kanemitsu Y, Hamajima H, Tajima K (2006) A gene-gene interaction between ALDH2 Glu487Lys and ADH2 His47Arg polymorphisms regarding the risk of colorectal cancer in Japan. Carcinogenesis 27:1018–1023

Miyasaka K, Hosoya H, Tanaka Y, Uegaki S, Kino K, Shimokata H, Kawanami T, Funakoshi A (2010) Association of aldehyde dehydrogenase 2 gene polymorphism with pancreatic cancer but not colon cancer. Geriatr Gerontol Int 10(Suppl 1):S120–S126

Mizoue T, Inoue M, Wakai K, Nagata C, Shimazu T, Tsuji I, Otani T, Tanaka K, Matsuo K, Tamakoshi A, Sasazuki S, Tsugane S (2008) Alcohol drinking and colorectal cancer in Japanese: a pooled analysis of results from five cohort studies. Am J Epidemiol 167:1397–1406

Moher D, Liberati A, Tetzlaff J, Altman DG (2009) Preferred reporting items for systematic reviews and meta-analyses: the PRISMA statement. Ann Intern Med 151:264–269. w264

Murata M, Tagawa M, Watanabe S, Kimura H, Takeshita T, Morimoto K (1999) Genotype difference of aldehyde dehydrogenase 2 gene in alcohol drinkers influences the incidence of Japanese colorectal cancer patients. Jpn J Cancer Res: Gann 90:711–719

Na H-K, Lee JY (2017) Molecular basis of alcohol-related gastric and colon cancer. Int J Mol Sci 18:1116

Naing C, Lai PK, Mak JW (2017) Immediately modifiable risk factors attributable to colorectal cancer in Malaysia. BMC Public Health 17:637–637

Otani T, Iwasaki M, Hanaoka T, Kobayashi M, Ishihara J, Natsukawa S, Shaura K, Koizumi Y, Kasuga Y, Yoshimura K, Yoshida T, Tsugane S (2005) Folate, vitamin B6, vitamin B12, and vitamin B2 intake, genetic polymorphisms of related enzymes, and risk of colorectal cancer in a hospital-based case-control study in Japan. Nutr Cancer 53:42–50

Park SY, Wilkens LR, Setiawan VW, Monroe KR, Haiman CA, Le Marchand L (2019) Alcohol intake and colorectal cancer risk in the multiethnic cohort study. Am J Epidemiol 188:67–76

Peng Y, Shi H, Qi X-b, C-j X, Zhong H, Ma R-lZ, Su B (2010) The ADH1B Arg47His polymorphism in east Asian populations and expansion of rice domestication in history. BMC Evol Biol 10:15

Pronko P, Bardina L, Satanovskaya V, Kuzmich A, Zimatkin S (2002) Effect of chronic alcohol consumption on the ethanol- and acetaldehyde-metabolizing systems in the rat gastrointestinal tract. Alcohol Alcohol (Oxford, Oxfordshire) 37:229–235

Ramzi NH, Chahil JK, Lye SH, Munretnam K, Sahadevappa KI, Velapasamy S, Hashim NAN, Cheah SK, Lim GCC, Hussein H, Haron MR, Alex L, Ler LW (2014) Role of genetic & environment risk factors in the aetiology of colorectal cancer in Malaysia. Indian J Med Res 139:873–882

Rao VR, Bhaskar LV, Annapurna C, Reddy AG, Thangaraj K, Rao AP, Singh L (2007) Single nucleotide polymorphisms in alcohol dehydrogenase genes among some Indian populations. Am J Hum Biol 19:338–344

Rossi M, Jahanzaib Anwar M, Usman A, Keshavarzian A, Bishehsari F (2018) Colorectal Cancer and alcohol consumption-populations to molecules. Cancers (Basel) 10:38

Salaspuro M (1997) Microbial metabolism of ethanol and acetaldehyde and clinical consequences. Addict Biol 2:35–46

Seitz HK, Stickel F (2007) Molecular mechanisms of alcohol-mediated carcinogenesis. Nat Rev Cancer 7:599–612

Seitz HK, Simanowski UA, Garzon FT, Rideout JM, Peters TJ, Koch A, Berger MR, Einecke H, Maiwald M (1990) Possible role of acetaldehyde in ethanol-related rectal cocarcinogenesis in the rat. Gastroenterology 98:406–413

Shin A, Joo J, Bak J, Yang HR, Kim J, Park S, Nam BH (2011) Site-specific risk factors for colorectal cancer in a Korean population. PLoS One 6:e23196

Shiotani A, Ishikawa H, Mutoh M, Takeshita T, Nakamura T, Morimoto K, Sakai T, Wakabayashi K, Matsuura N (2015) Genetic polymorphisms in ADH1B and ALDH2 are associated with colorectal tumors in Japan: a case-control study. J Cancer Ther 6:1054

Svensson T, Yamaji T, Budhathoki S, Hidaka A, Iwasaki M, Sawada N, Inoue M, Sasazuki S, Shimazu T, Tsugane S (2016) Alcohol consumption, genetic variants in the alcohol- and folate metabolic pathways and colorectal cancer risk: the JPHC study. Sci Rep 6:36607–36607

Tuan J, Chen YX (2016) Dietary and lifestyle factors associated with colorectal cancer risk and interactions with microbiota: fiber, red or processed meat and alcoholic drinks. Gastrointest Tumors 3:17–24

Wang Y, Duan H, Yang H, Lin J (2015) A pooled analysis of alcohol intake and colorectal cancer. Int J Clin Exp Med 8:6878–6889

Yang H, Zhou Y, Zhou Z, Liu J, Yuan X, Matsuo K, Takezaki T, Tajima K, Cao J (2009) A novel polymorphism rs1329149 of CYP2E1 and a known polymorphism rs671 of ALDH2 of alcohol metabolizing enzymes are associated with colorectal cancer in a southwestern Chinese population. Cancer epidemiology, biomarkers & prevention: a publication of the American Association for Cancer Research, cosponsored by the American society of preventive. Oncology 18:2522–2527

Yin G, Kono S, Toyomura K, Moore MA, Nagano J, Mizoue T, Mibu R, Tanaka M, Kakeji Y, Maehara Y, Okamura T, Ikejiri K, Futami K, Yasunami Y, Maekawa T, Takenaka K, Ichimiya H, Imaizumi N (2007) Alcohol dehydrogenase and aldehyde dehydrogenase polymorphisms and colorectal cancer: the Fukuoka colorectal cancer study. Cancer Sci 98:1248–1253

Yin G, Hamajima N, Morita M, Tajima O, Tabata S, Kono S (2011) Lack of influence of the ADH1B Arg47His genetic polymorphism on risk of colorectal adenoma in middle-aged Japanese men. Asian Pac J Cancer Prev: APJCP 12:297–302

Yokoyama A, Muramatsu T, Ohmori T, Yokoyama T, Okuyama K, Takahashi H, Hasegawa Y, Higuchi S, Maruyama K, Shirakura K, Ishii H (1998) Alcohol-related cancers and aldehyde dehydrogenase-2 in Japanese alcoholics. Carcinogenesis 19:1383–1387

Zhao H, Liu KJ, Lei ZD, Lei SL, Tian YQ (2014) Meta-analysis of the aldehyde dehydrogenases-2 (ALDH2) Glu487Lys polymorphism and colorectal cancer risk. PLoS One 9:e88656

Zhong Q, Wu RR, Zeng ZM (2016) Association of ADH1B Arg47His and ALDH2 Glu487Lys polymorphisms with risk of colorectal cancer and their interaction with environmental factors in a Chinese population. Genet Mol Res: GMR 15

Zuo W, Zhan Z, Ma L, Bai W, Zeng S (2019) Effect of ALDH2 polymorphism on cancer risk in Asians: a meta-analysis. Medicine (Baltimore) 98:e14855–e14855

Chapter 11
Gut Microbiota as Signatures in Non-communicable Diseases and Mucosal Immunity

Santosh Kumar Behera, Ardhendu Bhusan Praharaj, Gayathri Chalikonda, Gowru Srivani, and Namita Mahapatra

Abstract The human gut microbiota encompasses a complex and dynamic ecosystem that provides crucial signals for host development and physiology. The altering composition of the human microbiota is associated with modifications in human behavior and the rising prevalence of pathogenesis of late onset diseases such as life style non-communicable diseases, metabolic and neurological disorders etc. The factors that trigger modifications in the composition and function of the gut microbiota will aid in understanding and designing of therapies that target it, which may be quite formidable. Though several studies have been reported on altered gut microbial composition and its association with diseases, but information on mucin layer degradation, production and immune cell interactions is scanty. The alignment of the gut microbiota can be an emerging indicator or marker of health, as it could be a sensitive tool for identifying various risks. This review unzips the role and type of gut microbiota associated with different non-communicable diseases and their impact on intestinal mucosal immunity which could be the signatures required to promote the human health by the clinicians.

Keywords Gastrointestinal microbiome · Late onset disorders · Intestinal mucosa · Mucosal immunity · Mucins

---

S. K. Behera · N. Mahapatra (✉)
Biomedical Informatics Centre, ICMR-Regional Medical Research Centre, Chandrasekharpur, Bhubaneswar, Odisha, India

A. B. Praharaj
Department of Human Genetics, ICMR-Regional Medical Research Centre, Chandrasekharpur, Bhubaneswar, Odisha, India

G. Chalikonda
Department of Bioscience and Biotechnology, Banasthali University, Vanasthali, Rajasthan, India

G. Srivani
Department of Hematology and Medical Oncology, Winship Cancer Institute, Emory University, Atlanta, GA, USA

G. S. R. Raju, L.V.K.S. Bhaskar (eds.), *Theranostics Approaches to Gastric and Colon Cancer*, Diagnostics and Therapeutic Advances in GI Malignancies,
https://doi.org/10.1007/978-981-15-2017-4_11

Abbreviation

AD	Alzheimer's disease
AD	Auto immune disease
ADD	Antibiotic associated diarrhoea
ALS	Amyotrophic lateral sclerosis
ASD	Autism spectrum disorder
BBDP	Bio breeding diabetes prone
BMI	Body mass index
CD	Chron's disease
CHF	Chronic heart failure
CKD	Chronic kidney disease
CNS	Central nervous system
CRC	Colorectal Cancer
CVD	Cardio vascular disease
EU	Endotoxin Unit
GFD	Gluten free diet
GH	Glycoside hydrolases
GIT	Gastro intestinal tract
GM	Germ free
GPR	G protein coupled receptor
GWAS	Genome wide association studies
HBV	Hepatitis B Virus
HCC	Hepatocellular carcinoma
HCV	Hepatitis C Virus
HP	Helicobacter pylori
HPV	Human Papilloma Virus
IAP	Inhibitor of apoptosis protein
IBD	Inflammatory bowel diseases
IBS	Irritable bowel syndrome
Ig	Immunoglobulin
LDL	Low density lipoprotein
LPS	Lipopolysaccharide
NAFLD	Nonalcoholic fatty liver diseases
NASH	Nonalcoholic steatohepatitis
NCD	Non communicable disease
NFT	Neuro fibrillary tangles
NOD	Non obese diabetic
PAMP	Pathogen associated molecular patterns
PD	Parkinson's diseases
PDD	Pervasive developmental disorder
PPAR	Peroxisome proliferator activated receptor
PRR	Pattern recognition receptor
PSE	Plant sterol esters

RA	Rheumatoid arthritis
SCFA	Short chain fatty acid
SLE	Systemic lupus erythematosus
SNP	Single nucleotide polymorphism
T_1D	Type 1 diabetes
T_2D	Type 2 diabetes
TLR	Toll like receptor
TMA	Trimethylamine
TRL	Triglyceride rich lipoprotein
UC	Ulcerative colitis
WHO	World health organization

11.1 Gut Microbiota and Human Body

Humans are super-organisms who have learned to live and co-evolve with existing microbes on earth. The compilation of microorganisms that live in peaceful coexistence with their hosts has been referred as microbiota. "Microbiota" indicates a microbial system at a particular anatomical cite while "microbiome" indicates collective genes encoded by such microbiota. It is speculated that human gastrointestinal microbiota contains as many as 10^{14} bacteria, which is (Clarke and Daniell 2011; Qin et al. 2012) tenfold higher than the number of human cells present. Overall, bacterial cell biomass amounts to about 2 kg. The microbiota colonizes naturally at every surface of human body that interacts with external environment like our skin, genitourinary, gastrointestinal and respiratory tracts. Gastrointestinal tract (GIT) is most heavily colonized organ compared to others. It is estimated that highest no of human microbes reside in the colon (Ley et al. 2006).

Microbial colonization prefers the GIT, as it is rich in nutrients necessary for the survival of the microbes. Therefore, GIT is a complex microbial ecosystem that plays a pivotal role in the progression of diseases (Sommer and Bäckhed 2013). Gut microbiota also plays a vital role in various metabolic activities, including digestion, energy salvaging, supplying nutrients, as well as xenobiotic transformation. Overall, symbiotic gut microbiome provides several benefits, while microbial dysbiosis correlates with metabolic disorders (Bekkering et al. 2013; McPhee and Schertzer 2015; Winer et al. 2016).

The intestinal microbial genome is thought to have more genetic potential when compared to the human genome. However, studies evaluating mucosal microbiota are restricted by the invasive nature of collecting a sample from the area. The mucosa associated microbiota is reported to differ significantly from distal GIT luminal content (Eckburg et al. 2005). About 500–1000 species are estimated to be residing in the human gut that varies widely between different studies (Xu and Gordon 2003). Bacterial phyla namely *Bacteroidetes*, *Firmicutes, Actinobacteria*, and *Proteobacteria* are found to be dominated in healthy human gut with *Bacteroidetes* and *Firmicutes* found in many endemic gut bacteria (Tap et al. 2009). Various studies report that the majority of the intestinal bacteria are members of either *Bacteroidetes* (16–23%) or *Firmicutes* (49–76%) phyla (Mariat et al. 2009). To a lesser extent, the phyla *Proteobacteria* and *Actinobacteria* also contribute to the intestinal microbiome (Eckburg et al. 2005; Frank et al. 2007a). *Bacteroidetes* is comprised of *Bacteroidia* class as well as the *Bacteroides* and *Prevotella* genrae (Krieg et al. 2010). *Firmicutes*, on the other hand, is the largest bacterial phylum within the gut, as it has over 200 genera, which generally fall into two groups: *Clostridium coccoides* cluster XIVa and *C. leptum* cluster IV (Mariat et al. 2009). These two groups contain members of *Eubacterium, Clostridium*, and *Ruminococcus* genera. In addition to the two *Firmicutes and Bacteroidetes*, bacteria belonging to *Actinobacteria*, *Proteobacteria*, *Fusobacteria*, *Lentisphaerae*, *Verrucomicrobia,* and *Spirochaetes* phyla and have been detected as well.

It is an established fact that Gut microbial population may define disease encountered by individual including non-communicable diseases (NCDs) like obesity, celiac disease, diabetes mellitus type 1 (T1D), diabetes mellitus type 2 (T2D), Chron's disease, ulcerative colitis, autism, Alzheimer's (AD), Parkinson's (PD), asthma, food allergies, heart disease, and cancer. The gut microbial population has difference in prevalence of bacterial phylum in various NCD's as summarized in Table 11.1, which may occur due to dysbiosis of microbiota. The interrelationships between nutrition, microbiome, and host metabolism are important players in homeostasis (Sommer and Bäckhed 2013). Nutrition in early life is a significant risk factor for development of NCDs in adulthood. Various studies have provided clues regarding mechanisms that link nutritional deficiencies early on in life to disease in adulthood (Langley-Evans 2015). Intestinal microflora and food metabolites, especially butyrate, was found to be an important constituent, regulate energy balance by preventing excess energy accumulation and promoting fat catabolism, thereby altering energy metabolism (Gao et al. 2009). Short Chain Fatty Acids (SCFAs), including acetate, butyrate, and propionate are produced as complex dietary carbohydrates are cleaved by bacterial glycoside hydrolases (GH); these affect host's metabolism in several ways, as represented in Fig. 11.1. Based on in vitro studies, it was illustrated that acetate preferentially activates G-protein coupled receptor (GPR) 43, propionate activates GPR43 and GPR41 and butyrate preferentially activates GPR41 of the gut which in turn induces the intestinal hormone peptide YY (PYY), and thereby decreasing gut motility, inducing rate of intestinal transit, and decreasing the amount of energy harvested from diet. Engagement of GPR43 by SCFAs has been shown to trigger the incretin hormone as well as glucagon-like peptide (GLP-1) to induce insulin sensitivity. Insulin-mediated fat accumulation is suppressed by SCFA-mediated GPR43 activation and thus regulates energy balance by reducing excess energy accumulation and inducing fat consumption (Kimura et al. 2013).

The human gut microbiota influences and also get influenced by aging process whose cause and effect is unclear (Mariat et al. 2009; Yatsunenko et al. 2012). The means of feeding mechanism is said to be another factor which influences the microbiota (Yatsunenko et al. 2012). Gut microbiota's role in host physiology and

Table 11.1 Over and under representation of bacterial phyla in human gut associated with various non-communicable diseases (NCDs)

Disease category	Disease	More-Prevalent Bacterial phyla	Less-Prevalent Bacterial phyla	References
Metabolic disorder	Obesity	Phylum: Firmicutes:- *Lactobacillus reuteri, L. ductionin, L. paracaseiand L. plantarum*	Phylum: Bacteroidetes Genus: *Prevotella*	(Collado et al. 2008; Million et al. 2012b; Duncan et al. 2008; Jumpertz et al. 2011)
	Type 2 diabetes	Phylum: Firmicutes:- *Lactobacillus gasseri, Lactobacillus reuteri* and *Lactobacillus plantarum* Phylum: Proteobacteria:- Sulphate-reducing bacteria, *Desulfo vibrio*	Phylum: Actinobacteria: -Bifidobacterium phylum: Firmicutes:- Faecalibacterium prausnitzi, lactobacillus, and clostridium phylum: Verrucomicrobia:- Akkermansia muciniphila	(Furet et al. 2010; Sato et al. 2014)
	Type 1 diabetes	*Phylum: Bacteroidetes:-* *B. dorei, B. vulgatus*	Phylum: Actinobacteria: -Bifidobacterium Phylum: Firmicutes:- Lactobacillus and clostridium (phylum: Firmicutes)	(Murri et al. 2013; Sato et al. 2014)
	Dyslipidemia	*Phylum: Firmicutes:-* *Erysipelotrichaceae* *Phylum: Actinobacteria:-* *Coriobacteriaceae*	*Phylum: Firmicutes:-* Lactobacillus Phylum: Actinobacteria: - Bifidobacterium spp.	(Banjoko et al. 2012; Devkota et al. 2012; Wang et al. 2012, 2014)
	Non-alcoholic fatty liver diseases (NAFLD)	Phylum: Firmicutes:-genus: *Lactobacillus*	Phylum: Firmicutes:- Clostridium coccoides Faecalibacterium prausnitzi Oscillibacter valericigenes	(Zhu et al. 2013; Jiang et al. 2015; Raman et al. 2013)
Auto-immune/ immunological disorder	Inflammatory bowel disease (IBD)	Phylum: Proteobacteria:- *E. coli*	Phylum: Firmicutes:- Clostridium leptum groups, *Faecalibacterium prausnitzii*, Lachnospiraceae, Ruminococcaceae, Clostridiales	(Manichanh et al. 2006; Sokol et al. 2008)

(continued)

Table 11.1 (continued)

Disease category	Disease	More-Prevalent Bacterial phyla	Less-Prevalent Bacterial phyla	References
	Celiac disease	Phylum: Proteobacteria:- *E. coli*	Phylum:-Actinobacteria Bifidobacterium Phylum: Firmicutes Lactobacillus	(Sanz 2010)
	Rheumatoid arthritis (RA)	Phylum: Bacteroidetes:- Prevotella spp.	Phylum:-Actinobacteria Genus: *Bifidobacterium* Phylum: Bacteroidetes:- *Bacteroides fragilis*	(Yeoh et al. 2013; Scher et al. 2013; Vaahtovuo et al. 2008)
	Systemic lupus erythematosus (SLE)	Phylum: Firmicutes Lachnospiraceae	Phylum: Firmicutes Lactobacillaceae	(Collins et al. 1994)
Neurological disorder	Autism	Phylum:Bacteroidetes:- *Bacteroides vulgates* Phylum: Proteobacteria *Desulfovibrio species*	Phylum: Bacteroidetes Genus: *Prevotella*	(Tomova et al. 2015; Finegold 2011)
	Schizophrenia	Phylum: Firmicutes:- *Lactobacillus casei, Lactobacillus salivarias, Lactobacillus lactis, and Streptococcus thermophilius, S mitis* and *S mutans*	Phylum: Proteobacteria:- *Neisseria, Haemophilus* Prochlorococcus	(Yolken and Dickerson 2014)
	Alzheimer	Proteobacteria	*Phylum:-Actinobacteria* *Genus: Bifidobacterium* Phylum: Bacteroidetes Genus: *Prevotella*	(Cheng et al. 2011; Scheperjans et al. 2015)
	Parkinson's disease	Phylum: Proteobacteria:- *Helicobacter pylori*	Phylum: Bacteroidetes:- *Prevotellaceae*	(Scheperjans et al. 2015; Dobbs et al. 2000; Álvarez-Arellano and Maldonado-Bernal 2014; Nielsen et al. 2012)

Others	Cancer	Phylum: Firmicutes:- Phascolarctobacterium, Phylum: Actinobacteria:- Citrobacter farmeri Phylum: Verrucomicrobia:- Akkermansia muciniphila	Phylum: Firmicutes:- *Ruminococcus spp.* and *Pseudobutyrivibrio ruminis*	(Zhang et al. 2015; Weir et al. 2013)
	Chronic heart disease	Phylum: Proteobacteria:- Escherichia coli Phylum: Firmicutes:-*Lactobacillus plantarum* (CECT 7527,7528 &7529)	N/A	(Sandek et al. 2007)
	Chronic kidney disease (CKD)	*Phylum: Proteobacteria:-* *Escherichia coli*, *Klebsiella pneumoniae* Phylum: Firmicutes:- *Enterococcus*, *Clostridium perfringens*	Phylum: Actinobacteria:- *Bifidobacterium species,* *Bifidobacterium species,* *B. catenulatum, B. longum, B. bifidum* phylum:Firmicutes:- *Lactobacillus plantarum, lactobacillus paracasei*	(Hida et al. 1996; Wang et al. 2012)

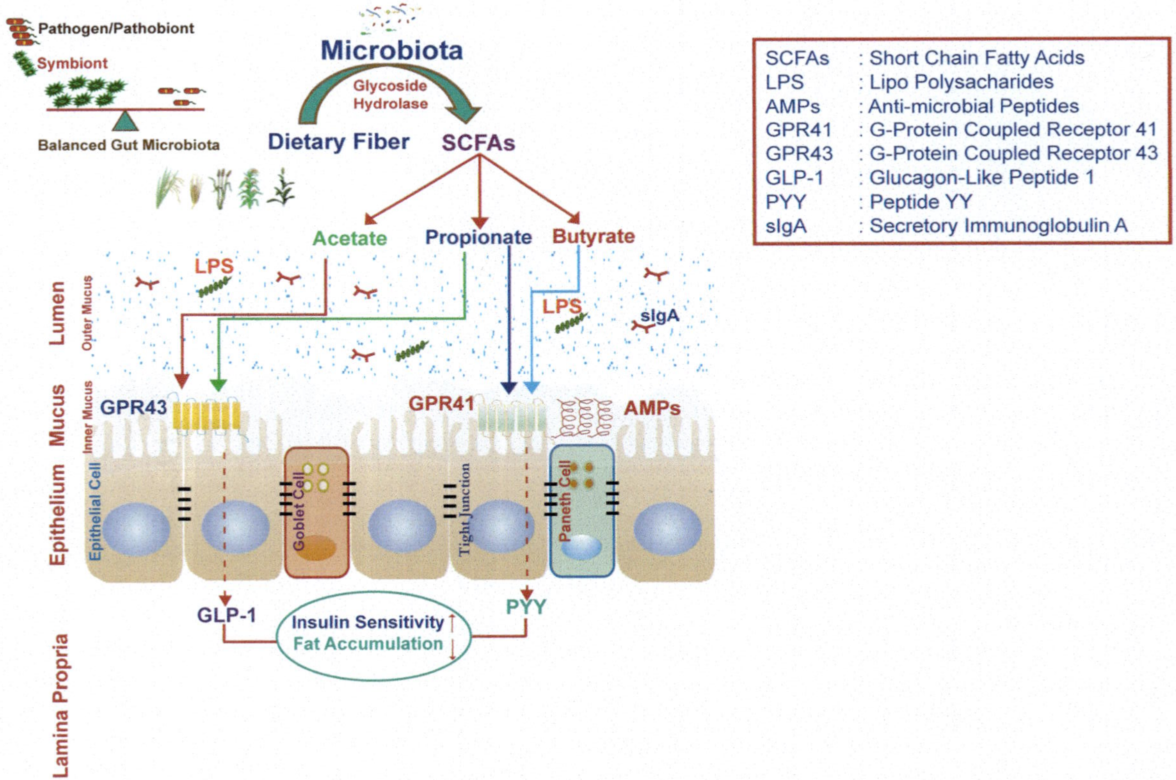

Fig. 11.1 The healthy gut microbiota leads to development of a functional barrier, with normal amounts of mucus, secreted IgA (sIgA), antimicrobial peptides (AMPs) and pattern recognition receptors (PRRs) which in turn holds the microbiota in the intestinal lumen and away from the intestinal epithelial cells. The

bacterial glycoside hydrolases (GH) breaks the complex carbohydrates derived from dietary fiber to short-chain fatty acids (SCFAs) i.e., acetate, propionate and butyrate. These SCFAs controls the gut metabolism through various methods. The SCFAs-dependent (i.e. propionate and butyrate) activation of G-Protein Coupled Receptor 41 (GPR41) induces the expression of peptide YY (PYY), an intestinal hormone that regulates gut motility, increases intestinal transit rate and lowers the harvest of energy from the diet. The G-Protein Coupled Receptor 43 and 41 (GPR43 and GPR41) by propionate triggers the incretin hormone glucagon-like peptide 1 (GLP-1) to increase insulin sensitivity. In adipocytes the SCFAs-mediated activation of GPR43 suppresses insulin signaling and inhibits fat accumulation in adipose tissue

health status is found to be crucial with age-related differences in its composition which may be responsible in progression of disease and fatality in the elderly (Biagi et al. 2010). This induces age related frailty, including innate-immunity, dementia, and sarcopaenia (O'Toole and Jeffery 2015). Therefore, healthy gut microbial composition is influenced by host genotype, age, diet, sex, medical history, and medications (Nicholson et al. 2005). It plays a crucial role in late onset of diseases by effecting host energy and drug metabolism, immunity and leading to inflammation by introduction of new metabolites (Eckburg et al. 2005; Elinav et al. 2011). Hence, a check on dysbiosis could be an important component of late onset disease management and controls, for which a fast and for most clear knowledge of specific bacteria involved in the disease needs to be identified.

Thus, this review is an effort to compile the role and type of gut microbiota associated with various late onset NCDs such as diseases related to obesity, cancers, cardiac, GIT and autoimmunity. Further, an attempt has been made to collate the pathophysiology of gut microbiota in late onset diseases and emphasise on its role through integrity of associated factors like mucin and mucosal immune system. The finding from this study is expected to help in formulating better clinical practices and bio medicines for the therapeutics of the diseases in a prognostic and diagnostic approach.

11.2 Role of Gut Microbiota in NCDs

The collective human microbiome composition is dramatically affected from alteration in the environment and societal norms (Sun and Chang 2014). Microorganisms are thought to be part of human body as they play a crucial role in human physiology. The human gut flora is a key player in functions like host nutrition, drug metabolism, epithelial homeostasis regulation, fat storage regulation, stimulation of intestinal angiogenesis, pathogen resistance and inflammatory responses (Bäckhed et al. 2004). Therefore, any perturbations in host intestine-microbe interactions lead to host responses and miscues that worsen various pathogenic processes. NCDs primarily studied in relation to gut microbiota include, IBD, cancers, obesity, diabetes and neurodegenerative diseases.

The knowledge of gut flora that reside in the mucosa and their part in health and disease progression is increasing at a remarkable rate. Host intestine and microbiome interacts through intestinal mucosal surface in human body. Variation in the intestinal mucin glycosylation has been linked with a various diseases including colitis, colorectal cancer, and IBD (Larsson et al. 2011; Behera et al. 2015). In colon, mucosa is divided into two layers, the nutrient-rich outer layer and an inner layer firmly attached to the surface of the epithelium. While the former provides habitat for microbiota, the second is largely devoid of bacteria. Access of bacteria to epithelium may cause inflammatory and immune responses. Evidence suggests that apart from contributing to the weight, insulin sensitivity, fat distribution, and metabolism, gut bacteria seem to have a role in systemic inflammation as well. Therefore, along with qualitative and quantitative variations in gut microbiota, differences among non-diabetic and diabetic individuals (Diamant et al. 2011) and in many other NCDs as listed have also been observed.

11.3 Gut Microbiota and Metabolic Disorder

The association between metabolic pathology and gut microbiota has grown into an attractive field of bio medical research, with recent studies focusing on the fundamental mechanisms of intestinal microbiota related pathogenesis and regulation of metabolic dysfunction. The underlying molecular mechanism of gut microbial association with metabolic disorders is poorly understood because of microbial complexity (Yang and Kweon 2016). Gut microbiota maintains energy homeostasis and regulates local as well as systemic immunity (Hur and Lee 2015). Dysbiosis of gut microbiota is associated with pathogenesis of diseases such as diabetes, obesity, dyslipidaemia and non-alcoholic fatty liver disease (NAFLD). Gut microbiota modulates the lipid metabolism, lipopolysaccharide accumulation, and SCFA production, thereby regulating food intake, inflammatory tone, and insulin signaling. The first part of this review summarizes the modulation of microbiota in various metabolic disorders.

11.3.1 Obesity

Obesity and associated disorders are a major modern day public health concern. Due to the increase in food intake and decreased energy expenditure, obesity is projected to affect greater than half a billion individuals in 10 years (Kelly et al. 2008). Obesity is the mother of many other deadly diseases, particularly diabetes, cardiovascular, NAFLD and some form of cancers. The aetiology of obesity is complex; various environmental factors, including gut microbiota, play critical roles (Aguirre and Venema 2015). Furthermore, the exact function of microbes is not yet established by scientific research. Although microbes are one of many factors that contribute to obesity, characteristic variations in gut microbiota have been documented in obese individuals. In fact, a system using microbial differences to categorize individuals as obese versus lean demonstrated 90% accuracy in mouse experiments. This established obesity as a somewhat transmissible phenotype (Ley et al. 2006; Turnbaugh et al. 2009; Knights et al. 2011).

On the other hand, the dysbiosis' interaction with obesity is less clear when compared to its link with IBD. There are many contradicting reports regarding phyla-level gut microbiota shifts in obese patients (Walters et al. 2014). Several studies have shown a large variation of the *Firmicutes*: *Bacteroidetes* bacteria ratio among infants versus adults (0.4 vs 10.9, respectively) as well as adults versus elderly (10.9 vs 0.6, respectively) (Mariat et al. 2009). However, other studies evaluated this same ratio and noted either the opposite or even no trend in microbiota (Mahowald et al. 2009; Verdam et al. 2013). The mechanisms of gut microbiota effects adiposity are largely unstudied. The proposed mechanisms suggest increased calorie harvest, adipogenesis induced by elevated SCFAs, over expression of the adiposity-associated genes, and higher lipopolysaccharides (LPS) production by gut bacteria. Excessive SCFAs produced by obese gut microbiota, present an additional energy which causes an imbalance in energy regulation (Murugesan et al. 2017). In a study on obese Indian individuals the molecular analysis of the gut microbiota revealed predominance in *Bacteroides* genus (Patil et al. 2012) with high archaeal densities together with SCFA levels in these subjects.

Fat-rich diet has been documented with higher proportion of Gram-negative bacteria; those are responsible for increasing concentrations of LPS in the blood stream. *Bacteroides fragilis* has a unique composition of LPS that allows it to be up to 1000 times less toxic than *Escherichia coli (E. coli)* (Wexler 2007). A higher concentration of LPS was found in the obese people, likely due to the LPS transport from the GI tract into the bloodstream. For example, increased chylomicron-led LPS transport and disruption of the gut's barrier leads to increased gut permeability and a decreased LPS degradation. *E.coli* depletion was linked to LPS levels of 1–1.3 EU/ml (Endotoxin Unit/millilitre) (Radilla-Vázquez et al. 2016), a component of Gram-negative bacterial cell walls (Amar et al. 2008). Higher levels of LPS (defined as metabolic endotoxemia) are associated with gut inflammation, adipose tissue, as well as several disease mechanisms like diabetes (Serre et al. 2010). Also high fat diets trigger characteristic changes of intestinal microbiota, which decrease inhibition of apoptosis proteins (IAP) and increase Toll-like receptor 4 (TLR-4) activation in the gut mucosa, thereby leading to increased gut permeability and inflammation (Serre et al. 2010). Enhanced GI permeability plays a role in inflammation, as increased plasma LPS levels lead to hyperglycemia and obesity, as depicted in Fig. 11.2.

Turnbaugh et al. (2009) and Furet et al. (2010) evaluated the low *Bacteroidetes* and unvarying trend of *Firmicutes* (*Bacteroides/Prevotella*) in obese individuals. On the other hand, Collado et al. (2008) reported that increased *Firmicutes* (*Staphylococcus aureus*) and *Bacteroidetes* (*Bacteroides/Prevotella*) species correlate with increased body mass index (BMI) in pregnant women and high *Bacteroides* concentrations were observed to be associated with excessive weight gain during pregnancy. Million et al. (2012a) detailed changes in *Firmicutes* composition in obese subjects, including an increase in *Lactobacillus reuteri, L. ductionin, L. paracasei,* and *L. plantarum*. However, other studies have failed to confirm etthese differences in *Firmicutes* and *Bacteroidetes* (Patil et al. 2012) in gut microbiota (Jumpertz et al. 2011). As such studies were done with a small sample size, further confirmation by large scale studies is needed to establish these gut microbiota findings.

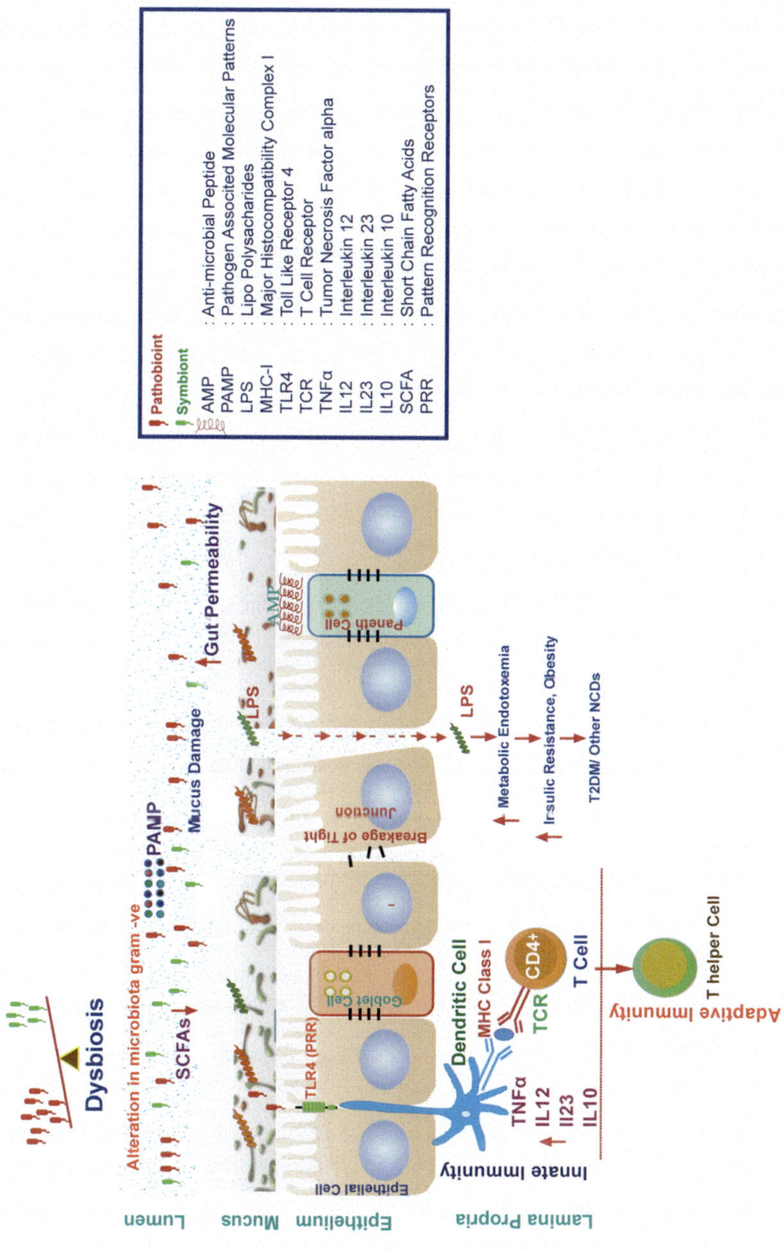

Fig. 11.2 Microbial dysbiosis leads to overgrowth of pathobionts (*e.g.* mainly gram –ve bacteria) which in turn leads to the loss of barrier integrity and breach in the epithelial barrier. Translocation of bacteria and bacterial components to epithelial cells triggers the intestinal immune system to direct a potentially harmful

11.3.2 Type 2 Diabetes (T2D)

The prevalence of T2D, per World Health Organization records, is about 10–33% depending on the region evaluated. As it is multifactorial, both genetic and environmental factors account for this non-communicable and chronic disease. Various research studies suggest that increased inflammation in relation leads to molecular dysregulation and insulin resistance (Gomes et al. 2014). The intestinal microbiota interacts with diet (may be due to geographical conditions) and susceptible genetic factors, thereby playing important role in maintenance of inflammatory response to these factors and in energy homeostasis. Therefore, dysbiosis in gut microbiota contributes to the development of inflammatory stress and imbalance in energy homeostasis, leading to diabetes (Cani et al. 2005).

Diabetic individuals less *Bifidobacterium* and *F.prausnitzii* Gram +ve bacteria (Furet et al. 2010). This group of bacteria associated with anti-inflammatory properties increase with by attenuation by nuclear factor kappa B (NF-κB) response (O'Mahony et al. 2010). Despite perturbations observed in gut microbiota of T2D subjects, it is still necessary to elucidate whether the variations in the microbiota, intestinal barrier, and metabolic endotoxemia are causes or consequences of diabetes. Nevertheless, studies indicate that the gut microbiome may contribute to T2D development. T2D subjects showed characteristic reduction in the number of *Clostridiales* bacteria (*Roseburia* and *F. prausnitzii*) and *Eubacterium rectal* (Qin et al. 2012). Decreased levels of *Akkermansia muciniphila* have been reported in diabetic and glucose-intolerant patients (Zhang et al. 2013). An additional study

---

Fig. 11.2 (continued) proinflammatory response to clear invading bacteria by secreting IL-1 and IL-6 from intestinal epithelial cells, promoting a TH1 and TH17 response by dendritic cells (DCs) and producing higher levels of commensal-specific IgG by B cells. The LPS translocation results in metabolic endotoxemia and insulin resistance resulting in obesity. LPS may also bind to receptor complex on macrophages and results in enhanced production of inflammatory cytokines including IFN-b, IFN-g, IL-1b, IL-6, TNFα, and IL-12. Therefore, changes in microbiota composition and related microbial products exert influences through complex mechanisms on host immune and metabolic functions resulting in Type 2 Diabetes, obesity and other NCDs involving autoimmunity and inflammation. Microbial dysbiosis may also increases gut permeability by reducing the expression of tight junction protein, in the intestinal epithelial cells, which results in the passage of lipopolysaccharide (LPS) into the portal blood circulation. When the harmful bacteria slip through the gut lining, and are recognized by Pattern Recognition Receptors (PRRs), TLRs gets activated. These TLRs signal immune cells, including T cells and macrophages, to produce inflammatory cytokines. The cytokines can weaken the tight junctions between epithelial cells, setting off an inflammatory cycle that might lead to various NCDs. The balance between proinflammatory and protective cytokines is crucial for the maintenance of gut homeostasis. Damage to the epithelium and induction of pro-inflammatory stimuli, including PAMPs derived from luminal antigens and local intestinal microflora, induce the expression of IL-1 family members that are subsequently released by necrotic IECs as potential alarmins (e.g., IL-33 and IL-1α). The endotoxins (toxic compounds) viz., lipopolysaccharides released from the cell walls of gram-negative bacteria are found to be immunogenic and inflammatory. A large influx of endotoxins into the body is termed as endotoxemia which in turn leads to other gut-dysbiosis related disorders

found different gut microbiota in diabetic and insulin-resistant patients compared to insulin-sensitive patients, while accounting for differences in carbohydrate metabolism (Serino et al. 2013).

Bacteroidetes ratios within T1D children have also been recently reported. Studies indicate that bacteria responsible for upholding gut mucosal integrity exist in much lower levels in diabetics, compared to healthy individuals (Murri et al. 2013). Sulfate-reducing microbes, including *Desulfovibrio, L. gasseri, L. reuteri,* and *L. plantarum* are increased T2D patients (Sato et al. 2014). Interestingly, treatment with α-glucosidase inhibitors increased *Lactobacillus spp* levels in Japanese diabetics (Sato et al. 2014). These differences in gut microbiota have been reported in relation to T2D patient who have undergone Roux-en-Y gastric bypass as well. Probiotic as well as prebiotic treatments have been suggested to control metabolic diseases (Cani et al. 2007). Overall, several observational studies have reported links between gut microbiota and T2D, but their conclusions are often conflicting.

11.3.3 Type 1 Diabetes (T1D) Mellitus

Type 1 Diabetes (T1D), an autoimmune chronic disease with several multisystem affects, occurs due to destruction of β islet cells. Some researchers have reported the contribution of gut in development of T1D, which is important to understand its effect on microbiome and digestive system (Pellegrini et al. 2017a). Animal models first demonstrated the role of microbiome in T1D pathogenesis; the non-obese diabetic (NOD) mouse strain and Bio-Breeding diabetes-prone (BB-DP) rats were carefully studied (Wen et al. 2008). Initially, germ free (GF) environments increased the risk of T1D while bacteria had a protective effect against T1D development (Wen et al. 2008; Pozzilli et al. 1993). When germ free non-obese diabetic (NOD) mice with T1D induced by a deletion of the innate immune adaptor myeloid differentiation primary response gene 88 (MyD88) were examined, a microbiota-dependent protection from T1D was noted (Burrows et al. 2015). Myeloid differentiation factor 88 (MyD88), an adaptor enzyme downstream to several Pattern recognition receptors (PRRs), recognizes microbes via the TLR response. Overall, microbes promote as well as inhibit autoimmunity through various signaling pathways. Changes in the amount of *Bifidobacterium, Lactobacillus, Clostridium, Firmicutes, and Bacteroidetes* have been described in association with T1D pathogenesis (Murri et al. 2013). Intestinal microbiota is now recognized as a critical epigenetic factor in modification of T1D predisposition. Also, in increased numbers of *Bacteroides (B. dorei, B. vulgatus)* have been reported Caucasian in Latin American children with beta-cell autoimmunity and T1D (Mejía-León and Barca 2015). Studies indicate that T1D-associated auto antibodies and *Bacteroides* are directly proportional to each other (de Goffau et al. 2013).

Restoration of gut flora, was proposed to reduce oxidative stress and inflammation within mucosa as well as deter T1D development and maintain barrier function

(Musso et al. 2011). Therefore, the pathogenesis of T1D may be linked intestinal with both impaired mucosal barrier and immune function. Patients with T1D demonstrated significantly more signs of inflammation in the duodenal mucosa linked to 10 specific genes namely CCL13, 19, 22, COX2, CCR2, CD68, IL4R, TNFα, PTX3, and VEGFA (Pellegrini et al. 2017b). The contributions of gut microbiota to human health and disease, as per the information collected from studies of humans and animal models for T1D, has been reported after evaluating supporting evidences (Atkinson and Chervonsky 2012). Examined the cohort longitudinal studies in humans and suggested that individuals progressing to T1D exhibit a decrease in microbiota diversity and intestinal dysbiosis.

11.3.4 Dyslipidemia

Dyslipidemia has developed into a prevalent condition due to increased rates of obesity. Dyslipidemia is associated with various metabolic syndromes including elevated fasting lipids, elevated postprandial TRLs (triglyceride-rich lipoproteins), reduced HDL (high-density lipoprotein) levels, elevated small dense LDL (low-density lipoprotein) (Ruotolo and Howard 2002). Variations in the microbiome are associated with adiposity, T2D and dyslipidemia. Further, obesity is associated with dyslipidemia, hypertension, and impaired glucose homeostasis (Woting and Blaut 2016). With oral supplementation of Lactobacillus strains, *curvatus* and *plantarum*, obese mice's weight and inflammatory markers improved significantly. On the other hand, supplementation with only *Lactobacillus plantarum* further attenuated body weight gain and dyslipidemia in such mice (Wu et al. 2015). *Lactobacillus* and *Bifidobacterium spp.* lower cholesterol and ameliorate dyslipidemia (Banjoko et al. 2012). In obese patients, LAB (lactic acid bacteria) probiotics decreased total cholesterol and LDL levels (Tuohy et al. 2014). Alternatively, *Coriobacteriaceae* and *Erysipelotrichaceae* are linked to lipid metabolism and dyslipidemic phenotypes. The alterations in microbiome may by potential pharmaceutical targets for improved metabolic function in the host. Analogous interactions have been demonstrated in mice and human models, particularly with altered cholesterol metabolism disorders, including obesity and metabolic syndrome. Most investigations indicate modulation of gut microbiota via dietary changes, which thereby alter the host's bile acid (Devkota et al. 2012). Gut microbiome changes associated host metabolism have various functional consequences (Vijay-Kumar et al. 2010; Cani et al. 2008).

Various studies reported that selective bacterial taxa might be used to lower cholesterol markers. With *Bifidobacteria* probiotic supplementation, HDL cholesterol increases, thereby alleviating dyslipidemia and insulin resistance (Kondo et al. 2010). As gut microbiota decreases overall cholesterol levels, *Coriobacteriaceae* and *Erysipelotrichaceae*'s cholesterol-lowering contribution cannot be ruled out. This study supports the theory that GI microbiota and lipid metabolism are linked;

therefore, dysbiosis may be a consequence, rather than a cause, of metabolic dysfunction (Martínez et al. 2013).

11.3.5 Non-alcoholic Fatty Liver Disease (NAFLD)

Non-alcoholic fatty liver disease (NAFLD), is a liver disease associated with metabolic syndrome; it is characterized by abnormalities in liver enzymes. NAFLD includes a pathological spectrum, ranging from simple hepatic steatosis and steatohepatitis to liver cirrhosis and liver cancer (Toshikuni et al. 2014). There are ambiguous and controversial statements from different reports over NAFLD. One of the reports represented a significant alcohol consumption in NAFLD, as alcohol use ranged from >1 EtOH drink (10 gm of EtOH per drink) to >40 gm daily (Chalasani et al. 2012). Various studies characterize NAFLD as fat infiltration (steatosis), NASH, and cirrhosis in the absence of alcohol use disorder (Anstee et al. 2011). Emerging data demonstrate dysbiosis to be linked with NAFLD that represented a spectrum of hepatic diseases including simple steatosis, fibrosis, cirrhosis, and NASH without alcohol use disorder. The primary etiology of liver disease is NAFLD, although its development to progression to advanced hepatic failure is poorly understood (Henao-Mejia et al. 2012). Studies have reported that NAFLD patients have increased risk of developing T2D and vascular diseases.

Animal studies have elucidated gut microbiota's role in NAFLD pathogenesis. It is not surprising that dysbiosis of intestinal microbiota is associated with fatty liver disease and NAFLD, as the liver and GI tract are closely linked (Boursier and Diehl 2015). Gut microbiota also plays a key role in fatty liver development via ethanol production (Cope et al. 2000). Interestingly, obese mice that were breath tested for ethanol levels showed much higher blood alcohol ratings than lean mice; also, antibiotic treatment reduced ethanol production by 50% (Dumas et al. 2006). Also, gut microbiota may contribute to NAFLD development via alteration of the choline and the bile acid metabolisms (Corbin and Zeisel 2012). Bacterial metabolism of choline in gut microbiota directly affects NAFLD pathogenesis, as the consumption of choline and production of methylamines leads to decreased synthesis of phosphatidylcholine, thereby adding to triglycerides accumulation in the liver. Similarly, choline-deficient diets are linked to hepatic steatosis. Recent studies evaluating gut microbiota in with fatty and healthy livers revealed differences within the phylum and genus levels. *Firmicutes* and *Bacteroidetes* are predominant in NAFLD patients, although their exact proportions vary among studies. On the other hand, *A. muciniphila* appears to deter development of metabolic syndrome.

NASH patients present with specific changes regarding gut microbial composition: depleted *Bacteroidetes* and high *C. coccoides* levels. Higher percentages of ethanol-producing bacteria, like *Enterobacteriaceae* and *E. coli*, occur in NASH patients (Zhu et al. 2013). Also, recent studies evaluating the effect of choline deficient diets found that this contributed to higher levels of *Gammaproteobacteria* and *Erysipelotrichia* gut bacteria and fatty liver fat (Spencer et al. 2011). Gut microbiota responsible for metabolic pathway regulation act differently in NASH patients. Such altered metabolic patterns can be useful biomarkers. Specific metabolites, such as ethanol, can be useful to track disease progression (Brandl and Schnabl 2017). Therefore, identifying metabolic profiles for individualized therapy may lead future microbiome-based approaches.

11.4 Gut Microbiota in Immunological Disorder

Immunologic disorder is the principal cause of various diseases, including autoimmune diseases, allergies, and cancer. The host's gut immune system and microbes interact in the GI tract, both symbiotic and pathogenic. Studies reported that adaptive immunity is primarily influenced by intestinal bacteria their subsequent molecular pathways. Studies suggest that dysbiosis of intestinal microbiota cause immunological diseases such as rheumatoid arthritis, inflammatory bowel syndrome, T1D, T2D, as well as obesity (Carding et al. 2015). It seems that the human immune system is controlled by microbes, rather than the immune system controlling microbes.

The microbiome is crucial for induction, training as well as immune system function, which maintains a symbiotic relationship between the host and microbes. Host cell pathogen-associated molecular patterns (PAMPs) are identified by PRRs, such as TLRs and nucleotide-binding oligomerization domain-containing protein (NOD), which allow interaction between the host and microbiome.

Gut flora provide a peptidoglycan source that allows innate immunity identify PRRs and NOD1/2, which are generally expressed on gut epithelial cells, macrophages, paneth and dendric cells (DC) (Artis 2008).

11.4.1 Autoimmune Disorder (AD)

Autoimmunity refers to a self destructive condition where immune cells mistakenly target their own cells via antibodies. Finally, following multiple attacks and ongoing inflammation, the organism may note tissue destruction and dysfunction (Tlaskalová-Hogenová et al. 2011). Such diseases include type 1 diabetes, RA, Graves' disease, systemic lupus erythematosus (SLE), and IBD (Bach 2002).

Studies have shown that allergic as well as autoimmune diseases are associated with microbial alternation, in addition to the well-known genetic components. The mechanisms that contribute to microbiota's effect on diseases are not yet understood, although they do appear to involve changes in gut commensals.

11.4.2 Inflammatory Bowel Disease (IBD)

IBD consists of Crohn's disease (CD) and ulcerative colitis (UC). Although the root causes of these chronic autoimmune disorders are still unclear, but microorganisms play a significant role in disease initiation and propagation.

Alterations in GI microbiota have been consistently linked with IBD. Most studies report decreased gut microbiota diversity in individuals affected by IBD (Willing et al. 2010; Manichanh et al. 2006; Tong et al. 2013). Fewer *Firmicutes,* increased *Proteobacteria,* and increased *Bacteroidetes* have been observed (Manichanh et al. 2006; Peterson et al. 2008). Such reduced diversity observed with IBD consistently reports reduced *Firmicutes* levels, as well as reduced *C. leptum* and *F.prausnitzii* (Manichanh et al. 2006; Wang et al. 2014).

The normal microbial load of gut bacteria is ecologically diversified, which is said to be reduced in IBD patients (Tong et al. 2013; Nishikawa et al. 2009). Unlike obesity, IBD has microbial taxonomic shifts; these include lower levels *Firmicutes* and *Bacteroides* and increased levels of *Proteobacteria* and *Actinobacteria*. CD patients see increased *Bacteroides,* (Andoh et al. 2012) and *Enterobacteriales,* which include genera *Shigella, Salmonella,* and *Escherichia* (Willing et al. 2010).

Mycobacterium and *Actinobacteria* bacteria (Bull et al. 2003) increase ___?__. Microbes associated with intestinal epithelium are key players in IBD pathogenesis, as opposed to fecal microbes, as such intestinal bacteria directly interact with mucosal immune tissues (Eckburg et al. 2005; Frank et al. 2007b). IBD have exhibit lower than normal levels of *Firmicutes* phylum (Walker et al. 2011) and *Clostridiales* order (Baumgart et al. 2007) bacteria. Various studies showed a reduction in *Lachnospiraceae* and *Ruminococcaceae* families, *Eubacterium, Faecalibacterium, Roseburia* genera (Hedin et al. 2014), as well as *Clostridium* cluster *IV* (Hedin et al. 2014) and *Clostridium* cluster *XIVa* (Nishikawa et al. 2009). *Lactobacillus* genus bacteria, which are categorized within phylum *Firmicutes,* are decreased in IBD patients (Ott et al. 2004). Despite differences between fecal and intestinal epithelial microbiota, IBD patients do present with fecal microbiota changes. Vindigni et al. (2016) discussed IBD, including the intestinal microbiota, barrier function, and immunity and reported a mutual interdependent contribution of dysbiosis, leaky gut and inflammation for IBD onset and exacerbation.

11.4.3 Celiac Disease

Celiac disease is an autoimmune disorder with systemic symptoms. Reactivity is triggered with ingestion of ingested gluten and prolamines, which are found in wheat, barley, and rye. Helper T cells (Th1) in combination with innate response leads to this pathology. Within the gut, CD causes lymphocyte infiltration into intestinal mucosa, atrophy of small intestine villous structures, as well as extra-intestinal manifestations, such as dermatitis herpetiformis, anemia, alopecia, and neuropathy. Treatment is following a gluten-free diet. In fact, several studies have concluded that by following such a diet, patients shift towards a healthier microbiome (Cenit et al. 2015).

Decreased Gram positive bacteria, such as *Firmicutes*, and increased Gram negative bacteria, such as *Bacteroides* and *E. coli*, were noted in Celiac fecal and duodenal bacteria (Sellitto et al. 2012). Alternatively, a gluten-free diet can reduce healthy bacteria as well, such as *Bifidobacterium*, *Lactobacillus,* and *B. longus*, as well as increase unhealthy bacteria, such as *Enterobacteriaceae* and *E. coli*. Additionally, altered microbiota reduces pro-inflammatory cytokines like TNF-α, IFN-γ, and IL-8, and anti-inflammatory IL-10 (Sanz 2010).

11.4.4 Rheumatoid Arthritis (RA)

RA is an autoimmune inflammatory disease that has systemic symptoms. The relationship between RA and microbiome is well studied and the disease may be a product of genetic and environmental factors, including included gut microbiome. Rheumatoid arthritis patients have reduced fecal *Bifidobacteria* as well as *Bacteroides fragilis* (Yeoh et al. 2013). There are several reports evaluating this microbial interaction. Despite evidence that *Porphyromonas, Prevotella,* and *Leptotricha* species are more prevalent in RA patients, the mechanisms leading to pathology remain enigmatic. The presence of prevotella genus was reported in the patients diagnosed with rheumatoid arthritis. Additionally, a study recognized that *Prevotella copri* may be a potential trigger for RA (Scher et al. 2013). Gut microbiota alterations in newly diagnosed RA cases were characterized by high levels of *P. copri* and low levels of *Bacteroidetes* and *ClostridiumXIV*. Such data suggests that these commensal bacteria have either a pathogenic or protective role in RA development.

Prevention and management of RA using microbial modifications has also been suggested (Scher et al. 2013; Sandhya et al. 2016). Certain gastrointestinal and urogenital infections have been found to reduce risk of RA (Sandberg et al. 2015), indicating a link between RA and the gut. It is also proposed that TLR triggered by intestinal microbiota may be involved in RA development (Rogier et al. 2015). High LPS uptake via gut lumen has been observed in obese mice models; exposure to LPS is thought to increase risk of RA development (Yeoh et al. 2013). Additionally it is also said that obesity may be a factor in the aetiology of RA. A study evaluating fecal samples of RA patients and fibromyalgia patients showed that *Bifidobacteria* and *Bacteroides Fragilis* levels are low RA patients. This suggests that changes in two commensal species could contribute to RA pathogenesis.

11.4.5 Systemic Lupus Erythematosus (SLE)

Systemic lupus erythematosus (SLE) is a autoimmune illness as well. It affects several systems, including the skin, kidneys, joints, brain, blood cells and more. Risk factors for SLE pathogenesis include genetic and environmental components. Dysbiosis has been described in the SLE; high *Bacteroidetes* levels and low *Firmicutes* have been noted (Hevia et al. 2014). Using metagenome functional inference methods, risky metabolic processes associated with SLE were identified: over representation of glycan metabolism and oxidative phosphorylation (Rojo et al. 2015). Removal of certain gut commensals with amoxicillin treatment, facilitates lupus progression via decreased bacterial metabolites, including homoserine lactone, *N*-acetylmuramic acid, and *N*-acetylglucosamine, thereby facilitating immune suppression (Rojo et al. 2015). Factors effecting gut microbiota, such as, diet (Cuervo et al. 2015; Mulak and Bonaz 2015) and pH of water has been observed to be involved in SLE.

11.5 Role of Gut Microbiota on Neurological Disorders

Various studies reported that gut microbiota directly affect neurotransmitter by altering its levels, which may enable them to communicate with neurons. Neuropsychiatric disorders are globally rapidly, and have led to significant disabilities (Moos et al. 2016). Evidences associates microbiota dysbiosis with various mental illnesses and neuropsychiatric conditions (Silberberg et al. 2015), including autism and schizophrenia.

11.5.1 Autism

Autism spectrum disorders (ASDs) neuro-developmental disorders, characterized by cognitive impairment, particular behaviors, and impaired social skills. ASDs include autism (AD), Asperger's syndrome, and PDD-NOS (pervasive developmental disorder not otherwise specified).

The influence of gut bacteria that might control brain function and behavior by interacting with the main nervous system via neural, endocrine and immune systems. Children with autistic (AD) reportedly have less diverse gut microbiomes; they have low *Coprococcus, Prevotella,* and *Veillonellaceae* genera bacteria (Kang et al. 2013). *Bacteroides vulgatus* and the *Desulfovibrio* species were more commonly found in fecal samples from AD children (Tomova et al. 2015). Also, another difference was that *Actinobacteria* and *Firmicutes* accounted for less of the whole flora of AD children's stools than the normal children's stools (Finegold 2011).

All these diseases are linked by a common disease mechanism, autoimmunity and inflammation. It is a matter of research that the autoimmunity leading to these diseases is to a self-antigen or to a non-self-molecule present in commensal bacteria or microbiome of the individual. LPS of human gut bacteria has been linked to be causing systemic inflammation leading to obesity and diabetes (Cani et al. 2007; Brown et al. 2012). Human microbiota is isolated from individuals immune system by mucosal barrier, breaching of this barrier may lead to infiltration of these molecules/substances to reach into the vascular system which depicts the role of mucosal immunity in non-communicable diseases.

11.5.2 Schizophrenia

Schizophrenia is a serious neuropsychiatric disorder, yet the mechanism of pathogenesis remains unclear. Recent genomic studies (GWAS) indicate that SNPs (small single nucleotide polymorphisms) may account for nearly 30% of genetic risk underlying schizophrenia (Kumar et al. 2014). Various studies have found increased serum proinflammatory cytokines in schizophrenia patients as compared with healthy individuals (Song et al. 2013) as well as a correlation linking inflammatory markers to clinical severity of the disease.

Studies examining gut microbiota's impact on neurological development show microbial alternation might contribute to brain development abnormalities, immune dysfunction, and metabolic disruption (Rogers et al. 2016). Nevertheless, defining the relationship between microbiota and neuropsychiatric dysfunction remains complex, as the connection may be nonspecific.

Higher lactic acid bacteria, such *Lactobacillus* species *casei*, *salivarias*, *lactis*, as well as *Streptococcus* species *thermophilius, mitis*, *mutans* had found in schizophrenics. Such bacteria linked with different Th2 immune responses (Yolken and Dickerson 2014). Alternatively, low levels of non-pathogenic bacteria, including *Neisseria, Prochlorococcus, Haemophilus, and Shwanella,* have been identified in schizophrenics (Yolken and Dickerson 2014). Schizophrenics have characteristic deficit syndrome and high gut immune activation, evidenced by increased antigens and antibodies to food. Also, they have increased pathogenic yeasts, including *Candida* species *glabrata* and *tropicalis*, but decreased levels of *albicans* (Yolken and Dickerson 2014). Probiotics helped normalize such altered microbiomes.

11.5.3 Alzheimer's Disease (AD)

Alzheimer's disease (AD) is identified by memory loss, disorientation, cognitive dysfunction and eventually bodily functions. In industrialized countries it affects up to 7% of the 60 and over population and 80% of the 80 and over population. Pathology of AD involves amyloid plaques accumulation and prion-like intracellular neurofibrillary tangles (NFTs) accumulation; these form amyloid (A) β and hyper-phosphorylated tau aggregates, respectively (Nisbet et al. 2015). In neurodegenerative diseases, there a disturbance of normal gut microbiota has been observed. Similar to diabetes, AD and PD are also associated decreased microbial diversity (Cheng et al. 2011) shifts in *Firmicutes* and *Bacteroides* microbial proportions.

Interestingly *Bacteroidetes*, approximately 48%, and *Firmicutes*, approximately 51%, make up the majority of GI microbiome. The other 1% is composed up *Fusobacteria, Cyanobacteria, Spirochaetes, Verrucomicrobia, Proteobacteria* as well as fungi, protozoa, and viruses.

With increased pathobionts, symbionts decrease; *Proteobacteria* and other pro-inflammatory species increase while *Bifidobacterium* (Cheng et al. 2011) and *Prevotella* decrease (Scheperjans et al. 2015). *Chlamydophila pneumoniae* is associated with pneumonia, CAD (coronary artery disease), arthritis, MS (multiple sclerosis), meningoencephalitis and AD (Balin and Hudson 2014). Recent studies confirm association between AD and *Helicobacter pylori, Borrelia, Treponema denticola* and *forsythia*, and *Porphyromonas gingivalis* (Miklossy 2011).

Chronic central nervous system (CNS) inflammation due to *Toxoplasma gondii* results in encephalitis and neurological dysfunction. Interestingly, AD has been linked to high anti-*T.gondii* antibodies, which suggests a link between the two pathologies (Prandota 2014).

11.5.4 Parkinson's Disease (PD)

Parkinson's disease (PD) is a condition where alfa-synuclein (α-syn) accumulates in the substantia nigra to cause neurodegenerative symptoms (Mulak and Bonaz 2015).

Some reports indicate that *H. pylori* can play a role in PD development (Dobbs et al. 2000). Nevertheless, the potential role of *H. pylori* in PD development remains controversial (Álvarez-Arellano and Maldonado-Bernal 2014; Rahne et al. 2013).

H. suis, which is a common zoonotic bacterium, is associated with higher mortality rates from PD (Blaecher et al. 2013). In fact, higher *H. suis* levels are observed in patients with parkinsonism symptoms. The study of microbiota regarding neurodegenerative diseases indicate clear differences (Scheperjans et al. 2015). Several bacterial species, including *Prevotellaceae,* are lower in PD subjects than in control subjects. Furthermore, microbiota related to the motor symptoms of PD suggest that gut microbiota could be a useful biomarker.

Scheperjans et al. (2015) showed reduced *Prevotellaceae* in as well as increased *Enterobacteriaceae* in patients with the postural and gait instability symptoms when compared to those with tremor-dominant PD phenotype.

11.6 Gut Microbiota in Other Disorders

11.6.1 Cancers

Cancer is a complex phenomenon identified as loss of control over growth of cells. Cancer formation is a result of defect in multiple gene including oncogenes, tumor suppressors and care-taker genes. Apart from genetic factors many environmental factors has a role in cancers of specific organs with inflammation being the underlining cause of these cancers. The relationship between microbiota and malignancy is complicated. Gut microbiota induces specific IgG antibody responses through T cells and TLR 4 on B cell (Zeng et al. 2016). Both distinct microbial species as well as dysregulation have been reported. This may lead to gastrointestinal cancer. Various studies reported that intestinal bacteria trigger macrophages to release diffusible clastogens, which are DNA-breaking factors. Colorectal cancer patients showed an under-representation of butyrate-forming bacterial genera (including *Pseudobutyrivibrio ruminis* and *Ruminococcus spp.*) in fecal samples, when compared to healthy individuals (Zhang et al. 2015). Butyrate is an vital nutrient for normal intestinal cell function, as, with propionate, it reduces cell proliferation and induces apoptosis of carcinomas (Zhang et al. 2015; Weir et al. 2013).

The complete absence of *Prevotella* species *copri*a and *oris* was seen in the colon carcinoma cells. *Prevotella* maximizes energy harvested from a plant-based diet; therefore, this finding reflects differences in the intake of fiber and various plant. At the same time, *Phascolarctobacterium, Acidaminobacter, Akkermansia muciniphila,* and *Citrobacter farmer* were very elevated in such patients (Weir et al. 2013). In another study a relatively lower abundance of *Clostridia* and increased abundance of *Fusobacterium* and *Porphyromonas* was found in CRC case than in control (Ahn et al. 2013). In CRC patients, the mucin-degrading species, *Akkermansia muciniphila*, was noted to be fourfold higher. All of these findings may influence the metabolites within the intestinal tract, thereby affecting the mucosal-epithelial barrier of the intestine.

Epidemiological studies indicate a positive association between various cancers and microbes; HPV is connected to cervical cancer pathogenesis; *H. pylori* is connected to gastric cancer pathogenesis; Hepatitis C virus (HCV) can develop into Hepatocellular carcinoma (HCC), and *Salmonella enterica* infection can lead to gallbladder cancer (Schwabe and Jobin 2013).

11.6.2 Chronic Heart Diseases

Studies so far reported that intestinal bacteria are a significant association risk factor for cardiovascular disease (CVD) development. Trimethylamine (TMA) from dietary choline acts as a proatherogenic compound following conversion into TMA

N-oxide in the liver. The higher level of invasive *E. coli*, known as the most common adherent bacteria were identified in chronic heart failure (CHF) patients. Altered intestinal barrier function and circulation can induce cytokine generation, further contributing to impaired cardiac function (Sandek et al. 2007). In a recent study, Lim (2015) have found out high prevalence of pathogenic bacteria, *viz Campylobacter, Salmonella, Shigella* and *Yersinia enterocolitica* (*Y.enterocolitica)* in patients with chronic heart failure as compared with healthy individuals.

11.6.3 Chronic Kidney Disease (CKD)

Studies have demonstrated that like obesity, chronic kidney disease (CKD) can be associated with microbiota changes, as oxidative stress and inflammation from the gastrointestinal tract can lead to progression of kidney disease (Mafra et al. 2014; Kasubuchi et al. 2015). Diets low plant fiber and symbiotic organisms can disturb the normal gut microbiome, causing overgrowth of harmful bacteria and overproduction of uremic toxins, including cresyl and indoxyl molecules. With translocation into circulation, such uremic toxins further systemic inflammation as well as CVD and CKD (Wing et al. 2016).

CKD patients were found to have higher levels of the aerobic *E. coli*, *Enterococcus,* and *Klebsiella pneumoniae*, as well as anaerobic *C.perfringens,* and lower levels anaerobic *Bifidobacterium* species (Hida et al. 1996). Gut bacteria, including *Bifidobacterium* species *longum*, *catenulatum*, *bifidum*, and *Lactobacillus* species *paracasei* and *plantarum*, are decreased in peritoneal dialysis patients (Wang et al. 2012). Also, certain intestinal bacteria produce major uremic toxins, including p-cresol sulfate, indoxylsulfate, and trimethylamine N-oxide, thereby contributing to CKD and CVD progression. A prebiotic containing fiber-high amylose and maize resistant starch type 2 is proven to increase *Bacteroidetes* to *Firmicutes* ratio as well as improved kidney function in mice, thereby suggesting that gut microbiome has an important role in kidney function (Kieffer et al. 2016). Therefore, targeting the gut microbiome to reduce inflammation and uremic toxin production may be a useful treatment method (Wing et al. 2016).

11.7 Mucosal Immunity and Gut-Microbiota

Dysregulation of immunity is the major cause of most of the age related NCD's like diabetes, arthritis, autism as well as cancer. Gut environment being continuously exposed to external environment, including environmental microbes, makes it the prime site to develop immune dysregulation due to change in its microbiota or dietary allergens. Gut mucosal barrier is known to prevent interaction between host immune system and microbiota. Depletion in mucosal layer is also found to be associated with phenomenon of ageing (Sipos et al. 2011). Development of noncommunicable age related diseases and mucosal; thinning could be a connected phenomenon which needs to be studied. Therefore, mucosal immunity may play an important role in these diseases.

Mucosal immunity is most abundantly expressed in the gut. The GIT helps prevent uptake of harmful agents and allows nutrient uptake from the small intestine lumen. Mucosal immunity keeps gut bacteria at bay using various mechanisms: epithelial barrier (Molloy et al. 2012), antimicrobial peptides (Honda and Littman 2012), mucosal innate, and adaptive immunity (Honda and Littman 2012). A coordinated functioning of all these mechanisms influences and is influenced by gut microbial population. Therefore, the host gut mucosal immune system not only protects from commensals but also influences microbiome composition.

Lining of gut with mucosa not only provides a strong physical barrier and contains anti-peptide antibodies but it also provides specific regulatory T cells in lymphoid tissue of the gut and secretory immunoglobulin A (sIgA); thus, providing innate as well as adaptive mucosal immunity to the organism (Frank et al. 2007a). These specific and well-coordinated immune mechanisms allow for microbiota to flourish within gut and limit microbiota-driven inflammation. At the same time, type and amount of microbiota also derives mucosal barrier functioning and thereby regulates immune response to commensal microbes. This function of gut may get impaired with ageing leading to many late onset disease triggers. Therefore, understanding host-microbiota interactions during intestinal homeostasis and gut inflammation may help in directing therapeutic approaches to above mentioned diseases.

11.8 Mucin-Desulfating Sulfatases and Gut Microbiota

The intestinal epithelium is covered with a mucus layer (Ermund et al. 2013) that protects it from microbial scavengers. Colonic mucosa is divided into two layers: an outer layer, which promotes microbiota growth, and an inner layer, which is devoid of bacteria (Johansson et al. 2011). Mucins are classified into two groups as per their pH value, then further classified into sulfated or nonsulfated groups.

Within the GI tract, certain strains of bacteria secrete sulfomucus, which are largely found in mouth and colon mucus (Levine et al. 1987). Studies indicate that mucin sulfation protects against bacterial degradation, thereby allowing for increased longevity of the mucin (Roberton and Corfield 1999). Mucin glycans are complex structures and difficult to study; bacterial sulfatases, proteases, and glycosidases are used to evaluate mucins (Table 11.2). *Bacteroidetes, Firmicutes, Actinobacteria, and Verrucomicrobia* bacteria are common mucin degrading bacteria (Tailford et al. 2015). Several glycoside hydrolases (GH) degrade mucins; fucosidases (GH29 and GH95), neuraminidases/sialidases (GH33), exo- and endo-b-N-acetylglucosaminidases (GH84 and GH85), a-N-acetylglucosaminidases (GH89), b-galactosidases (GH2, GH20 and GH42), endo-b1–4-galactosidases (GH98) and a-N-acetylgalactosaminidases (GH101, GH129) (Tailford et al. 2015). Interestingly, as most bacteria have incomplete GH enzymes, mucin degradation

Table 11.2 Gut bacterial species with mucin-degrading enzymes and their association with various diseases (if known)

Organisms	Enzymes	Apparently associated with dysbiosis	References
Bacteroides fragilis	Neuraminidase, sulphatase, Protease α-N-acetylgalactosaminidase, β-galactosidase, β -N-acetylglucosaminidase, α-fucosidases,	Rheumatoid arthritis	(Yeoh et al. 2013)
Bacteroides thetaiotaomicron	Sulphatase, neuraminidase α-fucosidase, β-galactosidase α-N-acetylgalactosaminidase β-N-acetylglucosaminidase	Predominate in intra-abdominal infections	(Xu and Gordon 2003)
Bacteroides vulgatus	Neuraminidase, α and β-galactosidases, α-fucosidase β-N-acetylglucosaminidase, α and β -Nacetylgalactosaminidase	Less diversity in patients with Crohn's disease compared to healthy patients & IBD	(Dicksved et al. 2008)
Bifidobacterium sp	α-L-Fucosidase, α-N-acetylgalactosaminidase	High fat diet reduced this spp. in the gut. Its reduction associated with development of T2D Milletus and obesity	(Boesten and de Vos 2008)
Bifidobacterium bifidum	Galactosyl-N-acetylhexosamine	Reduction associated with weight loss	(Boesten and de Vos 2008)
Clostridium cocleatum	Neuraminidase, β-galactosidases, β -glucosidase, β N acetylglucosaminidase, α-Nacetylgalactosaminidase	Unknown	(Cani et al. 2007)
Clostridium septicum	β-Galactosidase, β-N-acetylglucosaminidase, glycosulphatase, neuraminidase	Colorectal cancer	(Macfarlane et al. 2001)
Prevotella sp RS2	Glycosulphatase	Unknown	(Wright et al. 2000)
Ruminococcus torques	α-N-Acetylgalactosaminidase	IBS	(Hoskins et al. 1997; Malinen et al. 2010)
Streptomyces sp	α-L-Fucosidase	Unknown	(Goso et al. 2001)
Vibrio cholerae	Neuraminidase, endo-beta-Nacetylhexosaminidase, proteinases.	Diarrhoeal disease	(Stewart-Tull et al. 1986)

requires a consortium of bacteria and their GHs. Notably, mucin mucus associated microbes significantly impact immunity. *R. gnavus*, a mucin-degrading bacterium, induces dysbiosis and increases mucosa-associated bacteria in UC and Chron's (Crost et al. 2013). Also, the harmful bacteria *Pseudomonas aeruginosa* causes respiratory infections in CF (cystic fibrosis) patients; this bacterium uses the disulfated mucin a sulfur source (Robinson et al. 2012) (Fig. 11.3). With identification of keystone mucus-associated gastrointestinal bacteria can help develop diagnostic and therapeutic strategies.

11.9 Gut Microbiota and Therapies

A symbiotic relationship exists between both gut microbiota and its host. With an environment rich in SCFAs, including acetate, propionate and butyrate, and essential vitamins, the intestines help promote symbiotic bacterial proliferation (Kasubuchi et al. 2015). Recent advancements in sequencing techniques have helped researchers understand gut microbiota constituents as well as the pathophysiology of various NCDs.

11.9.1 Gut Microbiota as Present Drug Targets

A microbiome's adaptability allows for the possibility of therapeutic solutions focusing on manipulation of GI microbiota. In addition, such strategies may improve pharmaceutical efficacy. Future therapeutic and prophylactic disease management may likely involve pre- and probiotics targeting gut microbiota (Jia et al. 2008).

Certain diseases are derived from the growth of particular microorganisms, such as *Helicobacter pylori* induced chronic peptic ulcers. Therefore, therapy focus on antibiotics and dietary interventions. However, due to our limited knowledge of intestinal microbiota and lack of the flora induced pathogenesis, it may be premature to suggest probiotic based therapy. Note that *Bacteroides thetaiotaomicron,* which makes up a large portion of intestinal microbiota, a large portion of its genome is dedicated to metabolic processes that help the host organism by facilitating digestion of dietary polysaccharides (Xu and Gordon 2003). Also, the bacteria helps regulate mammalian peroxisome proliferator-activated receptor-γ (PPARγ), which is an receptor commonly expressed in adipose tissue, as it plays a role in insulin resistance (Kelly et al. 2004).

Revolutionary microbiota transplantations from a healthy donor's feces can help restore normal microbial composition in UC patients (Angelberger et al. 2013). Therefore, FMT (fecal microbiota transfer) is thought to help with IBD related dysbiosis (Matsuoka and Kanai 2015). Recent studies indicate that microbiota plays a large role in drug xenobiotic metabolism, stability, and oral bioavailability. So far, over thirty drugs have been isolated as bacterial substrates. Metformin, which

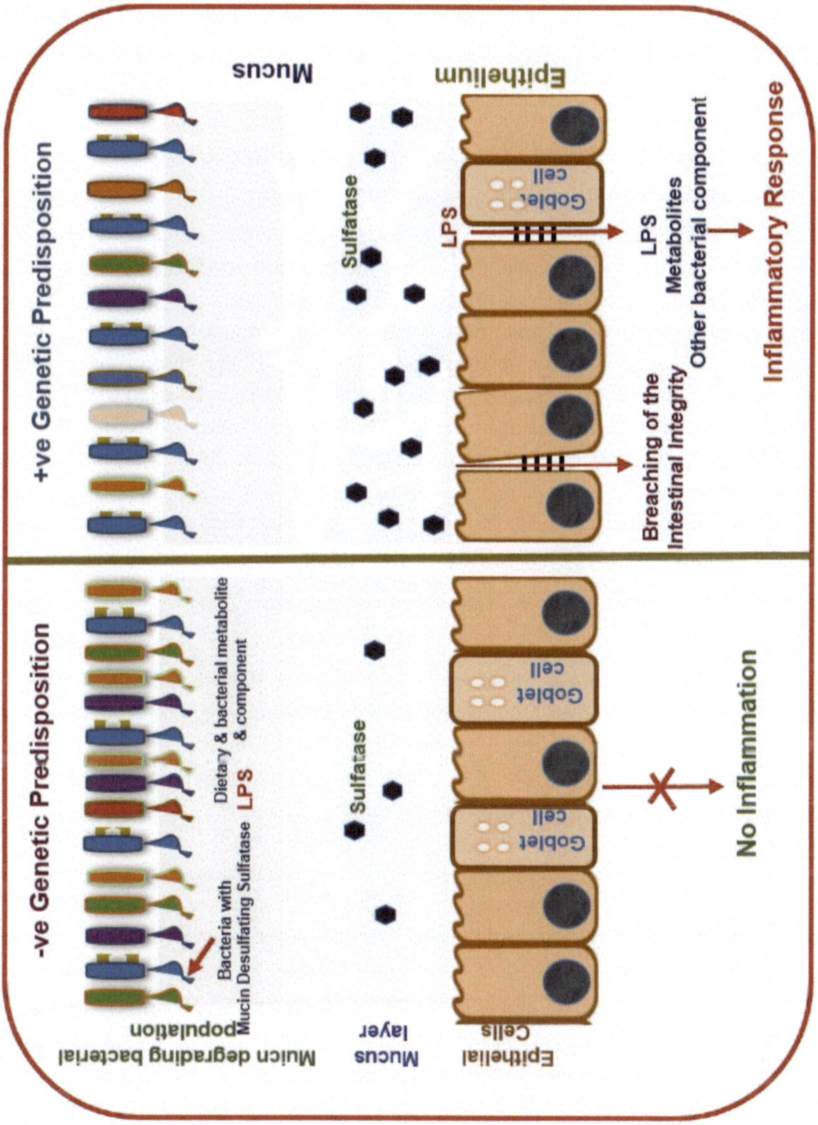

Fig. 11.3 The sulfation of mucins act as protective agent against degradation by bacteria and desulfation is one of the important rate-limiting steps in mucin degradation. Composition of microbiota and components from microbiota effects mucous layer integrity as well as integrity of epithelial cells. Several studies

is a commonly used T2D treatment, increases mucin-producing goblet cells (Tilg and Moschen 2014) and contributes to health maintenance by preventing microbial infiltration due to sustenance of gut mucosal system. It is also known that some species of gut bacteria feed on mucous, thereby, the drug in turn helps in maintenance of microbial ecosystem (intestinal microbiota). Subsequent studies indicated that *A. muciniphila* administration has beneficial effects on metabolism.

11.9.2 Mucosa Associated Gut Bacteria as Future Drug Targets

Although gut microbiota and disease are linked, there is no particular definition of healthy gut bacterial population or un-healthy. The gut bacteria are dynamic and complex. However, relative compositions of various bacteria can be interpreted to understand health status. Reports have stated that parts of the microbiome are "druggable targets," suggesting that the symbiotic relationship can be used to further human health (Wallace and Redinbo 2013). *Bacteroidetes* and *A.muciniphila's* ability to degrade mucin has been well studied (Eckburg et al. 2005; Derrien et al. 2004).

Commensal bacteria survive by using glycan-degrading enzymes to degrade. *Ruminococcus gnavus* is an example, as it exists in over 90% of human GI tracts and is linked to IBD development (Crost et al. 2013). Notably, with IBD, the host generally has increased mucosa-associated bacteria (Schultsz et al. 1999) and increased mucin-degrading bacteria (Png et al. 2010). Increased mucin-degrading bacteria may help sustain non-mucolytic mucosal bacteria in IBD. Also, *R.torques* is frequently increased in IBS cases (Malinen et al. 2010). By treating such bacteria, fat loss, lowered serum LPS levels, improved intestinal barrier function, and decreased insulin resistance can be achieved. *A.muciniphila* administration has shown to improve integrity of epithelial cell layer and reduce endotoximia (Reunanen et al. 2015), potentially underlying the improved glucose profiles (Shin et al. 2014). Even though mucin producing bacteria are also present in gut microbiome, that may also improve gut barrier, are found to be less studied. Genome sequencing of gut microbial population has started unravelling presence of mucin gene sequences

---

Fig. 11.3 (continued) have reported a potential involvement of mucin-degrading bacteria (like bacteria with mucin desulfating sulfatases) in pathogenesis of intestinal diseases. The relation between mucin and microbiota composition may vary depending on the age and host genetics. Higher representation of mucin degrading bacteria may be harmful to genetically predisposed individuals with low mucosa production (less number of goblet cells) and degrading intestinal integrity with age. In such individuals this may lead to inflammatory diseases due to infiltration of LPS or other bacterial components into host system. Individuals with higher microbial variability are also less predisposed to diseases as depicted in this figure. Microbial variability is again regulated by host genetics

among them. More studies are required on this front in order to study their role in disease and health so as to make therapeutic use of them. However, the association between mucosa degrading bacteria and a disease should be studied in relation with age as there are reports to suggest that under healthy gut conditions, some mucous degrading bacteria instead trigger goblet cell to produce more mucous.

11.9.3 Pre, Pro, Postbiotics and Feacal Microbiota Transplantation as Therapy

Generally, prebiotics consist of indigestible carbohydrates that beneficial bacterial growth. These include inulins and sugars, such as fructo-, galactic- and xylo-oligosaccharides (Saulnier et al. 2008). Via colonic fermentation, SCFAs form and exhibit an anti-inflammatory in IBS (Belmonte et al. 2012). With 12 weeks of therapy using *L. rhamnosus* and *Bifidobacterium lactis*, 16% increase of *Lactobacilli* and 18% increase of *Bifidobacteria* was observed as well as 31% decrease of *C. perfringens* (Cani et al. 2008).

Microbial composition helps perpetuate inflammatory diseases, including allergic disease, IBD, asthma, obesity, and neurological as well as psychologic mental health factors. As these diseases have multisystem effects, improved gut microbiota oriented strategies may reduce risk of NCDs. Further studies must evaluate specific prebiotics' ability to promote balanced microbiota decrease inflammation (West et al. 2015).

Probiotics are generally defined as active bacterial organisms, that produce non-specific health benefits for the host (Caselli et al. 2013). Probiotics help stabilize the intestinal mucosal barrier by targeting bacterial overgrowth, increasing mucin production, stimulating immune IgA, and increasing antioxidants (Mao et al. 1996; Oishi et al. 2008). The main probiotics in lactic acid bacterial samples have *Lactobacilli*, species *reuteri, plantarum, casei, para-casei fermentum, rhamnosus,* and *salivarius*, as well as *Bifidobacteria*, species *infantis, bifidum,* and *longum* (Ahrne and Johansson Hagslatt 2011).

Several comprehensive meta-analyses have shown that probiotics significantly decrease the incidence of Antibiotic-associated diarrhea (AAD) (RR 0.39–0.43) (Crow et al. 2015). In addition, probiotics help with CVD, osteoporosis, allergic reactions, and certain bacterial infections such as periodontal disease, halitosis, cavities, respiratory infections, urogenital infections, intestinal infections and *H. pylori* gastric infections (Martinez et al. 2015).

Also, microbiome studies can be used to identify disease risk for obesity and T2D. As 90% of cancer drugs induce GI side effects, pre- and probiotics may be treat such symptoms (Sheridan et al. 2014). A study evaluating probiotics use in UC found that probiotics are not statistically significantly different in safety from anti-inflammatory drugs (Veerappan et al. 2012; Zigra et al. 2008).

Postbiotics strengthen gut microbial diversity as they supplement the diet according to molecules depleted in various diseases (Guo et al. 2013). Additionally, gut microbiota metabolizes amino acid byproducts to create "postbiotics".

Recent studies suggest that the success rate of fecal transplants therapy for *C. difficile* infection results in success in 90% of patients (Bakken et al. 2011). Although, since such therapy is not commonly practiced due to risk of potentially introducing a pathogen to the recipient. With further research, a refined microbial inoculum can be established to produce desired mechanisms in the recipient. Although there is not sufficient evidence currently regarding the best therapeutic approach for certain clinical symptoms, emerging scientific concepts will guide probiotic developments.

This review promotes understanding of the relationship between microbiota and health, which in turn help devise new pre- and probiotic strategies for treatment of various diseases. This, study has unzipped the contradictory observations in the field of gut microbiome associated with diseases and explored its key factors in human health. It stated that not only the gut microbiome of a diseased but also its associates play a key role in its expression. In addition, as gut microbiome is influenced by several environmental factors like diet, physical activity along with individual's genetic factors and heterogeneity in the study samples needs to be taken care of before making any conclusions. Nevertheless, observation from a robust study on interaction between mucosal layer and microbial composition may have potential to therapeutic usage of gut microbiota in non-communicable disease and their impact on intestinal mucosal immunity which could be the signatures required for promoting the human health by the clinicians.

Acknowledgements The work was supported by the financial assistance from the extramural research grant of Indian Council of Medical Research (ICMR), Second Phase of Biomedical Informatics Centres of ICMR Grant Number: BIC/12(19)/2013. The Authors acknowledge Department of Science and Technology, Government of India, New Delhi, for granting inspire fellowship to Mr. Ardhendu Bhusan Praharaj. The authors also acknowledge ICMR-Regional Medical Research Centre, Bhubaneswar for providing infrastructure facilities to carry out this work.

Conflict of Interest Disclosure The authors declare no conflict of interest exists.

Declaration of Funding Sources The work was supported by the financial assistance from the extramural research grant of Indian Council of Medical Research (ICMR), Second Phase of Biomedical Informatics Centres of ICMR Grant Number: BIC/12(19)/2013 and DST, Govt. of India.

References

Aguirre M, Venema K (2015) Does the gut microbiota contribute to obesity? Going beyond the gut feeling. Microorganisms 3(2):213–235

Ahn J, Sinha R, Pei Z, Dominianni C, Wu J, Shi J, Goedert JJ, Hayes RB, Yang L (2013) Human gut microbiome and risk for colorectal cancer. J Natl Cancer Inst 105(24):1907–1911

Ahrne S, Johansson Hagslatt M-L (2011) Effect of lactobacilli on paracellular permeability in the gut. Nutrients 3(1):104–117

Álvarez-Arellano L, Maldonado-Bernal C (2014) Helicobacter pylori and neurological diseases: married by the laws of inflammation. World J Gastrointest Pathophysiol 5(4):400

Amar J, Burcelin R, Ruidavets JB, Cani PD, Fauvel J, Alessi MC, Chamontin B, Ferriéres J (2008) Energy intake is associated with endotoxemia in apparently healthy men. Am J Clin Nutr 87 (5):1219–1223

Andoh A, Kuzuoka H, Tsujikawa T, Nakamura S, Hirai F, Suzuki Y, Matsui T, Fujiyama Y, Matsumoto T (2012) Multicenter analysis of fecal microbiota profiles in Japanese patients with Crohn's disease. J Gastroenterol 47(12):1298–1307

Angelberger S, Reinisch W, Makristathis A, Lichtenberger C, Dejaco C, Papay P, Novacek G, Trauner M, Loy A, Berry D (2013) Temporal bacterial community dynamics vary among ulcerative colitis patients after fecal microbiota transplantation. Am J Gastroenterol 108 (10):1620

Anstee QM, McPherson S, Day CP (2011) How big a problem is non-alcoholic fatty liver disease? BMJ 343:d3897

Artis D (2008) Epithelial-cell recognition of commensal bacteria and maintenance of immune homeostasis in the gut. Nat Rev Immunol 8(6):411–420

Atkinson M, Chervonsky A (2012) Does the gut microbiota have a role in type 1 diabetes? Early evidence from humans and animal models of the disease. Diabetologia 55(11):2868–2877

Bach J-F (2002) The effect of infections on susceptibility to autoimmune and allergic diseases. N Engl J Med 347(12):911–920

Bäckhed F, Ding H, Wang T, Hooper LV, Koh GY, Nagy A, Semenkovich CF, Gordon JI (2004) The gut microbiota as an environmental factor that regulates fat storage. Proc Natl Acad Sci 101 (44):15718–15723

Bakken J, Borody T, Brandt L, Brill J, Demarco D, Franzos M, Kelly C, Khoruts A, Louie T, Martinelli L et al (2011) Treating Clostridium difficile infection with fecal microbiota transplantation. Clinical Gastroenterol hepatol: official Clin Pract J Am Gastroenterol Assoc 9:1044–1049

Balin BJ, Hudson AP (2014) Etiology and pathogenesis of late-onset Alzheimer's disease. Curr Allergy Asthma Rep 14(3):417

Banjoko IO, Adeyanju MM, Ademuyiwa O, Adebawo OO, Olalere RA, Kolawole MO, Adegbola IA, Adesanmi TA, Oladunjoye TO, Ogunnowo AA (2012) Hypolipidemic effects of lactic acid bacteria fermented cereal in rats. Lipids Health Dis 11(1):170

Baumgart M, Dogan B, Rishniw M, Weitzman G, Bosworth B, Yantiss R, Orsi RH, Wiedmann M, McDonough P, Kim SG et al (2007) Culture independent analysis of ileal mucosa reveals a selective increase in invasive Escherichia coli of novel phylogeny relative to depletion of clostridiales in Crohn's disease involving the ileum. ISME J 1(5):403–418

Behera SK, Praharaj AB, Dehury B, Negi S (2015) Exploring the role and diversity of mucins in health and disease with special insight into non-communicable diseases. Glycoconj J 32 (8):575–613

Bekkering P, Jafri I, Van Overveld FJ, Rijkers GT (2013) The intricate association between gut microbiota and development of type 1, type 2 and type 3 diabetes. Expert Rev Clin Immunol 9 (11):1031–1041

Belmonte L, Youmba SB, Bertiaux-Vandaele N, Antonietti M, Lecleire S, Zalar A, Gourcerol G, Leroi A-M, Déchelotte P, Coëffier M (2012) Role of toll like receptors in irritable bowel syndrome: differential mucosal immune activation according to the disease subtype. PLoS One 7(8):e42777

Biagi E, Nylund L, Candela M, Ostan R, Bucci L, Pini E, Nikkïla J, Monti D, Satokari R, Franceschi C (2010) Through ageing, and beyond: gut microbiota and inflammatory status in seniors and centenarians. PLoS One 5(5):e10667

Blaecher C, Smet A, Flahou B, Pasmans F, Ducatelle R, Taylor D, Weller C, Bjarnason I, Charlett A, Lawson A (2013) Significantly higher frequency of H elicobacter suis in patients with idiopathic parkinsonism than in control patients. Aliment Pharmacol Ther 38 (11–12):1347–1353

Boesten RJ, de Vos WM (2008) Interactomics in the human intestine: lactobacilli and bifidobacteria make a difference. J Clin Gastroenterol 42:S163–S167

Boursier J, Diehl AM (2015) Implication of gut microbiota in nonalcoholic fatty liver disease. PLoS Pathog 11(1):e1004559

Brandl K, Schnabl B (2017) Intestinal microbiota and nonalcoholic steatohepatitis. Curr Opin Gastroenterol 33(3):128–133

Brown K, DeCoffe D, Molcan E, Gibson DL (2012) Diet-induced dysbiosis of the intestinal microbiota and the effects on immunity and disease. Nutrients 4(8):1095–1119

Bull TJ, McMinn EJ, Sidi-Boumedine K, Skull A, Durkin D, Neild P, Rhodes G, Pickup R, Hermon-Taylor J (2003) Detection and verification of Mycobacterium avium subsp. paratuberculosis in fresh ileocolonic mucosal biopsy specimens from individuals with and without Crohn's disease. J Clin Microbiol 41(7):2915–2923

Burrows MP, Volchkov P, Kobayashi KS, Chervonsky AV (2015) Microbiota regulates type 1 diabetes through toll-like receptors. Proc Natl Acad Sci U S A 112(32):9973–9977

Cani PD, Amar J, Iglesias MA, Poggi M, Knauf C, Bastelica D, Neyrinck AM, Fava F, Tuohy KM, Chabo C et al (2007) Metabolic endotoxemia initiates obesity and insulin resistance. Diabetes 56(7):1761–1772

Cani PD, Bibiloni R, Knauf C, Waget A, Neyrinck AM, Delzenne NM, Burcelin R (2008) Changes in gut microbiota control metabolic endotoxemia-induced inflammation in high-fat diet–induced obesity and diabetes in mice. Diabetes 57(6):1470–1481

Cani PD, Daubioul CA, Reusens B, Remacle C, Catillon GG, Delzenne NM (2005) Involvement of endogenous glucagon-like peptide-1(7–36) amide on glycaemia-lowering effect of oligofructose in streptozotocin-treated rats. J Endocrinol 185(3):457

Carding S, Verbeke K, Vipond DT, Corfe BM, Owen LJ (2015) Dysbiosis of the gut microbiota in disease. Microb Ecol Health Dis 26:26191–26191

Caselli M, Cassol F, Calò G, Holton J, Zuliani G, Gasbarrini A (2013) Actual concept of "probiotics": is it more functional to science or business? World J Gastroenterol: WJG 19 (10):1527

Cenit MC, Olivares M, Codoñer-Franch P, Sanz Y (2015) Intestinal microbiota and celiac disease: cause, consequence or co-evolution? Nutrients 7(8):6900–6923

Chalasani N, Younossi Z, Lavine JE, Diehl AM, Brunt EM, Cusi K, Charlton M, Sanyal AJ (2012) The diagnosis and management of non-alcoholic fatty liver disease: practice guideline by the American Association for the Study of Liver Diseases, American College of Gastroenterology, and the American Gastroenterological Association. Hepatology 55(6):2005–2023

Cheng J, Palva AM, de Vos WM, Satokari R (2011) Contribution of the intestinal microbiota to human health: from birth to 100 years of age. In: Between pathogenicity and commensalism. Springer, Berlin, Heidelberg, pp 323–346

Clarke JL, Daniell H (2011) Plastid biotechnology for crop production: present status and future perspectives. Plant Mol Biol 76(3–5):211–220

Collado MC, Isolauri E, Laitinen K, Salminen S (2008) Distinct composition of gut microbiota during pregnancy in overweight and normal-weight women. Am J Clin Nutr 88(4):894–899

Collins M, Lawson P, Willems A, Cordoba J, Fernandez-Garayzabal J, Garcia P, Cai J, Hippe H, Farrow J (1994) The phylogeny of the genus clostridium: proposal of five new genera and eleven new species combinations. Int J Syst Evol Microbiol 44(4):812–826

Cope K, Risby T, Diehl AM (2000) Increased gastrointestinal ethanol production in obese mice: implications for fatty liver disease pathogenesis. Gastroenterology 119(5):1340–1347

Corbin KD, Zeisel SH (2012) Choline metabolism provides novel insights into non-alcoholic fatty liver disease and its progression. Curr Opin Gastroenterol 28(2):159

Crost EH, Tailford LE, Le Gall G, Fons M, Henrissat B, Juge N (2013) Utilisation of mucin glycans by the human gut symbiont ruminococcus gnavus is strain-dependent. PLoS One 8(10):e76341

Crow JR, Davis SL, Chaykosky DM, Smith TT, Smith JM (2015) Probiotics and fecal microbiota transplant for primary and secondary prevention of C lostridium difficile infection. Pharmacother: J Hum Pharmacol Drug Ther 35(11):1016–1025

Cuervo A, Hevia A, López P, Suárez A, Sánchez B, Margolles A, González S (2015) Association of polyphenols from oranges and apples with specific intestinal microorganisms in systemic lupus erythematosus patients. Nutrients 7(2):1301–1317

de Goffau MC, Luopajärvi K, Knip M, Ilonen J, Ruohtula T, Härkönen T, Orivuori L, Hakala S, Welling GW, Harmsen HJ (2013) Fecal microbiota composition differs between children with β-cell autoimmunity and those without. Diabetes 62(4):1238–1244

Derrien M, Vaughan EE, Plugge CM, de Vos WM (2004) Akkermansia muciniphila gen. nov., sp. nov., a human intestinal mucin-degrading bacterium. Int J Syst Evol Microbiol 54 (5):1469–1476

Devkota S, Wang Y, Musch MW, Leone V, Fehlner-Peach H, Nadimpalli A, Antonopoulos DA, Jabri B, Chang EB (2012) Dietary-fat-induced taurocholic acid promotes pathobiont expansion and colitis in Il10−/− mice. Nature 487(7405):104

Diamant M, Blaak E, De Vos W (2011) Do nutrient–gut–microbiota interactions play a role in human obesity, insulin resistance and type 2 diabetes? Obes Rev 12(4):272–281

Dicksved J, Halfvarson J, Rosenquist M, Järnerot G, Tysk C, Apajalahti J, Engstrand L, Jansson JK (2008) Molecular analysis of the gut microbiota of identical twins with Crohn's disease. ISME J 2(7):716

Dobbs S, Dobbs R, Weller C, Charlett A (2000) Link between helicobacter pylori infection and idiopathic parkinsonism. Med Hypotheses 55(2):93–98

Dumas M-E, Barton RH, Toye A, Cloarec O, Blancher C, Rothwell A, Fearnside J, Tatoud R, Blanc V, Lindon JC (2006) Metabolic profiling reveals a contribution of gut microbiota to fatty liver phenotype in insulin-resistant mice. Proc Natl Acad Sci 103(33):12511–12516

Duncan SH, Lobley G, Holtrop G, Ince J, Johnstone A, Louis P, Flint HJ (2008) Human colonic microbiota associated with diet, obesity and weight loss. Int J Obes 32(11):1720

Eckburg PB, Bik EM, Bernstein CN, Purdom E, Dethlefsen L, Sargent M, Gill SR, Nelson KE, Relman DA (2005) Diversity of the human intestinal microbial flora. Science 308 (5728):1635–1638

Elinav E, Strowig T, Kau AL, Henao-Mejia J, Thaiss CA, Booth CJ, Peaper DR, Bertin J, Eisenbarth SC, Gordon JI (2011) NLRP6 inflammasome regulates colonic microbial ecology and risk for colitis. Cell 145(5):745–757

Ermund A, Schütte A, Johansson MEV, Gustafsson JK, Hansson GC (2013) Studies of mucus in mouse stomach, small intestine, and colon. I. Gastrointestinal mucus layers have different properties depending on location as well as over the Peyer's patches. Am J Physiol Gastrointest Liver Physiol 305(5):G341–G347

Finegold SM (2011) State of the art; microbiology in health and disease. Intestinal bacterial flora in autism. Anaerobe 17(6):367–368

Frank DN, Amand ALS, Feldman RA, Boedeker EC, Harpaz N, Pace NR (2007a) Molecular-phylogenetic characterization of microbial community imbalances in human inflammatory bowel diseases. Proc Natl Acad Sci 104(34):13780–13785

Frank DN, St Amand AL, Feldman RA, Boedeker EC, Harpaz N, Pace NR (2007b) Molecular-phylogenetic characterization of microbial community imbalances in human inflammatory bowel diseases. Proc Natl Acad Sci U S A 104(34):13780–13785

Furet J-P, Kong L-C, Tap J, Poitou C, Basdevant A, Bouillot J-L, Mariat D, Corthier G, Doré J, Henegar C et al (2010) Differential adaptation of human gut microbiota to bariatric surgery–induced weight loss. Links with metabolic and low-grade inflammation markers. Diabetes 59 (12):3049–3057

Gao Z, Yin J, Zhang J, Ward RE, Martin RJ, Lefevre M, Cefalu WT, Ye J (2009) Butyrate improves insulin sensitivity and increases energy expenditure in mice. Diabetes 58(7):1509–1517

Gomes AC, Bueno AA, de Souza RGM, Mota JF (2014) Gut microbiota, probiotics and diabetes. Nutr J 13(1):60

Goso Y, Ishihara K, Sugawara S, Hotta K (2001) Purification and characterization of α-l-fucosidases from Streptomyces sp. OH11242. Comp Biochem Physiol B: Biochem Mol Biol 130(3):375–383

Guo Y, Mah E, Davis C, Jalili T, Ferruzzi M, Chun O, Bruno R (2013) Dietary fat increases quercetin bioavailability in overweight adults. Mol Nutr Food Res 57:896–905

Hedin CR, McCarthy NE, Louis P, Farquharson FM, McCartney S, Taylor K, Prescott NJ, Murrells T, Stagg AJ, Whelan K et al (2014) Altered intestinal microbiota and blood T cell phenotype are shared by patients with Crohn's disease and their unaffected siblings. Gut 63 (10):1578–1586

Henao-Mejia J, Elinav E, Jin C, Hao L, Mehal WZ, Strowig T, Thaiss CA, Kau AL, Eisenbarth SC, Jurczak MJ (2012) Inflammasome-mediated dysbiosis regulates progression of NAFLD and obesity. Nature 482(7384):179

Hevia A, Milani C, López P, Cuervo A, Arboleya S, Duranti S, Turroni F, González S, Suárez A, Gueimonde M et al (2014) Intestinal dysbiosis associated with systemic lupus erythematosus. MBio 5(5):e01548–e01514

Hida M, Aiba Y, Sawamura S, Suzuki N, Satoh T, Koga Y (1996) Inhibition of the accumulation of uremic toxins in the blood and their precursors in the feces after oral administration of Lebenin®, a lactic acid bacteria preparation, to uremic patients undergoing hemodialysis. Nephron 74(2):349–355

Honda K, Littman DR (2012) The microbiome in infectious disease and inflammation. Annu Rev Immunol 30:759–795

Hoskins LC, Boulding ET, Larson G (1997) Purification and characterization of blood group A-degrading isoforms of α-N-acetylgalactosaminidase from ruminococcus torques strain IX-70. J Biol Chem 272(12):7932–7939

Hur KY, Lee M-S (2015) Gut microbiota and metabolic disorders. Diabetes Metab J 39(3):198–203

Jia W, Li H, Zhao L, Nicholson JK (2008) Gut microbiota: a potential new territory for drug targeting. Nat Rev Drug Discov 7(2):123

Jiang W, Wu N, Wang X, Chi Y, Zhang Y, Qiu X, Hu Y, Li J, Liu Y (2015) Dysbiosis gut microbiota associated with inflammation and impaired mucosal immune function in intestine of humans with non-alcoholic fatty liver disease. Sci Rep 5:8096

Johansson ME, Larsson JMH, Hansson GC (2011) The two mucus layers of colon are organized by the MUC2 mucin, whereas the outer layer is a legislator of host–microbial interactions. Proc Natl Acad Sci 108(Supplement 1):4659–4665

Jumpertz R, Le DS, Turnbaugh PJ, Trinidad C, Bogardus C, Gordon JI, Krakoff J (2011) Energy-balance studies reveal associations between gut microbes, caloric load, and nutrient absorption in humans. Am J Clin Nutr 94(1):58–65

Kang D-W, Park JG, Ilhan ZE, Wallstrom G, Labaer J, Adams JB, Krajmalnik-Brown R (2013) Reduced incidence of prevotella and other fermenters in intestinal microflora of autistic children. PLoS One 8(7):e68322–e68322

Kasubuchi M, Hasegawa S, Hiramatsu T, Ichimura A, Kimura I (2015) Dietary gut microbial metabolites, short-chain fatty acids, and host metabolic regulation. Nutrients 7(4):2839–2849

Kelly D, Campbell JI, King TP, Grant G, Jansson EA, Coutts AG, Pettersson S, Conway S (2004) Commensal anaerobic gut bacteria attenuate inflammation by regulating nuclear-cytoplasmic shuttling of PPAR-γ and RelA. Nat Immunol 5(1):104

Kelly T, Yang W, Chen C-S, Reynolds K, He J (2008) Global burden of obesity in 2005 and projections to 2030. Int J Obes 32(9):1431

Kieffer DA, Piccolo BD, Vaziri ND, Liu S, Lau WL, Khazaeli M, Nazertehrani S, Moore ME, Marco ML, Martin RJ (2016) Resistant starch alters gut microbiome and metabolomic profiles concurrent with amelioration of chronic kidney disease in rats. Am J Physiol Renal Physiol 310 (9):F857–F871

Kimura I, Ozawa K, Inoue D, Imamura T, Kimura K, Maeda T, Terasawa K, Kashihara D, Hirano K, Tani T (2013) The gut microbiota suppresses insulin-mediated fat accumulation via the short-chain fatty acid receptor GPR43. Nat Commun 4:1829

Knights D, Parfrey LW, Zaneveld J, Lozupone C, Knight R (2011) Human-associated microbial signatures: examining their predictive value. Cell Host Microbe 10(4):292–296

Kondo S, Xiao J-Z, Satoh T, Odamaki T, Takahashi S, Sugahara H, Yaeshima T, Iwatsuki K, Kamei A, Abe K (2010) Antiobesity effects of bifidobacterium breve strain B-3 supplementation in a mouse model with high-fat diet-induced obesity. Biosci Biotechnol Biochem 74 (8):1656–1661

Krieg N, Staley J, Brown D, Hedlund B, Paster B, Ward N, Ludwig W, Whitman W (2010) Bergey's manual of systematic bacteriology, vol 4. Springer Science+ Business Media, New York, NY

Kumar A, Agarwal S, Phadke SR, Jaiswal Y (2014) Genetic insight of schizophrenia: past and future perspectives. Gene 535(2):97–100

Langley-Evans S (2015) Nutrition in early life and the programming of adult disease: a review. J Hum Nutr Diet 28:1–14

Larsson JMH, Karlsson H, Crespo JG, Johansson ME, Eklund L, Sjövall H, Hansson GC (2011) Altered O-glycosylation profile of MUC2 mucin occurs in active ulcerative colitis and is associated with increased inflammation. Inflamm Bowel Dis 17(11):2299–2307

Levine MJ, Reddy MS, Tabak LA, Loomis RE, Bergey EJ, Jones PC, Cohen RE, Stinson MW, Al-Hashimi I (1987) Structural aspects of salivary glycoproteins. J Dent Res 66(2):436–441

Ley RE, Turnbaugh PJ, Klein S, Gordon JI (2006) Microbial ecology: human gut microbes associated with obesity. Nature 444(7122):1022

Lim GB (2015) Heart failure: gut flora—pathogenic role in chronic heart failure. Nat Rev Cardiol 13(2):61

Macfarlane S, Hopkins MJ, Macfarlane GT (2001) Toxin synthesis and mucin breakdown are related to swarming phenomenon in clostridium septicum. Infect Immun 69(2):1120–1126

Mafra D, Lobo JC, Barros AF, Koppe L, Vaziri ND, Fouque D (2014) Role of altered intestinal microbiota in systemic inflammation and cardiovascular disease in chronic kidney disease. Future Microbiol 9(3):399–410

Mahowald MA, Rey FE, Seedorf H, Turnbaugh PJ, Fulton RS, Wollam A, Shah N, Wang C, Magrini V, Wilson RK (2009) Characterizing a model human gut microbiota composed of members of its two dominant bacterial phyla. Proc Natl Acad Sci 106(14):5859–5864

Malinen E, Krogius-Kurikka L, Lyra A, Nikkilä J, Jääskeläinen A, Rinttilä T, Vilpponen-Salmela T, von Wright AJ, Palva A (2010) Association of symptoms with gastrointestinal microbiota in irritable bowel syndrome. World J Gastroenterol: WJG 16(36):4532

Manichanh C, Rigottier-Gois L, Bonnaud E, Gloux K, Pelletier E, Frangeul L, Nalin R, Jarrin C, Chardon P, Marteau P et al (2006) Reduced diversity of faecal microbiota in Crohn's disease revealed by a metagenomic approach. Gut 55(2):205–211

Mao Y, Nobaek S, Kasravi B, Adawi D, Stenram U, Molin G, Jeppsson B (1996) The effects of lactobacillus strains and oat fiber on methotrexate-induced enterocolitis in rats. Gastroenterology 111(2):334–344

Mariat D, Firmesse O, Levenez F, Guimarães V, Sokol H, Doré J, Corthier G, Furet J (2009) The Firmicutes/Bacteroidetes ratio of the human microbiota changes with age. BMC Microbiol 9 (1):123

Martínez I, Perdicaro DJ, Brown AW, Hammons S, Carden TJ, Carr TP, Eskridge KM, Walter J (2013) Diet-induced alterations of host cholesterol metabolism are likely to affect the gut microbiota composition in hamsters. Appl Environ Microbiol 79(2):516–524

Martinez RCR, Bedani R, Saad SMI (2015) Scientific evidence for health effects attributed to the consumption of probiotics and prebiotics: an update for current perspectives and future challenges. Br J Nutr 114(12):1993–2015

Matsuoka K, Kanai T (2015) The gut microbiota and inflammatory bowel disease. In: Seminars in immunopathology: 2015. Springer, Berlin, pp 47–55

McPhee JB, Schertzer JD (2015) Immunometabolism of obesity and diabetes: microbiota link compartmentalized immunity in the gut to metabolic tissue inflammation. Clin Sci 129 (12):1083–1096

Mejía-León ME, Barca AM (2015) Diet, microbiota and immune system in type 1 diabetes development and evolution. Nutrients 7(11):9171–9184

Miklossy J (2011, Sep 20) Emerging roles of pathogens in Alzheimer disease. Expert Rev Mol Med 13:e30. https://doi.org/10.1017/S1462399411002006

Million M, Maraninchi M, Henry M, Armougom F, Richet H, Carrieri P, Valero R, Raccah D, Vialettes B, Raoult D (2012a) Obesity-associated gut microbiota is enriched in lactobacillus reuteri and depleted in Bifidobacterium animalis and Methanobrevibacter smithii. Int J Obes 36 (6):817–825

Million M, Maraninchi M, Henry M, Armougom F, Richet H, Carrieri P, Valero R, Raccah D, Vialettes B, Raoult D (2012b) Obesity-associated gut microbiota is enriched in lactobacillus reuteri and depleted in Bifidobacterium animalis and Methanobrevibacter smithii. Int J Obes 36 (6):817

Molloy MJ, Bouladoux N, Belkaid Y (2012) Intestinal microbiota: shaping local and systemic immune responses. In: Seminars in immunology: 2012. Elsevier, pp 58–66

Moos WH, Maneta E, Pinkert CA, Irwin MH, Hoffman ME, Faller DV, Steliou K (2016) Epigenetic treatment of neuropsychiatric disorders: autism and schizophrenia. Drug Dev Res 77(2):53–72

Mulak A, Bonaz B (2015) Brain-gut-microbiota axis in Parkinson's disease. World J Gastroenterol 21(37):10609–10620

Murri M, Leiva I, Gomez-Zumaquero JM, Tinahones FJ, Cardona F, Soriguer F, Queipo-Ortuño MI (2013) Gut microbiota in children with type 1 diabetes differs from that in healthy children: a case-control study. BMC Med 11(1):46

Murugesan S, Nirmalkar K, García-Espitia M, Pizano-Zárate ML, Hoyo-Vadillo C, Garcia-Mena J (2017) Current insight into the role of gut microbiota in Mexican childhood obesity. SOJ Pharm Pharm Sci 4:1–5

Musso G, Gambino R, Cassader M (2011) Interactions between gut microbiota and host metabolism predisposing to obesity and diabetes. Annu Rev Med 62:361–380

Nicholson JK, Holmes E, Wilson ID (2005) Gut microorganisms, mammalian metabolism and personalized health care. Nat Rev Microbiol 3(5):431

Nielsen HH, Qiu J, Friis S, Wermuth L, Ritz B (2012) Treatment for helicobacter pylori infection and risk of Parkinson's disease in Denmark. Eur J Neurol 19(6):864–869

Nisbet RM, Polanco J-C, Ittner LM, Götz J (2015) Tau aggregation and its interplay with amyloid-β. Acta Neuropathol 129(2):207–220

Nishikawa J, Kudo T, Sakata S, Benno Y, Sugiyama T (2009) Diversity of mucosa-associated microbiota in active and inactive ulcerative colitis. Scand J Gastroenterol 44(2):180–186

O'Mahony D, Murphy S, Boileau T, Park J, O'Brien F, Groeger D, Konieczna P, Ziegler M, Scully P, Shanahan F et al (2010) Bifidobacterium animalis AHC7 protects against pathogen-induced NF-κB activation in vivo. BMC Immunol 11:63–63

O'Toole PW, Jeffery IB (2015) Gut microbiota and aging. Science 350(6265):1214–1215

Oishi K, Sato T, Yokoi W, Yoshida Y, Ito M, Sawada H (2008) Effect of probiotics, Bifidobacterium breve and lactobacillus casei, on bisphenol a exposure in rats. Biosci Biotechnol Biochem 72(6):1409–1415

Ott SJ, Musfeldt M, Wenderoth DF, Hampe J, Brant O, Fölsch UR, Timmis KN, Schreiber S (2004) Reduction in diversity of the colonic mucosa associated bacterial microflora in patients with active inflammatory bowel disease. Gut 53(5):685–693

Patil DP, Dhotre DP, Chavan SG, Sultan A, Jain DS, Lanjekar VB, Gangawani J, Shah PS, Todkar JS, Shah S (2012) Molecular analysis of gut microbiota in obesity among Indian individuals. J Biosci 37(4):647–657

Pellegrini S, Sordi V, Bolla AM, Saita D, Ferrarese R, Canducci F, Clementi M, Invernizzi F, Mariani A, Bonfanti R et al (2017a) Duodenal mucosa of patients with type 1 diabetes shows distinctive inflammatory profile and microbiota. J Clin Endocrinol Metabol 102(5):1468–1477

Pellegrini S, Sordi V, Bolla AM, Saita D, Ferrarese R, Canducci F, Clementi M, Invernizzi F, Mariani A, Bonfanti R (2017b) Duodenal mucosa of patients with type 1 diabetes shows distinctive inflammatory profile and microbiota. J Clin Endocrinol Metabol 102(5):1468–1477

Peterson DA, Frank DN, Pace NR, Gordon JI (2008) Metagenomic approaches for defining the pathogenesis of inflammatory bowel diseases. Cell Host Microbe 3(6):417–427

Png CW, Lindén SK, Gilshenan KS, Zoetendal EG, McSweeney CS, Sly LI, McGuckin MA, Florin TH (2010) Mucolytic bacteria with increased prevalence in IBD mucosa augment in vitro utilization of mucin by other bacteria. Am J Gastroenterol 105(11):2420

Pozzilli P, Signore A, Williams AJK, Beales PE (1993) NOD mouse colonies around the world-recent facts and figures. Immunol Today 14(5):193–196

Prandota J (2014) Possible link between toxoplasma gondii and the anosmia associated with neurodegenerative diseases. Am J Alzheimers Dis Other Dement 29(3):205–214

Qin J, Li Y, Cai Z, Li S, Zhu J, Zhang F, Liang S, Zhang W, Guan Y, Shen D (2012) A metagenome-wide association study of gut microbiota in type 2 diabetes. Nature 490(7418):55

Radilla-Vázquez RB, Parra-Rojas I, Martínez-Hernández NE, Márquez-Sandoval YF, Illades-Aguiar B, Castro-Alarcón N (2016) Gut microbiota and metabolic endotoxemia in young obese Mexican subjects. Obes Facts 9(1):1–11

Rahne K-E, Tagesson C, Nyholm D (2013) Motor fluctuations and helicobacter pylori in Parkinson's disease. J Neurol 260(12):2974–2980

Raman M, Ahmed I, Gillevet PM, Probert CS, Ratcliffe NM, Smith S, Greenwood R, Sikaroodi M, Lam V, Crotty P (2013) Fecal microbiome and volatile organic compound metabolome in obese humans with nonalcoholic fatty liver disease. Clin Gastroenterol Hepatol 11(7):868–875. e863

Reunanen J, Kainulainen V, Huuskonen L, Ottman N, Belzer C, Huhtinen H, de Vos WM, Satokari R (2015) Akkermansia muciniphila adheres to enterocytes and strengthens the integrity of the epithelial cell layer. Appl Environ Microbiol 81(11):3655–3662

Roberton AM, Corfield AP (1999) Mucin degradation and its significance in inflammatory conditions of the gastrointestinal tract. In: Medical importance of the normal microflora. Springer, pp 222–261

Robinson CV, Elkins MR, Bialkowski KM, Thornton DJ, Kertesz MA (2012) Desulfurization of mucin by Pseudomonas aeruginosa: influence of sulfate in the lungs of cystic fibrosis patients. J Med Microbiol 61(12):1644–1653

Rogers G, Keating D, Young R, Wong M, Licinio J, Wesselingh S (2016) From gut dysbiosis to altered brain function and mental illness: mechanisms and pathways. Mol Psychiatry 21(6):738

Rogier R, Koenders MI, Abdollahi-Roodsaz S (2015) Toll-like receptor mediated modulation of T cell response by commensal intestinal microbiota as a trigger for autoimmune arthritis. J Immunol Res 2015:527696–527696

Rojo D, Hevia A, Bargiela R, López P, Cuervo A, González S, Suárez A, Sánchez B, Martínez-Martínez M, Milani C et al (2015) Ranking the impact of human health disorders on gut metabolism: systemic lupus erythematosus and obesity as study cases. Sci Rep 5:8310–8310

Ruotolo G, Howard BV (2002) Dyslipidemia of the metabolic syndrome. Curr Cardiol Rep 4 (6):494–500

Sandberg MEC, Bengtsson C, Klareskog L, Alfredsson L, Saevarsdottir S (2015) Recent infections are associated with decreased risk of rheumatoid arthritis: a population-based case-control study. Ann Rheum Dis 74(5):904–907

Sandek A, Bauditz J, Swidsinski A, Buhner S, Weber-Eibel J, von Haehling S, Schroedl W, Karhausen T, Doehner W, Rauchhaus M (2007) Altered intestinal function in patients with chronic heart failure. J Am Coll Cardiol 50(16):1561–1569

Sandhya P, Danda D, Sharma D, Scaria V (2016) Does the buck stop with the bugs?: an overview of microbial dysbiosis in rheumatoid arthritis. Int J Rheum Dis 19(1):8–20

Sanz Y (2010) Effects of a gluten-free diet on gut microbiota and immune function in healthy adult humans. Gut Microbes 1(3):135–137

Sato J, Kanazawa A, Ikeda F, Yoshihara T, Goto H, Abe H, Komiya K, Kawaguchi M, Shimizu T, Ogihara T et al (2014) Gut dysbiosis and detection of "live gut bacteria" in blood of Japanese patients with type 2 diabetes. Diabetes Care 37(8):2343–2350

Saulnier DM, Gibson GR, Kolida S (2008) In vitro effects of selected synbiotics on the human faecal microbiota composition. FEMS Microbiol Ecol 66(3):516–527

Scheperjans F, Aho V, Pereira PA, Koskinen K, Paulin L, Pekkonen E, Haapaniemi E, Kaakkola S, Eerola-Rautio J, Pohja M (2015) Gut microbiota are related to Parkinson's disease and clinical phenotype. Mov Disord 30(3):350–358

Scher JU, Sczesnak A, Longman RS, Segata N, Ubeda C, Bielski C, Rostron T, Cerundolo V, Pamer EG, Abramson SB et al (2013) Expansion of intestinal Prevotella copri correlates with enhanced susceptibility to arthritis. elife 2:e01202–e01202

Schultsz C, van den Berg FM, Fiebo W, Tytgat GN, Dankert J (1999) The intestinal mucus layer from patients with inflammatory bowel disease harbors high numbers of bacteria compared with controls. Gastroenterology 117(5):1089–1097

Schwabe RF, Jobin C (2013) The microbiome and cancer. Nat Rev Cancer 13(11):800

Sellitto M, Bai G, Serena G, Fricke WF, Sturgeon C, Gajer P, White JR, Koenig SSK, Sakamoto J, Boothe D et al (2012) Proof of concept of microbiome-metabolome analysis and delayed gluten exposure on celiac disease autoimmunity in genetically at-risk infants. PLoS One 7(3):e33387–e33387

Serino M, Fernández-Real JM, García-Fuentes E, Queipo-Ortuño M, Moreno-Navarrete JM, Sánchez A, Burcelin R, Tinahones F (2013) The gut microbiota profile is associated with insulin action in humans. Acta Diabetol 50(5):753–761

Serre CBL, Ellis CL, Lee J, Hartman AL, Rutledge JC, Raybould HE (2010) Propensity to high-fat diet-induced obesity in rats is associated with changes in the gut microbiota and gut inflammation. Am J Physiol Gastrointest Liver Physiol 299(2):G440–G448

Sheridan PO, Bindels LB, Saulnier DM, Reid G, Nova E, Holmgren K, O'Toole PW, Bunn J, Delzenne N, Scott KP: Can prebiotics and probiotics improve therapeutic outcomes for undernourished individuals?; 201474-82

Shin N-R, Lee J-C, Lee H-Y, Kim M-S, Whon TW, Lee M-S, Bae J-W (2014) An increase in the Akkermansia spp. population induced by metformin treatment improves glucose homeostasis in diet-induced obese mice. Gut 63(5):727–735

Silberberg D, Anand NP, Michels K, Kalaria RN (2015) Brain and other nervous system disorders across the lifespan — global challenges and opportunities. Nature 527:S151

Sipos F, Leiszter K, Tulassay Z (2011) Effect of ageing on colonic mucosal regeneration. World J Gastroenterol: WJG 17(25):2981

Sokol H, Cosnes J, Chazouilleres O, Beaugerie L, Tiret E, Poupon R, Seksik P (2008) Disease activity and cancer risk in inflammatory bowel disease associated with primary sclerosing cholangitis. World J Gastroenterol: WJG 14(22):3497

Sommer F, Bäckhed F (2013) The gut microbiota—masters of host development and physiology. Nat Rev Microbiol 11(4):227

Song X, Fan X, Song X, Zhang J, Zhang W, Li X, Gao J, Harrington A, Ziedonis D, Lv L (2013) Elevated levels of adiponectin and other cytokines in drug naive, first episode schizophrenia patients with normal weight. Schizophr Res 150(1):269–273

Spencer MD, Hamp TJ, Reid RW, Fischer LM, Zeisel SH, Fodor AA (2011) Association between composition of the human gastrointestinal microbiome and development of fatty liver with choline deficiency. Gastroenterology 140(3):976–986

Stewart-Tull D, Ollar R, SCOBIE TS (1986) Studies on the vibrio cholerae mucinase complex. I. Enzymic activities associated with the complex325. J Med Microbiol 22 (4):325–333

Sun J, Chang EB (2014) Exploring gut microbes in human health and disease: pushing the envelope. Genes Dis 1(2):132–139

Tailford LE, Crost EH, Kavanaugh D, Juge N (2015) Mucin glycan foraging in the human gut microbiome. Front Genet 6:81

Tap J, Mondot S, Levenez F, Pelletier E, Caron C, Furet JP, Ugarte E, Muñoz-Tamayo R, Paslier DL, Nalin R (2009) Towards the human intestinal microbiota phylogenetic core. Environ Microbiol 11(10):2574–2584

Tilg H, Moschen AR (2014) Microbiota and diabetes: an evolving relationship. Gut 63 (9):1513–1521

Tlaskalová-Hogenová H, Stěpánková R, Kozáková H, Hudcovic T, Vannucci L, Tučková L, Rossmann P, Hrnčíř T, Kverka M, Zákostelská Z et al (2011) The role of gut microbiota (commensal bacteria) and the mucosal barrier in the pathogenesis of inflammatory and autoimmune diseases and cancer: contribution of germ-free and gnotobiotic animal models of human diseases. Cell Mol Immunol 8(2):110–120

Tomova A, Husarova V, Lakatosova S, Bakos J, Vlkova B, Babinska K, Ostatnikova D (2015) Gastrointestinal microbiota in children with autism in Slovakia. Physiol Behav 138:179–187

Tong M, Li X, Wegener Parfrey L, Roth B, Ippoliti A, Wei B, Borneman J, McGovern DPB, Frank DN, Li E et al (2013) A modular organization of the human intestinal mucosal microbiota and its association with inflammatory bowel disease. PLoS One 8(11):e80702–e80702

Toshikuni N, Tsutsumi M, Arisawa T (2014) Clinical differences between alcoholic liver disease and nonalcoholic fatty liver disease. World J Gastroenterol: WJG 20(26):8393

Tuohy KM, Fava F, Viola R (2014) 'The way to a man's heart is through his gut microbiota'–dietary pro-and prebiotics for the management of cardiovascular risk. Proc Nutr Soc 73(2):172–185

Turnbaugh PJ, Hamady M, Yatsunenko T, Cantarel BL, Duncan A, Ley RE, Sogin ML, Jones WJ, Roe BA, Affourtit JP (2009) A core gut microbiome in obese and lean twins. Nature 457 (7228):480

Vaahtovuo J, Munukka E, Korkeamäki M, Luukkainen R, Toivanen P (2008) Fecal microbiota in early rheumatoid arthritis. J Rheumatol 35(8):1500–1505

Veerappan G, Betteridge J, Young P (2012) Probiotics for the treatment of inflammatory bowel disease. Curr Gastroenterol Rep 14:324–333

Verdam FJ, Fuentes S, de Jonge C, Zoetendal EG, Erbil R, Greve JW, Buurman WA, de Vos WM, Rensen SS (2013) Human intestinal microbiota composition is associated with local and systemic inflammation in obesity. Obesity 21(12):E607–E615

Vijay-Kumar M, Aitken JD, Carvalho FA, Cullender TC, Mwangi S, Srinivasan S, Sitaraman SV, Knight R, Ley RE, Gewirtz AT (2010) Metabolic syndrome and altered gut microbiota in mice lacking toll-like receptor 5. Science 328(5975):228–231

Vindigni SM, Zisman TL, Suskind DL, Damman CJ (2016) The intestinal microbiome, barrier function, and immune system in inflammatory bowel disease: a tripartite pathophysiological circuit with implications for new therapeutic directions. Ther Adv Gastroenterol 9(4):606–625

Walker AW, Sanderson JD, Churcher C, Parkes GC, Hudspith BN, Rayment N, Brostoff J, Parkhill J, Dougan G, Petrovska L (2011) High-throughput clone library analysis of the mucosa-associated microbiota reveals dysbiosis and differences between inflamed and non-inflamed regions of the intestine in inflammatory bowel disease. BMC Microbiol 11:7–7

Wallace BD, Redinbo MR (2013) The human microbiome is a source of therapeutic drug targets. Curr Opin Chem Biol 17(3):379–384

Walters WA, Xu Z, Knight R (2014) Meta-analyses of human gut microbes associated with obesity and IBD. FEBS Lett 588(22):4223–4233

Wang I-K, Lai H-C, Yu C-J, Liang C-C, Chang C-T, Kuo H-L, Yang Y-F, Lin C-C, Lin H-H, Liu Y-L (2012) Real-time PCR analysis of the intestinal microbiotas in peritoneal dialysis patients. Appl Environ Microbiol 78(4):1107–1112

Wang W, Chen L, Zhou R, Wang X, Song L, Huang S, Wang G, Xia B (2014) Increased proportions of bifidobacterium and the lactobacillus group and loss of butyrate-producing bacteria in inflammatory bowel disease. J Clin Microbiol 52(2):398–406

Weir TL, Manter DK, Sheflin AM, Barnett BA, Heuberger AL, Ryan EP (2013) Stool microbiome and metabolome differences between colorectal cancer patients and healthy adults. PLoS One 8 (8):e70803

Wen L, Ley RE, Volchkov PY, Stranges PB, Avanesyan L, Stonebraker AC, Hu C, Wong FS, Szot GL, Bluestone JA et al (2008) Innate immunity and intestinal microbiota in the development of type 1 diabetes. Nature 455(7216):1109–1113

West CE, Jenmalm M, Prescott S (2015) The gut microbiota and its role in the development of allergic disease: a wider perspective. Clin Exp Allergy 45(1):43–53

Wexler HM (2007) Bacteroides: the good, the bad, and the nitty-gritty. Clin Microbiol Rev 20 (4):593–621

Willing BP, Dicksved J, Halfvarson J, Andersson AF, Lucio M, Zheng Z, Järnerot G, Tysk C, Jansson JK, Engstrand L (2010) A pyrosequencing study in twins shows that gastrointestinal microbial profiles vary with inflammatory bowel disease phenotypes. Commentary. Gastroenterology 139(6):1844–1854. e1

Winer DA, Luck H, Tsai S, Winer S (2016) The intestinal immune system in obesity and insulin resistance. Cell Metab 23(3):413–426

Wing MR, Patel SS, Ramezani A, Raj DS (2016) Gut microbiome in chronic kidney disease. Exp Physiol 101(4):471–477

Woting A, Blaut M (2016) The intestinal microbiota in metabolic disease. Nutrients 8(4):202

Wright IC, Rabe-Hesketh S, Woodruff PW, David AS, Murray RM, Bullmore ET (2000) Meta-analysis of regional brain volumes in schizophrenia. Am J Psychiatr 157(1):16–25

Wu C-C, Weng W-L, Lai W-L, Tsai H-P, Liu W-H, Lee M-H, Tsai Y-C (2015) Effect of lactobacillus plantarum strain K21 on high-fat diet-fed obese mice. Evid Based Complement Alternat Med 2015

Xu J, Gordon JI (2003) Honor thy symbionts. Proc Natl Acad Sci 100(18):10452–10459

Yang J-Y, Kweon M-N (2016) The gut microbiota: a key regulator of metabolic diseases. BMB Rep 49(10):536

Yatsunenko T, Rey FE, Manary MJ, Trehan I, Dominguez-Bello MG, Contreras M, Magris M, Hidalgo G, Baldassano RN, Anokhin AP (2012) Human gut microbiome viewed across age and geography. Nature 486(7402):222

Yeoh N, Burton J, Suppiah P, Reid G, Stebbings S (2013) The role of the microbiome in rheumatic diseases. Curr Rheumatol Rep 15:314

Yolken R, Dickerson F (2014) The microbiome-the missing link in the pathogenesis of schizophrenia. Schizophr Res 153:S16

Zeng MY, Cisalpino D, Varadarajan S, Hellman J, Warren HS, Cascalho M, Inohara N, Núñez G (2016) Gut microbiota-induced immunoglobulin G controls systemic infection by symbiotic bacteria and pathogens. Immunity 44(3):647–658

Zhang X, Shen D, Fang Z, Jie Z, Qiu X, Zhang C, Chen Y, Ji L (2013) Human gut microbiota changes reveal the progression of glucose intolerance. PLoS One 8(8):e71108–e71108

Zhang M, Fan X, Fang B, Zhu C, Zhu J, Ren F (2015) Effects of lactobacillus salivarius Ren on cancer prevention and intestinal microbiota in 1, 2-dimethylhydrazine-induced rat model. J Microbiol 53(6):398–405

Zhu L, Baker SS, Gill C, Liu W, Alkhouri R, Baker RD, Gill SR (2013) Characterization of gut microbiomes in nonalcoholic steatohepatitis (NASH) patients: a connection between endogenous alcohol and NASH. Hepatology 57(2):601–609

Zigra P, Maipa V, Alamanos Y (2008) Probiotics and remission of ulcerative colitis: a systematic review. Neth J Med 65:411–418

Chapter 12
Immunotherapy for Colon Cancer: Recent Perspectives

Christoffer B. Lambring, Chloe Smith, Sohail Siraj, Krishna Patel, and Riyaz Basha

Abstract Colorectal cancer (CRC) is third in cancer-related deaths among male and female populations. It is a slowly progressing malignancy and therefore, early detection of ulcers or the non-malignant polyps can prevent the disease and result in better patient outcomes. There are several treatment regimens, including chemotherapy and radiation that have been used for decades in treating CRC. Morbidity caused by intensive therapy and limitations in treatment due to late stage diagnoses impact the survival of CRC patients and dampen the quality of life of surviving patients. Alternative approaches such as immunotherapy have been tested for some time to address the issues associated with intensive chemotherapy. Several strategies that have been studied in pre-clinical models for CRC are recently gaining attention due to promising results, leading to clinical trials. Strategies including the use of monoclonal antibodies, vaccines, T cell stimulators, and immune checkpoint inhibitors are currently in clinical trials. The results of these clinical trials are crucial for immunotherapy usage as a part of standard care. Less toxic and more specific immunotherapeutic approaches are urgently required to address the morbidity of CRC patients, these approaches have the ability to reduce metastasis and increase the survival rate in CRC patients.

Keywords Colorectal cancer · Immunotherapy · PD-1 inhibitors · Cancer vaccines

---

C. B. Lambring · S. Siraj · K. Patel · R. Basha (✉)
Texas College of Osteopathic Medicine, UNT Health Science Center, Fort Worth, TX, USA
e-mail: Riyaz.Basha@unthsc.edu

C. Smith
Old Dominion University, Norfolk, VA, USA

G. S. R. Raju, L.V.K.S. Bhaskar (eds.), *Theranostics Approaches to Gastric and Colon Cancer*, Diagnostics and Therapeutic Advances in GI Malignancies,
https://doi.org/10.1007/978-981-15-2017-4_12

12.1 Introduction

Colon cancer is a slow, progressive malignancy originating in the large intestine. If the cancer is initiated in rectum, the last part of the large intestine, it is termed as rectal cancer. Morbidity varies for colorectal cancer (CRC) based on stage of the disease, patient health, and other factors, but among men and women it is expected to cause 51,020 deaths during 2019 (Siegel et al. 2019). It is predicted that 145,600 new cases of colorectal (colon: 101,420; rectal: 44,180) cancer will be diagnosed among men and women in 2019 in the United States (Siegel et al. 2019). The chances for developing colorectal cancer during life is 1:22 for men, and 1:24 for women. An increase in screenings and early detection are key factors to minimize the occurrence of CRC deaths. In patients diagnosed with the localized disease the 5-year survival rate is about 90% in the United States. The survival rates of the patients who are diagnosed at later stages are poor, ranging from 71% for the patients diagnosed with the regional disease or 14% for the patients diagnosed with distant metastasis (Siegel et al. 2017). It is predicted that the global CRC burden may grow 60% by 2030 leading to 2.2 million new cases as well as 1.1 million deaths (Arnold et al. 2017).

There are several types of CRC, the most common being adenocarcinoma. CRC begins as benign polyps in the colon. Most polyps are hyperplastic and inflammatory, which are not generally a pre-cancerous condition. While colonoscopy is the most common technique used for the screening of CRC, other tests including blood based tests, stool, and radiological tests are also employed (Bond 2002; Carroll et al. 2014; Force et al. 2016; Issa and Noureddine 2017; Levin et al. 2008a). Polyps are usually removed when a patient undergoes a regular screening, such as a colonoscopy.

12.1.1 Adenocarcinoma

The most prevalent colon cancer is adenocarcinoma. "Adeno" meaning gland, which describes the cellular properties of adenocarcinomas that form within the lining of internal organs. Adenocarcinomas can form in many different organs including the lungs and breasts. Early tumors start as polyps that continue to grow into malignant tumors. These early polyps, known as adenomatous polyps, are also considered a pre-cancerous condition due to their tendency to turn into malignant tumors (Shinya and Wolff 1979; Muto et al. 1975).

12.1.2 Gastrointestinal Stromal Tumors

Gastrointestinal stromal tumors (GISTs) are relatively atypical tumors (0.1–3%) of the gastrointestinal (GI) tract (Mazzola et al. 2008; Raut et al. 2007). GISTs originate in the Interstitial cells of Cajal (ICCs) and are associated with the autonomic nervous system tract (Mazzola et al. 2008; Raut et al. 2007; Min and Leabu 2006). GISTs can begin at any location in the GI tract, but rarely appear in the colon. GISTs seem to occur more frequently in patients that have a history with neurofibromatosis. Mutations in tyrosine protein kinase or CD117 and platelet-derived growth factor receptor-alpha are associated with the majority of GISTs (Mazzola et al. 2008; Raut et al. 2007; El-Menyar et al. 2017).

12.1.3 Gastrointestinal Carcinoid Tumors

GI tract neuroendocrine cells are termed carcinoids. Neuroendocrine cells control the release of digestive enzymes and the rate at which food moves through the GI tract. Gastrointestinal carcinoid tumors are classified by tumor grade, which describes how fast the tumor is predicted to grow and spread. Tumors classified as grades 1 and 2 are slow-growing, slow-spreading, and are called GI neuroendocrine tumors (NETs). Tumors classified as grade 3 grow and spread quickly and are called GI Neuroendocrine Carcinomas (NECs) (Ilgenfritz and Mathews 1955; Altman and Mann 1948; Weng et al. 1996).

12.1.4 Colorectal Lymphoma

Non-Hodgkin lymphoma (NHL) is a malignant tumor that stems from immune system cells. In 2016, the World Health Organization revised the classification of lymphoid neoplasms recognizing $>$60 types of lymphomas (Swerdlow et al. 2016). The GI tract is not a viewed as a classic lymphoid organ, however the largest number of lymphocytes are present in gut lymphoid tissues. Up to 40% of extra-nodal lymphoma cases are found in the GI tract (d'Amore et al. 1994; Herrmann et al. 1980). The least common primary lymphoma location is the colon and rectum only accounting for $\sim$10% of GI lymphomas and $<$ 1% of large bowel malignancies (Blackledge et al. 1979; Contreary et al. 1980; Henry and Berry 1988; Koch et al. 2001). Colorectal lymphoma is a NHL of the GI tract. Lymphoma affects the body's lymph tissue and starts in locations such as the digestive tract. Lymphomas most commonly affect B-cells and occasionally T-cells. NHLs can also be organized by their growth rate and metastasis. Indolent lymphomas are slow growing. The most frequent indolent NHL in the United States is follicular lymphoma. The most prevalent aggressive NHL, high growth and metastasis rate, in the United States is diffuse large B-cell lymphoma.

12.1.5 Hereditary Colorectal Cancers

There are mutated genes that family members can inherit which will increase an individual's risk to colon cancer (Li et al. 2016). Heritable mutations are responsible for about 5–10% of CRC (Valle et al. 2019). The major types of hereditary colon cancer are Lynch syndrome and familial adenomatous polyposis (FAP), they are designated hereditary nonpolyposis colorectal cancer (HNPCC) (Carethers and Stoffel 2015). There are also rare conditions that have been associated with a predisposition for colon cancer such as MUTYH-associated polyposis (MAP) and attenuated familial adenomatous polyposis (AFAP) (DeRycke et al. 2017).

12.2 Symptoms

Many people, especially younger populations, with early stage CRC do not show symptoms until the cancer has advanced. Older patients are likely to present with some symptoms ranging from fatigue that does not get better with rest, certain bowel conditions like constipation or diarrhea, bright or dark red stool, unexplained anemia, and unexplainable weight loss (Bat et al. 1992; Trombold et al. 2013). People with symptoms of the disease are recommended for full screening for CRC through standard diagnostic tools.

12.3 Diagnosis

12.3.1 Blood Tests and Imaging

Doctors will consider risk factors for a patient in order to determine the best diagnostic test to use. Depending on a patient's symptoms, general health, family history, and medical test results, a doctor can determine what laboratory test will benefit the patient for a precise diagnosis. Typically, diagnosis is carried out by performing a colonoscopy and taking a biopsy of a suspicious polyp. Colonoscopy allows gastroenterologists to look closely at the rectum or colon of the patient (Segnan et al. 2005; Force 2008; Levin et al. 2008b). Biopsies can be taken during colonoscopies or surgeries and are used to confirm whether a patient has colorectal cancer and are helpful for investigating the type of disease and determining the treatment plan (Koelzer et al. 2016; Lugli et al. 2017).

Blood tests and imaging techniques including Computed tomography (CT), Magnetic resonance imaging (MRI), Computed tomography (CT), Positron emission tomography (PET), and Ultrasound are used for the diagnosis of CRC (Deveney and Way 1984; Winawer and Sherlock 1976; Gothlin 1988; Matsui et al. 1987; Machi et al. 1987; Lee and Lee 2017). MRI is useful in identifying the location, PET and CT scans are used to measure tumor size and are useful to determine the location of metastasis, and endorectal ultrasounds are used for detecting how deep colorectal cancer has grown (Lee and Lee 2017; Nasseri and Langenfeld 2017).

12.3.2 Molecular Testing of the Tumor

Molecular testing of tumors includes identifying specific genes, proteins, and other characteristics found in tumors (Lee et al. 2017; Harada and Morlote 2020; Sepulveda et al. 2017). All CRCs need testing for mismatch repair protein problems denoted mismatch repair defects. This test can determine if a patient has Lynch-Syndrome and if immunotherapy should be considered as a treatment option in metastatic colorectal cancer (Liu et al. 1996; Lynch and Smyrk 1996; Cohen et al. 2019; Antelo et al. 2019).

12.4 Risk Factors

Risk factors that can contribute to the development of colon cancer include old age, African-American race, inflammatory intestinal conditions, personal or familial history with colon cancer, inherited syndromes like Lynch Syndrome and FAP, smoking, heavy use of alcohol, a sedentary lifestyle, obesity, diabetes, a diet low in fiber and high in fat, and radiation from previous cancer treatment (Northover and Murday 1989; Lynch et al. 1990a; Lynch et al. 1990b; Potter and McMichael 1986; Rossi et al. 2018; Anderson et al. 2019).

12.5 Prognosis

The prognosis for colorectal cancer is dependent on the stage of cancer at diagnosis, general health, colon blockage, and presence of carcinoembryonic antigens. The earlier the cancer is detected, the better a patient's prognosis will be, as later stage CRC is highly aggressive (Mahar et al. 2017). When colon cancer has a late stage diagnosis, the breadth of lymph nodes involved and number of organs the cancer has spread to will lengthen the treatment plan, and negatively impact prognosis (Wu et al. 2019; Kim et al. 1976). A patient's general health affects their ability to tolerate treatment, and in many cases healthier patients are able to deal with treatment side effects better. Colon cancer can cause a blockage of the colon and impact a patient's outlook. Carcinoembryonic antigen (CEA) is a protein molecule in the blood that increases when colon cancer is present and may affect how well a patient responds to treatment. The prognosis for colorectal cancer is still poor despite recent advancements in chemotherapeutic treatment options (Yasunaga et al. 2019).

12.6 Immunotherapy

Cancer immunotherapy are treatments designed to prompt a patient's immune system to recognize cancer cells. Major emerging cancer immunotherapies include cancer vaccines, immune checkpoint inhibitors, and adoptive immunotherapy. Immune checkpoints are often ignored by cancer cells, this increases their chances of evading the immune system. Immune checkpoint inhibitors can lead to the immune system regaining tumor immunity (Singh et al. 2015). Cancer vaccines involve sensitizing the immune system to assist in the defense against cancer cell invasion. In adoptive immunotherapy or adoptive T cell therapy (ACT), T cells are extracted, multiplied to larger quantities and transferred back into patients. Clinical attempts for various immunotherapies in relation to CRC are highlighted below.

12.6.1 Clinical Trials

Several clinical trials have been initiated and conducted for immunotherapy in CRC patients. A total of 190 studies were registered in *clinicalstrials.gov*. These trials are either completed (41 studies) or currently recruiting patients (77). In the United States alone, 110 studies were recorded (completed: 27; recruiting: 53). The details of the number of studies on vaccines and PD-1 inhibitors with status (completed or recruiting) and at various Phases (I to IV) are presented in the links below and in Table 12.1.

12.6.2 Total Studies

https://clinicaltrials.gov/ct2/results?cond=Colo-rectal+Cancer&term=immunotherapy&cntry=&state=&city=&dist=&Search=Search&flds=fk

12.6.3 Clinical Trials Registered in the United States

https://clinicaltrials.gov/ct2/results?cond=Colo-rectal+Cancer&term=immunotherapy&cntry=US&state=&city=&dist=&Search=Search&flds=fk

Table 12.1 Data and links for showing the details of clinical trials (data from November, 2019; clinicaltrials.gov)

Immunotherapy, all counties
Total: 190; Phase I: 108; Phase II: 104; Phase III: 17; Phase IV: 0
Recruiting: 77; Phase I: 41; Phase II: 40; Phase III: 5; Phase IV: 0
Completed: 41; Phase I: 27; Phase II: 21; Phase III: 4; Phase IV: 0
https://clinicaltrials.gov/ct2/results?cond=Colo-rectal+Cancer&term=immunotherapy&cntry=&state=&city=&dist=&Search=Search&flds=fk
Immunotherapy, United States
Total: 110; Phase I: 69; Phase II: 50; Phase III: 7; Phase IV: 0
Recruiting: 53; Phase I: 31; Phase II: 20; Phase III: 4; Phase IV: 0
Completed: 27; Phase I: 19; Phase II: 13; Phase III: 2; Phase IV: 0
https://clinicaltrials.gov/ct2/results?cond=Colo-rectal+Cancer&term=immunotherapy&cntry=US&state=&city=&dist=&Search=Search&flds=fk
Vaccine, all counties
Total: 126; Phase I: 76; Phase II: 63; Phase III: 6; Phase IV: 0
Recruiting: 23; Phase I: 15; Phase II: 11; Phase III: 1; Phase IV: 0
Completed: 56; Phase I: 35; Phase II: 26; Phase III: 2; Phase IV: 0
https://clinicaltrials.gov/ct2/results?cond=Colo-rectal+Cancer&term=vaccine&cntry=&state=&city=&dist=&Search=Search&flds=fk
Vaccine, United States
Total: 80; Phase I: 52; Phase II: 32; Phase III: 2; Phase IV: 0
Recruiting: 16; Phase I: 11; Phase II: 9; Phase III: 1; Phase IV: 0
Completed: 39; Phase I: 27; Phase II: 15; Phase III: 0; Phase IV: 0
https://clinicaltrials.gov/ct2/results?cond=Colo-rectal+Cancer&term=vaccine&cntry=US&state=&city=&dist=&Search=Search&flds=fk
PD-1, all counties
Total: 111; Phase I: 70; Phase II: 63; Phase III: 5; Phase IV: 0
Recruiting: 67; Phase I: 42; Phase II: 36; Phase III: 4; Phase IV: 0
Completed: 6; Phase I: 4; Phase II: 4; Phase III: 0; Phase IV: 0
https://clinicaltrials.gov/ct2/results?cond=Colo-rectal+Cancer&term=PD-1&cntry=&state=&city=&dist=&Search=Search&flds=fk
PD-1; United States
Total: 58; Phase I: 45; Phase II: 33; Phase III: 1; Phase IV: 0
Recruiting: 37; Phase I: 30; Phase II: 18; Phase III: 1; Phase IV: 0
Completed: 5; Phase I: 4; Phase II: 4; Phase III: 0; Phase IV: 0
https://clinicaltrials.gov/ct2/results?cond=Colo-rectal+Cancer&term=PD-1&cntry=US&state=&city=&dist=&flds=fk

12.6.4 *Immune Checkpoint Inhibitors*

Immune system checkpoints serve regulatory functions in immune cells by differentiating self from non-self. Immune checkpoints on T cells ensure that it is active in the presence of a pathogen and inactivated in the absence of pathogens in order to prevent attack on normal healthy cells. Cancer cells can avoid the immune system

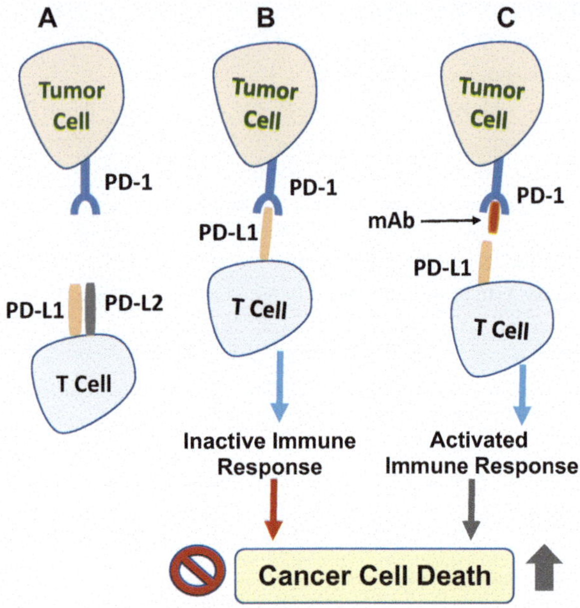

Fig. 12.1 PD-1/PD-L1 pathway and killing of cancer cells. (**a**) The interaction between PD-1 and it's legands (PD-L1 and PD-L2) impacts the T cell mediated immune response for killing cancer cells. (**b**) PD-1 ligands on tumor cells recognize and binds to PD-1 on the surface of T cells. This binding rules out the activation of T cell mediated immune response and fail to induce cancer cell killing. (**c**) monoclonal antibodies (mAbs) used as PD-1 or PD-L1 or PD-L2 inhibitors bind to respective receptor or ligand, activate T cell mediated immune response to kill cancer cells

response by using proteins to interact with the checkpoints, effectively masking themselves from T cell interference. Immune checkpoint inhibitors are drugs that are designed to disrupt the interaction between cancer cells and the checkpoints on T cells, allowing the T cell to recognize the cancer and proceed with the immune response (Fig. 12.1). Current investigational immune checkpoint inhibitors are PD-1/PD-L1 inhibitors and indoleamine 2, 3-dioxygenase (IDO). IDO represses the activity of both natural killer (NK) cells and CD8+ T effector cells (Prendergast et al. 2018). PD-1 and PD-L1 define a specific receptor-ligand interaction between T-cells and cancer cells.

12.6.5 Clinical Trials Using PD-1 Inhibitors

Multiple clinical trials have been initiated and conducted to test PD-1 inhibitors in CRC patients. A total of 111 studies were registered in *clinicalstrials.gov*. These trials are either completed (6 studies) or currently recruiting patients (67). In the United States alone, 58 studies were recorded (completed: 5, Table 12.2; recruiting: 37; Table 12.3). The details of the clinical trials recruiting or completed and the status in various Phases (I to IV) are presented in the links below and in Table 12.1.

Table 12.2 Completed clinical trials in United States with PD-1 inhibitor (data from November, 2019; clinicaltrials.gov)

	NCT Number	Clinical Trial Title	Phase
1	NCT02298946	AMP-224, a PD-1 Inhibitor, With Stereotactic Body Radiation Therapy in Metastatic Colorectal Cancer	Phase 1
2	NCT02335918	A Dose Escalation and Cohort Expansion Study of Anti-CD27 (Varlilumab) and Anti-PD-1 (Nivolumab) in Advanced Refrac tory Solid Tumors	Phase 1 Phase 2
3	NCT02860546	A Study Evaluating TAS-102 Plus Nivolumab in Patients With MSS CRC	Phase 2
4	NCT02280811	T Cell Receptor Immunotherapy Targeting HPV-16 E6 for HPV-Associated Cancers	Phase 1 Phase 2
5	NCT03361228	A Study to Evaluate the Safety, Tolerability, and Antitumor Activity of INCB001158 Plus Epacadostat, With or Without Pembrolizumab, in Advanced Solid Tumors	Phase 1 Phase 2

12.6.6 Total Studies

https://clinicaltrials.gov/ct2/results?cond=Colo-rectal+Cancer&term=PD-1&cntry=&state=&city=&dist=&Search=Search&flds=fk

12.6.7 Clinical Trials Registered in the United States

https://clinicaltrials.gov/ct2/results?cond=Colo-rectal+Cancer&term=PD-1&cntry=US&state=&city=&dist=&flds=fk

12.6.8 Cancer Vaccines

Vaccines work by introducing an attenuated or inactivated pathogenic molecule into the body to act as an antibody-stimulating antigen without inducing the disease itself. Vaccines also exist for viruses that can lead to cancers, for example, some head and neck cancers are associated with human papillomavirus (HPV) 16 infection (Marur et al. 2010; Perica et al. 2015). While these vaccines are useful in certain circumstances, most cancers are not infection related.

Vaccine cancer immunotherapies can include whole-tumor-cell vaccines or autologous vaccines, viral based vaccines, peptide vaccines, and dendritic cell vaccines (Singh et al. 2015). Autologous vaccines take tumor antigens from a patient and reintroduce it in order to better acclimate the body to the invading cells. Peptide

Table 12.3 Clinical trials recruiting CRC patients in the United States with PD-1 inhibitor (data from November, 2019; clinicaltrials.gov). The search on clinictrials.gov showed 37 studies registered, however the Table has the studies related to colorectal cancer and the studies in Phase 1 to Phase 4

	NCT Number	Clinical Trial Title	Phase
1	NCT03711058	Study of PI3Kinase Inhibition (Copanlisib) and Anti- PD-1 Antibody Nivolumab in Relapsed/Refractory Solid Tumors With Expansions in Mismatch-repair Proficient (MSS) Colo rectal Cancer	Phase 1 Phase 2
2	NCT03642067	Study of Nivolumab and Relatlimab in Patients With Micro satellite Stable (MSS) Advanced Colorectal Cancer	Phase 2
3	NCT03250832	Study of TSR-033 With an Anti-PD-1 in Patients With Advanced Solid Tumors	Phase 1
4	NCT03531632	MGD007 Combined With MGA012 in Relapsed/ Refractory Metastatic Colorectal Cancer	Phase 1 Phase 2
5	NCT03206073	A Phase I/II Study of Pexa-Vec Oncolytic Virus in Combina tion With Immune Checkpoint Inhibition in Refractory Colo rectal Cancer	Phase 1 Phase 2
6	NCT02837263	PI Pembro in Combination With Stereotactic Body Radiother apy for Liver Metastatic Colorectal Cancer	Phase 1
7	NCT03785210	Nivolumab (Anti-PD1), Tadalafil and Oral Vancomycin in People With Refractory Primary Hepatocellular Carcinoma or Liver Dominant Metastatic Cancer From Colorectal or Pancre atic Cancers	Phase 2
8	NCT03891953	Study of Safety and Efficacy of DKY709 Alone or in Combi nation With PDR001 in Patients With Advanced Solid Tumors	Phase 1
9	NCT03374254	Safety and Efficacy of Pembrolizumab (MK-3475) Plus Binimetinib Alone or Pembrolizumab Plus Chemotherapy With or Without Binimetinib in Metastatic Colorectal Cancer (mCRC) Participants (MK-3475-651)	Phase 1
10	NCT04046445	Phase 1b Study to Evaluate ATP128, With or Without BI 754091, in Patients With Stage IV Colorectal Cancer	Phase 1 Phase 2
11	NCT02636036	Phase I Study of Enadenotucirev and PD-1 Inhibitor in Sub jects With Metastatic or Advanced Epithelial Tumors	Phase 1
12	NCT03228667	QUILT-3.055: A Study of ALT-803 in Combination With PD-1/PD-L1 Checkpoint Inhibitor in Patients With Advanced Cancer	Phase 2
13	NCT03775850	A Study of EDP1503 in Patients With Colorectal Cancer, Breast Cancer, and Checkpoint Inhibitor Relapsed Tumors	Phase 1 Phase 2
14	NCT02890069	A Study of PDR001 in Combination With LCL161, Everolimus or Panobinostat	Phase 1
15	NCT03953235	A Study of a Personalized Cancer Vaccine Targeting Shared Neoantigens	Phase 1 Phase 2
16	NCT03829436	TPST-1120 as Monotherapy and in Combination With (Nivolumab, Docetaxel or Cetuximab) in Subjects With Advanced Cancers	Phase 1
17	NCT03948763	A Study of mRNA-5671/V941 as Monotherapy and in Com bination With Pembrolizumab (V941-001)	Phase 1

(continued)

Table 12.3 (continued)

	NCT Number	Clinical Trial Title	Phase
18	NCT03639714	A Study of a Personalized Neoantigen Cancer Vaccine	Phase 1 Phase 2
19	NCT02903914	Arginase Inhibitor INCB001158 as a Single Agent and in Combination With Immune Checkpoint Therapy in Patients With Advanced/Metastatic Solid Tumors	Phase 1 Phase 2
20	NCT02817633	A Phase 1 Study of TSR-022, an Anti-TIM-3 Monoclonal Antibody, in Patients With Advanced Solid Tumors (AMBER)	Phase 1
21	NCT03454451	CPI-006 Alone and in Combination With CPI-444 and With Pembrolizumab for Patients With Advanced Cancers	Phase 1
22	NCT02947165	Phase I/Ib Study of NIS793 in Combination With PDR001 in Patients With Advanced Malignancies	Phase 1
23	NCT03629756	A Study to Evaluate the Safety and Tolerability of Immuno therapy Combinations in Participants With Advanced Malignancies	Phase 1
24	NCT03628677	A Study to Evaluate the Safety and Tolerability of AB154 in Participants With Advanced Malignancies	Phase 1
25	NCT02997228	Combination Chemotherapy, Bevacizumab, and/ or Atezolizumab in Treating Patients With Deficient DNA Mismatch Repair Metastatic Colorectal Cancer	Phase 3
26	NCT02834052	Pembrolizumab + Poly-ICLC in MRP Colon Cancer	Phase 1 Phase 2
27	NCT03761914	Galinpepimut-S in Combination With Pembrolizumab in Patients With Selected Advanced Cancers	Phase 1 Phase 2
28	NCT03549000	A Phase I/Ib Study of NZV930 Alone and in Combination With PDR001 and /or NIR178 in Patients With Advanced Malignancies	Phase 1
29	NCT03170960	Study of Cabozantinib in Combination With Atezolizumab to Subjects With Locally Advanced or Metastatic Solid Tumors	Phase 1 Phase 2
30	NCT03797326	Efficacy and Safety of Pembrolizumab (MK-3475) Plus Lenvatinib (E7080/MK-7902) in Previously Treated Partici pants With Select Solid Tumors (MK-7902-005/E7080-G000-224/LEAP-005)	Phase 2

vaccines utilize sequences of amino acids as tumor antigens alongside a vaccine adjuvant (Singh et al. 2015). Viral based vaccines incorporate peptide vaccines into a viral vector to increase the immunogenicity of the vaccine and elicit an immunologic response (Singh et al. 2015). Dendritic cell vaccines aim to load dendritic cells (DCs) with antigens. DC vaccines use varying methods to deliver tumor associated antigens (TAAs) to DCs through tumor cell lysates (Nestle et al. 1998), tumor RNA (Koido et al. 2000), and synthetic peptides (Celluzzi et al. 1996). DCs recognize pathogenic antigens and work with T cells so that an immune response can be initiated.

12.7 Vaccine Immunotherapy for CRC Patients

Vaccines are used in cancer for both inhibiting tumor growth and protecting against recurrence of the disease. Depending on the source and characteristics, different types of vaccines are tested or are under testing for CRC (Kalyan et al. 2018).

12.7.1 Autologous Vaccines

These vaccines are developed from the autologous tumor cells extracted from the patient's tumor. The autologous cells are comprised of tumor associated antigens and induce adaptive immunity. These vaccines are designed to disrupt the tumor cells evasion of the immune response (Mosolits et al. 2005). One of the limitations of this type of vaccine is lack of specificity. Many of the targeted antigens in this process also present in non-malignant cells (Klebanoff et al. 2011). To overcome the shortfalls comprehended with autologous tumor cell vaccines, a few options have been attempted for CRC.

Bacillus Calmette–Guérin vaccine (BCG), has been tested in combination with an autologous vaccine in order to improve the specificity. A phase III clinical trial (randomized) was conducted at multi-institutions in order to gauge the efficacy of the BCG and patient-specific vaccines generated with autologous tumor cells (Vermorken et al. 1999). Even though this trial didn't show a significant response (in disease free or overall survival) in median follow up among CRC patients, a long follow-up analysis demonstrated a notable improvement in overall, event-free, recurrence-free interval as well as recurrence-free survival in stage II patients (Vermorken et al. 1999; Uyl-de Groot et al. 2005).

In a different approach, a non-lytic, Newcastle disease virus (NDV)'s low pathogenic strain was tested for augmentation of the immunogenicity of autologous vaccines. A phase II clinical trial involving metastatic CRC patients (Schlag et al. 1992), and a phase III (randomized) trial participated by liver metastatic patients with both colon and rectal cancers (Schulze et al. 2009) showed no remarkable improvement in either overall or event (metastasis)-free survival between control and vaccinated groups. Detailed analyses revealed some benefit in the patients with colon cancer in relation to patients with rectal cancer.

The clinical trials to determine benefits of autologous vaccines revealed only limited advantages in CRC. Detailed analyses establishing some benefit in stage II CRC patients versus the stage III patients participating in clinical tails. Furthermore, autologous vaccines seem to be beneficial in a sub-group of patients. The precise reasons for such results have not yet been evaluated in CRC. Due to such limitations, autologous vaccines are not currently popular in CRC and significant improvements in specificity and better application among all sub-groups will be needed.

Clinical trials: Multiple clinical trials have been initiated and conducted to test vaccines in CRC patients. A total of 126 studies were registered in *clinicalstrials.*

gov. These trials are either completed (56 studies) or currently recruiting patients (23). In the United States alone, 80 studies were recorded (completed: 39, Table 12.4; recruiting: 16; Table 12.5). The details of the clinical trials, recruiting or completed, and the status in various Phases (I to IV) are presented in the links below and in Table 12.1.

12.7.2 Total Studies

https://clinicaltrials.gov/ct2/results?cond=Colo-rectal+Cancer&term=vaccine&cntry=&state=&city=&dist=&Search=Search&flds=fk

12.7.3 Clinical Trials Registered in the United States

https://clinicaltrials.gov/ct2/results?cond=Colo-rectal+Cancer&term=vaccine&cntry=US&state=&city=&dist=&Search=Search&flds=fk

12.7.4 Adoptive Immunotherapy

Adoptive immunotherapy or adoptive T cell therapy (ACT) involves the utilization of cytotoxic T cells (CTLs), obtained from patients and grown ex-vivo, to promote a more effective immune response. This approach infuses CTLs that are tumor specific into patients so that they may target and destroy tumor cells (Perica et al. 2015). CTLs in the body bind to specific antigens and induce apoptosis of the target cell. In order for cancer cells to effectively elude the immune system they must mask themselves from CTLs and other immune components. The issue with ACT is specificity. Producing T cells in bulk that lack useful in vivo characteristics and then transferring them back into patients is not as effective as treatments that involve highly specific T cells (Carvalho et al. 2016). Chimeric antigen receptor (CAR) T cell therapy involves the use of synthetic T cells to target specific receptors. Another specific ACT is known as T cell receptor transfer (TCR). TCR is similar to CAR in that it involves the removal and modification of T cells. The hypervariable regions of the αβ TCR contact the peptide and the major histocompatibility complex (Perica et al. 2015). This allows the T cells to bind more accurately to cancer cells, potentially avoiding any off-target binding.

Table 12.4 Clinical trials completed in the United States with vaccines (data from November, 2019; clinicaltrials.gov): The search on clinictrials.gov showed 39 studies registered, however the Table has the studies related to colorectal cancer and the studies in Phase 1 to Phase 4

	NCT Number	Clinical Trial Title	Phase
1	NCT00558051	Alpha-type-1 Dendritic Cell-based Vaccines in Patients With Metastatic Colorectal Cancer	Phase 1
2	NCT00019591	Vaccine Therapy With or Without Interleukin-2 in Treating Patients With Locally Advanced or Metastatic Colorectal Cancer	Phase 1 Phase 2
3	NCT00773097	Study of the MUC1 Peptide-Poly-ICLC Adjuvant Vaccine in Individuals With Advanced Colorectal Adenoma	Phase 2
4	NCT00656123	Study of Colon GVAX and Cyclophosphamide in Patients With Metastatic Colorectal Cancer	Phase 1
5	NCT02981524	Study of GVAX (With CY) and Pembrolizumab in MMR-p Advanced Colorectal Cancer	Phase 2
6	NCT01671592	Safety of Labeled Dendritic Cell (DC) Vaccines and Feasibil ity of Tracking by Magnetic Resonance Imaging (MRI)	Phase 1
7	NCT01890213	Immunotherapy With CEA(6D) VRP Vaccine (AVX701) in Patients With Stage III Colorectal Cancer	Phase 1
8	NCT00016133	Vaccine Therapy in Treating Patients With Stage II or Stage III Colon Cancer That Has Been Removed During Surgery	Phase 1 Phase 2
9	NCT00924092	An Open Label Phase I Study to Eval the Safety and Tolera bility of a Vaccine (GI-6207) Consisting of Whole, Heat-killed Recombinant Saccharomyces Cerevisiae (Yeast) Genetically Modified to Express CEA Protein in Adults With Metastatic CEA- expressing	Phase 1
10	NCT00529984	A Phase I/II Study With CEA(6D) VRP Vaccine in Patients With Advanced or Metastatic CEA- Expressing Malignancies	Phase 1 Phase 2
11	NCT00088413	PANVAC-V and PANVAC-F Vaccines Plus Sargramostim to Treat Advanced Cancer	Phase: Phase 1 Phase 2
12	NCT00033748	Vaccine Therapy in Treating Patients With Colorectal Cancer Metastatic to the Liver	Phase 2
13	NCT00648102	Phase I Study of CDX-1307, hCG-B Vaccine, for Patients With Incurable, Locally Advanced or Metastatic Breast, Colorectal, Pancreatic, Bladder or Ovarian Cancer	Phase 1
14	NCT00003433	Immunotherapy in Treating Patients With Resected Liver Metastases From Colon Cancer	Phase 1 Phase 2
15	NCT00103142	Vaccine Therapy in Treating Patients With Liver or Lung Metastases From Colorectal Cancer	Phase 2
16	NCT00020267	Vaccine Therapy in Treating Patients With Metastatic Cancer	Phase 1
17	NCT00709462	A Study of CDX-1307, in Patients With Incurable Breast, Colorectal, Pancreatic, Ovarian or Bladder Cancer (CDX 1307-01)	Phase 1
18	NCT00128622	Denileukin Diftitox Followed by Vaccine Therapy in Treating Patients With Metastatic Cancer	Phase 1
19	NCT00054912	An Open Label Study of a Peptide Vaccine in Patients With Stage III Colon Cancer	Phase 1

(continued)

Table 12.4 (continued)

	NCT Number	Clinical Trial Title	Phase
20	NCT00019084	Vaccine Therapy and Biological Therapy in Treating Patients With Advanced Cancer	Phase 2
21	NCT00019331	Vaccine Therapy Plus Biological Therapy in Treating Adults With Metastatic Solid Tumors	Phase 2
22	NCT00027534	Vaccine Therapy in Treating Patients With Advanced or Met astatic Cancer	Phase 1
23	NCT00002475	Cyclophosphamide Plus Vaccine Therapy in Treating Patients With Advanced Cancer	Phase 2
24	NCT00019006	Vaccine Therapy in Treating Patients With Colon, Pancreatic, or Lung Cancer	Phase 1
25	NCT00004604	Biological Therapy in Treating Patients With Metastatic Cancer	Phase 1
26	NCT01191684	Vaccine Therapy in Treating Patients With Colorectal, Stom ach, or Pancreatic Cancer	Phase 1
27	NCT01519817	Cancer Vaccine Targeting Brachyury Protein in Tumors	Phase 1
28	NCT01972737	Phase I Study of Ad5-hGCC (Human Guanylyl Cyclase C)-PADRE in Stage I/II Colon Cancer	Phase 1
29	NCT01147965	Active Immunotherapy CEA Vaccine in Patients With Malig nancies Expressing CEA	Phase 1 Phase 2
30	NCT01522820	Vaccine Therapy With or Without Sirolimus in Treating Patients With NY-ESO-1 Expressing Solid Tumors	Phase 1
31	NCT00028496	Vaccine Therapy With or Without Sargramostim in Treating Patients With Advanced or Metastatic Cancer	Phase 1
32	NCT00861107	In-Situ Therapeutic Cancer Vaccine for Metastatic Cancer Combining AlloStim With Tumor Cryoablation	Phase 1 Phase 2

12.7.5 Non-specific Immunotherapies

Non-specific immunotherapies target cancer cells indirectly by activating the general immune system. Some non-specific immunotherapies include cytokines, which are secreted proteins that can help direct other immune system components to assist with the normal immune response depending on which cells bind to the cytokines. Cytokines are a part of the bodies normal innate and adaptive immune response and activate in response to recognition of a specific antigen.

Interferons (IFNs) are involved in the regulation of cancer cell growth and metastasis (Di Franco et al. 2017). Interferon-alpha (IFN-alpha) is used to treat some cancers, IFN-α exhibits high anti-proliferative ability on a variety of tumor cells (Maher et al. 2008; Palmer et al. 2000).

Immunotherapy involving interleukin (IL)-2 is being used more frequently as a treatment method for patients with different cancers (Eastman et al. 2001). Interleukins are cytokines that use chemical signaling to direct and regulate cellular behavior. Interleukins, namely IL-2, can be combined with other cancer treatments in order to enhance the immune systems actions against tumor antigens.

Table 12.5 Clinical recruiting CRC patients in the United States for clinical trials with vaccines (data from November, 2019; clinicaltrials.gov): Table has the studies related to colorectal cancer and the studies in Phase 1 to Phase 4

	NCT Number	Clinical Trial Title	Phase
1	NCT03730948	DC Vaccine in Colorectal Cancer	Phase 1
2	NCT01952730	GVAX for Colorectal Cancer	Phase 1
3	NCT03050814	Standard of Care Alone or in Combination With Ad-CEA Vaccine and Avelumab in People With Previously Untreated Metastatic Colorectal Cancer QUILT-2.004	Phase 2
4	NCT03547999	A Trial of Perioperative CV301 Vaccination in Combination With Nivolumab and Systemic Chemotherapy for Metastatic CRC	Phase 2
5	NCT03953235	A Study of a Personalized Cancer Vaccine Targeting Shared Neoantigens	Phase 1 Phase 2
6	NCT03552718	QUILT-2.025 NANT Neoepitope Yeast Vaccine (YE-NEO-001): Adjuvant Immunotherapy Using a Personal ized Neoepitope Yeast-Based Vaccine To Induce T-Cell Responses In Subjects W/ Previously Treated Cancers	Phase 1
7	NCT03639714	A Study of a Personalized Neoantigen Cancer Vaccine	Phase 1 Phase 2
8	NCT04046445	Phase 1b Study to Evaluate ATP128, With or Without BI 754091, in Patients With Stage IV Colorectal Cancer	Phase 1 Phase 2
9	NCT02865135	Trial To Test Safety And Efficacy Of Vaccination For Incur able HPV 16-Related Oropharyngeal, Cervical And Anal Cancer	Phase 1 Phase 2
10	NCT03827967	Trial of PalloV-CC in Colon Cancer	Phase 1
11	NCT03761914	Galinpepimut-S in Combination With Pembrolizumab in Patients With Selected Advanced Cancers	Phase 1 Phase 2
12	NCT02448173	A Multicenter Study of Active Specific Immunotherapy With OncoVax® in Patients With Stage II Colon Cancer	Phase 3
13	NCT03289962	A Study of RO7198457 as a Single Agent and in Combination With Atezolizumab in Participants With Locally Advanced or Metastatic Tumors	Phase 1
14	NCT03311334	A Study of DSP-7888 Dosing Emulsion in Combination With Immune Checkpoint Inhibitors in Adult Subjects With Advanced Solid Tumors	Phase 1 Phase 2

12.7.6 Monoclonal Antibodies (mAbs)

Monoclonal Antibodies can be used in cancer immunotherapy to activate the patient's immune response. This is done by designing antibodies that can target a specific cancer antigen. Antibodies recognize a specific epitope and bind to it, initiating the immune response and allowing the recruitment of other immune cells. In cancer immunotherapy monoclonal antibodies are specific to proteins and antigens of T cells located at the surface of cancer cells (Corraliza-Gorjon et al. 2017).

Monoclonal antibodies include naked, conjugated, and bispecific mAbs (BsmAbs). Naked mAbs work with no additional material attached to them. If the antibodies alone aren't effective against the cancer cells, conjugated mAbs are used in combination with chemotherapy drugs or in conjunction with radiation to increase their anti-tumor effects (Corraliza-Gorjon et al. 2017). Antibodies that contain two antigen binding sites on a single molecule are deemed bispecific (Sedykh et al. 2018). BsmAbs allow for the binding of receptors that activate CTLs on one antigen binding site, and cancer antigen binding on another.

12.8 Recent Developments and Future Perspectives

Recently, researchers have been investigating CRC cell antigens that are more specific to the tumor tissue in order to develop more effective mAbs while protecting the safety of the patient (Myers 2018; Terme et al. 2013). Expression analysis of colorectal tumor cells and their normal counterpart, mucoepithelial cells, is necessary to find specific tumor molecules. TMEM180, among others, is a recently discovered colorectal tumor specific antigen, and its antibodies were found in myeloma and lymph node cells from rats (Terme et al. 2013). The anti-TMEM180 mAb has shown to be highly effective in treatment while showing little to no toxicity to mucoepithelial cell lines that were given anti-TMEM180 mAb (Terme et al. 2013).

Another gene CLDN1 has been recently discovered which can serve as a biomarker in metastatic colorectal cancer (Cherradi et al. 2017). A mAb that inhibits expression of CLDN1 could have immunotherapeutic potential for metastatic colorectal cancer. To avoid negative side effects of immunotherapies such as checkpoint inhibition, researchers are interested in understanding the molecular composition of each patient's colorectal cancer in order to provide the most effective and safe immunotherapies.

Future perspectives of colorectal cancer immunotherapy include patient-tailored approaches that move away from non-specific engagement of the body's immune system (Myers 2018; Terme et al. 2013). Researchers are looking for more colorectal cancer biomarkers that are specific only to a patient's tumor tissue, so that prevention of off-site damage is reduced in the body. With the recent discoveries of some colorectal cancer biomarkers, clinical trials ensuring the safety of immunotherapies recently discovered, such as anti-TMEM180 mAbs, should be seen in the future (Arora et al. 2017).

12.9 Conclusion

The third most cancer related deaths in men and women can be attributed to colorectal cancer. The prognosis for colorectal cancer, regardless of recent advancements in treatment, is poor. Immunotherapy is a novel treatment option that utilizes the anti-tumor effects of the immune system in conjunction with other therapies against cancer. Cancer immunotherapy is an emerging state-of-the-art strategy to boost the immune response in order to target malignant cells. In an adoptive immunotherapy strategy, anti-tumor lymphocytes obtained from patients or healthy individuals are employed in adoptive cell transfer to trigger tumor cell killing (McCloskey et al. 2018). In another approach, vaccines are also used for immune system activation. T cell responses were activated through the antigen-presenting cells (Lohmueller and Finn 2017) providing anti-cancer activity to T cells. Immunotherapies can be dangerous if not managed to manipulate the immune system in such a way to target malignant cells. Therefore, researchers are trying to combat CRC by focusing on finding tumor-specific biomarkers to utilize in cancer immunotherapy. Some of the strategies attempted using immunotherapy approaches show promising results which has led to the advancement of several of these trials to further stages. Several immunotherapeutic approaches are generating excellent results in pre-clinical screening. The PD-1/PD-L1 immune checkpoint pathway that is responsible for negatively regulating T-cell mediated immune response has been exploited for developing therapeutic application in several cancers including CRC. The strategy to block the contact of PD-1/PD-L1 is showing only limited effects in CRC, inhibitors of ligands L1 and L2 may have better implications in CRC therapy. The discovery and testing of mAbs against CLDN1 and TMEM180 expands researcher's comprehension on CRC and its targets for therapy. While a number of approaches are in early to late phases in clinical trials, many clinical trials are in the pipeline. Future studies should warrant more clinical efficiency based off of how the gaps in knowledge concerning immunotherapy are addressed. More research is needed to develop safe but efficient immunotherapies for CRC.

References

Altman V, Mann N (1948) Metastisizing carcinoid tumor of the appendix and cecum. Am J Surg 76 (4):434–439

Anderson AS et al (2019) Lifestyle in patients at increased risk of colorectal cancer. J Hum Nutr Diet 32(5):570–577

Antelo M et al (2019) Lynch-like syndrome is as frequent as Lynch syndrome in early-onset nonfamilial nonpolyposis colorectal cancer. Int J Cancer 145(3):705–713

Arnold M et al (2017) Global patterns and trends in colorectal cancer incidence and mortality. Gut 66(4):683–691

Arora N, Gupta A, Singh PP (2017) Biological agents in gastrointestinal cancers: adverse effects and their management. J Gastrointest Oncol 8(3):485–498

Bat L et al (1992) Colonoscopy in patients aged 80 years or older and its contribution to the evaluation of rectal bleeding. Postgrad Med J 68(799):355–358
Blackledge G et al (1979) A study of gastro-intestinal lymphoma. Clin Oncol 5(3):209–219
Bond JH (2002) Fecal occult blood test screening for colorectal cancer. Gastrointest Endosc Clin N Am 12(1):11–21
Carethers JM, Stoffel EM (2015) Lynch syndrome and Lynch syndrome mimics: the growing complex landscape of hereditary colon cancer. World J Gastroenterol 21(31):9253–9261
Carroll MR, Seaman HE, Halloran SP (2014) Tests and investigations for colorectal cancer screening. Clin Biochem 47(10–11):921–939
Carvalho S et al (2016) Immunotherapy of cancer: from monoclonal to oligoclonal cocktails of anti-cancer antibodies: IUPHAR review 18. Br J Pharmacol 173(9):1407–1424
Celluzzi CM et al (1996) Peptide-pulsed dendritic cells induce antigen-specific CTL-mediated protective tumor immunity. J Exp Med 183(1):283–287
Cherradi S et al (2017) Antibody targeting of claudin-1 as a potential colorectal cancer therapy. J Exp Clin Cancer Res 36(1):89
Cohen SA, Pritchard CC, Jarvik GP (2019) Lynch syndrome: from screening to diagnosis to treatment in the era of modern molecular oncology. Annu Rev Genomics Hum Genet 20:293–307
Contreary K, Nance FC, Becker WF (1980) Primary lymphoma of the gastrointestinal tract. Ann Surg 191(5):593–598
Corraliza-Gorjon I et al (2017) New strategies using antibody combinations to increase cancer treatment effectiveness. Front Immunol 8:1804
d'Amore F et al (1994) Non-Hodgkin's lymphoma of the gastrointestinal tract: a population-based analysis of incidence, geographic distribution, clinicopathologic presentation features, and prognosis. Danish lymphoma study group. J Clin Oncol 12(8):1673–1684
DeRycke MS et al (2017) Targeted sequencing of 36 known or putative colorectal cancer susceptibility genes. Mol Genet Genomic Med 5(5):553–569
Deveney KE, Way LW (1984) Follow-up of patients with colorectal cancer. Am J Surg 148 (6):717–722
Di Franco S et al (2017) Role of type I and II interferons in colorectal cancer and melanoma. Front Immunol 8:878
Eastman ME et al (2001) Central venous device-related infection and thrombosis in patients treated with moderate dose continuous-infusion interleukin-2. Cancer 91(4):806–814
El-Menyar A, Mekkodathil A, Al-Thani H (2017) Diagnosis and management of gastrointestinal stromal tumors: an up-to-date literature review. J Cancer Res Ther 13(6):889–900
Force USPST (2008) Screening for colorectal cancer: U.S. preventive services task force recommendation statement. Ann Intern Med 149(9):627–637
Force USPST et al (2016) Screening for colorectal cancer: US preventive services task force recommendation statement. JAMA 315(23):2564–2575
Gothlin JH (1988) Modern radiological diagnostics in colorectal cancer. Scand J Gastroenterol Suppl 149:59–61
Harada S, Morlote D (2020) Molecular pathology of colorectal cancer. Adv Anat Pathol 7(1):20–26
Henry CA, Berry RE (1988) Primary lymphoma of the large intestine. Am Surg 54(5):262–266
Herrmann R et al (1980) Gastrointestinal involvement in non-Hodgkin's lymphoma. Cancer 46 (1):215–222
Ilgenfritz HC, Mathews WR (1955) Carcinoid tumors of the gastro-intestinal tract. Ann Surg 141 (6):807–818
Issa IA, Noureddine M (2017) Colorectal cancer screening: an updated review of the available options. World J Gastroenterol 23(28):5086–5096
Kalyan A et al (2018) Updates on immunotherapy for colorectal cancer. J Gastrointest Oncol 9 (1):160–169
Kim US, Papatestas AE, Aufses AH Jr (1976) Prognostic significance of peripheral lymphocyte counts and carcinoembryonic antigens in colorectal carcinoma. J Surg Oncol 8(3):257–262

Klebanoff CA et al (2011) Therapeutic cancer vaccines: are we there yet? Immunol Rev 239 (1):27–44

Koch P et al (2001) Primary gastrointestinal non-Hodgkin's lymphoma: I. anatomic and histologic distribution, clinical features, and survival data of 371 patients registered in the German multicenter study GIT NHL 01/92. J Clin Oncol 19(18):3861–3873

Koelzer VH, Zlobec I, Lugli A (2016) Tumor budding in colorectal cancer–ready for diagnostic practice? Hum Pathol 47(1):4–19

Koido S et al (2000) Induction of antitumor immunity by vaccination of dendritic cells transfected with MUC1 RNA. J Immunol 165(10):5713–5719

Lee DH, Lee JM (2017) Whole-body PET/MRI for colorectal cancer staging: is it the way forward? J Magn Reson Imaging 45(1):21–35

Lee HS et al (2017) Molecular testing for gastrointestinal cancer. J Pathol Transl Med 51 (2):103–121

Levin B et al (2008a) Screening and surveillance for the early detection of colorectal cancer and adenomatous polyps, 2008: a joint guideline from the American Cancer Society, the US multi-society task force on colorectal cancer, and the American College of Radiology. CA Cancer J Clin 58(3):130–160

Levin B et al (2008b) Screening and surveillance for the early detection of colorectal cancer and adenomatous polyps, 2008: a joint guideline from the American Cancer Society, the US multi-society task force on colorectal cancer, and the American College of Radiology. Gastroenterology 134(5):1570–1595

Li J et al (2016) Point mutations in exon 1B of APC reveal gastric adenocarcinoma and proximal polyposis of the stomach as a familial adenomatous polyposis variant. Am J Hum Genet 98 (5):830–842

Liu B et al (1996) Analysis of mismatch repair genes in hereditary non-polyposis colorectal cancer patients. Nat Med 2(2):169–174

Lohmueller J, Finn OJ (2017) Current modalities in cancer immunotherapy: Immunomodulatory antibodies, CARs and vaccines. Pharmacol Ther 178:31–47

Lugli A et al (2017) Recommendations for reporting tumor budding in colorectal cancer based on the international tumor budding consensus conference (ITBCC) 2016. Mod Pathol 30 (9):1299–1311

Lynch HT, Smyrk T (1996) Hereditary nonpolyposis colorectal cancer (Lynch syndrome). An updated review. Cancer 78(6):1149–1167

Lynch HT et al (1990a) Phenotypic variation in colorectal adenoma/cancer expression in two families. Hereditary flat adenoma syndrome. Cancer 66(5):909–915

Lynch HT et al (1990b) Genetic diagnosis of Lynch syndrome II in an extended colorectal cancer-prone family. Cancer 66(10):2233–2238

Machi J et al (1987) Intraoperative ultrasonography in screening for liver metastases from colorectal cancer: comparative accuracy with traditional procedures. Surgery 101(6):678–684

Mahar AL et al (2017) Personalizing prognosis in colorectal cancer: a systematic review of the quality and nature of clinical prognostic tools for survival outcomes. J Surg Oncol 116 (8):969–982

Maher SG et al (2008) IFNalpha and IFNlambda differ in their antiproliferative effects and duration of JAK/STAT signaling activity. Cancer Biol Ther 7(7):1109–1115

Marur S et al (2010) HPV-associated head and neck cancer: a virus-related cancer epidemic. Lancet Oncol 11(8):781–789

Matsui O et al (1987) Liver metastases from colorectal cancers: detection with CT during arterial portography. Radiology 165(1):65–69

Mazzola P et al (2008) Epidemiology and molecular biology of gastrointestinal stromal tumors (GISTs): a population-based study in the south of Switzerland, 1999-2005. Histol Histopathol 23(11):1379–1386

McCloskey CW et al (2018) Ovarian cancer immunotherapy: preclinical models and emerging therapeutics. Cancers (Basel) 10(8):244

Min KW, Leabu M (2006) Interstitial cells of Cajal (ICC) and gastrointestinal stromal tumor (GIST): facts, speculations, and myths. J Cell Mol Med 10(4):995–1013

Mosolits S, Nilsson B, Mellstedt H (2005) Towards therapeutic vaccines for colorectal carcinoma: a review of clinical trials. Expert Rev Vaccines 4(3):329–350

Muto T, Bussey HJ, Morson BC (1975) The evolution of cancer of the colon and rectum. Cancer 36 (6):2251–2270

Myers G (2018) Immune-related adverse events of immune checkpoint inhibitors: a brief review. Curr Oncol 25(5):342–347

Nasseri Y, Langenfeld SJ (2017) Imaging for colorectal cancer. Surg Clin North Am 97(3):503–513

Nestle FO et al (1998) Vaccination of melanoma patients with peptide- or tumor lysate-pulsed dendritic cells. Nat Med 4(3):328–332

Northover JM, Murday V (1989) Familial colorectal cancer and familial adenomatous polyposis. Baillieres Clin Gastroenterol 3(3):593–613

Palmer KJ et al (2000) Interferon-alpha (IFN-alpha) stimulates anti-melanoma cytotoxic T lymphocyte (CTL) generation in mixed lymphocyte tumour cultures (MLTC). Clin Exp Immunol 119(3):412–418

Perica K et al (2015) Adoptive T cell immunotherapy for cancer. Rambam Maimonides Med J 6(1): e0004

Potter JD, McMichael AJ (1986) Diet and cancer of the colon and rectum: a case-control study. J Natl Cancer Inst 76(4):557–569

Prendergast GC et al (2018) Indoleamine 2,3-Dioxygenase and its therapeutic inhibition in cancer. Int Rev Cell Mol Biol 336:175–203

Raut CP, Morgan JA, Ashley SW (2007) Current issues in gastrointestinal stromal tumors: incidence, molecular biology, and contemporary treatment of localized and advanced disease. Curr Opin Gastroenterol 23(2):149–158

Rossi M et al (2018) Colorectal cancer and alcohol consumption-populations to molecules. Cancers (Basel) 10(2):38

Schlag P et al (1992) Active specific immunotherapy with Newcastle-disease-virus-modified autologous tumor cells following resection of liver metastases in colorectal cancer. First evaluation of clinical response of a phase II-trial. Cancer Immunol Immunother 35(5):325–330

Schulze T et al (2009) Efficiency of adjuvant active specific immunization with Newcastle disease virus modified tumor cells in colorectal cancer patients following resection of liver metastases: results of a prospective randomized trial. Cancer Immunol Immunother 58(1):61–69

Sedykh SE et al (2018) Bispecific antibodies: design, therapy, perspectives. Drug Des Devel Ther 12:195–208

Segnan N et al (2005) Randomized trial of different screening strategies for colorectal cancer: patient response and detection rates. J Natl Cancer Inst 97(5):347–357

Sepulveda AR et al (2017) Molecular biomarkers for the evaluation of colorectal cancer: guideline from the American Society for Clinical Pathology, College of American Pathologists, Association for Molecular Pathology, and the American Society of Clinical Oncology. J Clin Oncol 35 (13):1453–1486

Shinya H, Wolff WI (1979) Morphology, anatomic distribution and cancer potential of colonic polyps. Ann Surg 190(6):679–683

Siegel RL et al (2017) Colorectal cancer statistics, 2017. CA Cancer J Clin 67(3):177–193

Siegel RL, Miller KD, Jemal A (2019) Cancer statistics, 2019. CA Cancer J Clin 69(1):7–34

Singh PP et al (2015) Immune checkpoints and immunotherapy for colorectal cancer. Gastroenterol Rep (Oxf) 3(4):289–297

Swerdlow SH et al (2016) The 2016 revision of the world health organization classification of lymphoid neoplasms. Blood 127(20):2375–2390

Terme M et al (2013) VEGFA-VEGFR pathway blockade inhibits tumor-induced regulatory T-cell proliferation in colorectal cancer. Cancer Res 73(2):539–549

Trombold J, Farmer RW, McCafferty M (2013) The impact of colorectal cancer screening in a veteran hospital population. Am Surg 79(3):296–300

Uyl-de Groot CA et al (2005) Immunotherapy with autologous tumor cell-BCG vaccine in patients with colon cancer: a prospective study of medical and economic benefits. Vaccine 23 (17–18):2379–2387

Valle L et al (2019) Genetic predisposition to colorectal cancer: syndromes, genes, classification of genetic variants and implications for precision medicine. J Pathol 247(5):574–588

Vermorken JB et al (1999) Active specific immunotherapy for stage II and stage III human colon cancer: a randomised trial. Lancet 353(9150):345–350

Weng YJ et al (1996) Carcinoid tumors of the gastrointestinal tract in Chinese of Taiwan: an analysis of fifty cases. Zhonghua Yi Xue Za Zhi (Taipei) 58(4):254–258

Winawer SJ, Sherlock P (1976) Approach to screening and diagnosis in colorectal cancer. Semin Oncol 3(4):387–397

Wu X, Lin H, Li S (2019) Prognoses of different pathological subtypes of colorectal cancer at different stages: a population-based retrospective cohort study. BMC Gastroenterol 19(1):164

Yasunaga M et al (2019) Significant antitumor effect of an antibody against TMEM180, a new colorectal cancer-specific molecule. Cancer Sci 110(2):761–770

Chapter 13
Nanotechnology Applications in Gastric Cancer

Begum Dariya, Eluri Pavitra, Saimila Momin, and Ganji Seeta Rama Raju

Abstract The Gastric cancer (GC) is the third deadliest malignant disease worldwide. The most common conventional therapies include the use of anticancer drugs; however, their use is ineffective and is limited due to the poor solubility of the drug and its ability to damage the immune system, since anticancer drugs have a small therapeutic drug, overdosing can lead to the toxicity in the systemic organs such as lung, liver and kidneys. Nanotechnology, which has been developed about a decade ago is considered to be a more advantageous therapy compared to the conventional methods for the therapy of cancers like GC. Varied type of nanoparticles that are multifunctional are now used to enhance the solubility and effectiveness of the drug for the treatment, prevention and diagnosis with the annual clinical trials. Moreover, they are biocompatible and less toxic to the normal healthy cells. The organic and inorganic nanoparticles that are used in the cancer diagnosis and drug delivery system include lipid, protein, metal and polymer-based materials. This chapter gives an overview on the types of nanoparticle used in the research for the GC therapy. Additionally, we have briefly described the combinational approaches, including gene therapy and immunotherapy, along with nanoparticles that includes antibody mediated, enzyme and ligand mediated strategies.

B. Dariya
Department of Bioscience and Biotechnology, Banasthali University, Vanasthali, Rajasthan, India

E. Pavitra
Department of Biological Engineering, Biohybrid Systems Research Center (BSRC), Inha University, Incheon, Republic of Korea

S. Momin
Department of Hematology and Medical Oncology, Winship Cancer Institute, Emory University, Atlanta, GA, USA

G. S. R. Raju (✉)
Department of Energy and Materials Engineering, Dongguk University-Seoul, Seoul, Republic of Korea
e-mail: gseetaramaraju7@dongguk.edu; gseetaramaraju@live.in

G. S. R. Raju, L.V.K.S. Bhaskar (eds.), *Theranostics Approaches to Gastric and Colon Cancer*, Diagnostics and Therapeutic Advances in GI Malignancies,
https://doi.org/10.1007/978-981-15-2017-4_13

Keywords Gastric cancer · Nanoparticles · Chemodrugs · Drug delivery system · Genetherapy

Abbreviations

AFM	Atomic force microscopy
AlPcS4	Aluminum phthalocyanine chloride tetrasulfonic acid
AuNC	Gold nanocages
BCRP	Breast cancer resistant protein
BRET	Bioluminescence resonance energy transfer
CIC	Cancer initiating cells
CL-NA	Cationic liposome nucleic acid
CSC	Cancer stem cells
CT	Computed tomography
DAC	Deoxycytidine
DFC	Farnesiferol C with dendrosome
DPA	Diphenyl amine
EBVaGC	Epstein-barr virus associated GC
ECM	Extracellular matrix
FA	Folic acid
FR	Folate receptor
FTIR	Fourier transform infrared spectroscopy
GC	Gastric cancer
His6	Histidine tag
ICG	Indocyanine green loaded mesoporous silica
LOX	Lysyl oxidase
mAb	Monoclonal antibody
MBA	Mercapto benzoic acid
MET	Methionine
MMP	Matric metalloproteins
MPA	Mercapto propionic acid
MRI	Magnetic resonance imaging
PAGA	Pluronic-poly[α-(4-aminobutyl)-1-glycolic acid]
PAMAM	Polyamidoamine
PBA	Phenylboronic acid
PDT	Photodynamic therapy
PEG	Polyethylene glycol
PEI	Polyethylenimine
PET	Positron emission tomography
PLGA	Poly(D, L-lactide-co-glycolide acid)
QD	Quantum dots
rBSA	Reduced bovine serum albumin
rLuc	Renilla luciferase
ROS	Reactive oxygen species

SATB1	AT rich sequence binding protein 1
SERS	Surface enhanced raman scatterings
SPION	Superparamagnetic iron oxide nanoparticle
TF	Tissue factor
TPA	Two photon absorption

13.1 Introduction

Gastric cancer (GC) is the fifth most common diagnosed malignant disease after colorectal cancer and the third most common cause for the cancer-related deaths globally with the annual death record of 783,000 and an estimated survival rate of 31% in the US, according to the analysis of GLOBOCAN in 2018 (Rawla and Barsouk 2019). The occurrence of the disease is due to risk factors including diet, urban life style, chemical exposure (Fortunato and Rushton 2015), genetically inherited mutations, GERD (Ye et al. 2001) and *Helicobacter pylori* infections (Ahmed 2005). The disease is highly symptomatic in its advanced stage and recurrent as a local or distant metastatic disease, especially after resection of gastric tumors that have rapid spree of disease in their early stages. Even though there are endless efforts made in the discovery and development of novel diagnostic and therapeutic tools to bring better results in the survival rates of these cancer patients, the prognosis rate is very low; thus, it requires more advanced therapies. The tremendous development in therapies include chemo- and radiation therapies, which are highly advantageous; however, there are various side effects and are less effective at the site of the tumor. Moreover, unlike other systemic tumors, GC is a localized tumor and possesses loco-regional metastasis and is generally associated with negative prognostic factors (Imano et al. 2012; Ishigami et al. 2009, 2010). Thus, the cancer therapies are higher risk for healthy tissue and are partial in terms of full recovery. In recent years nanotechnology has been highly researched since its exciting advances has got potential advantageous in the field of science including anticancer therapies, drug delivery mechanisms and imaging.

The nanotechnology is a multidisciplinary research determination that includes nanomedicine or nanoparticles that have unique physical and chemical properties, which are highly beneficial for the early and loco-regional diagnosis and therapy. Moreover, the traditional use of nanomedicine minimizes the toxicity via targeting therapies at the tumor site as well improving the pharmacokinetics. Researchers have already designed a multifunctional theragnostic nanoparticles that gained intensive consideration since it has an effective dual function—it can be used for both diagnosis and therapy (Chen et al. 2014; Muthu et al. 2014; Wang et al. 2013). In this chapter we highlight the emerging research advances and applications of nanoparticles for biomedical imaging, diagnosis and therapy.

13.2 Nanoparticles in Biomedical Imaging for GC Diagnosis and Therapy

The traditional imaging techniques used for diagnosis of various cancers like GC includes computed tomography (CT), positron emission computed tomography (PET) and magnetic resonance imaging (MRI) etc., which come with undesirable side effects and rapid clearance of the tracer used (Li et al. 2011, 2013; Mirkin et al. 2015). The current, advanced research technologies are developing a diverse set of nanoparticles and clearly describing their importance in molecular imaging. To enhance the imaging efficacy, the use of nanoparticles as contrast agents could be a promising strategy. The theragnosis or theragnostics, that are loaded with numerous components, simultaneously achieves disease detection as well therapy (Chen et al. 2014; Muthu et al. 2014; Wang et al. 2013). The molecular imaging is the process of developing molecular probes to view the characterization and functionalization at their molecular level without disturbing them. Furthermore, the multi component nanoparticles that have been developed have improved resolution to overcome the limitations like single imaging. It is also advantageous for better penetration into tissue and temporal resolution time (Key and Leary 2014). There are many nanoparticles, including organic and inorganic nanoparticles for cancer therapeutics and imaging; however, the nanoparticles with much specificity for the tumor cells and maximum circulation half life time are considered to be more beneficial. The nanoparticles are developed as such to show better affinity towards the over expressed receptor proteins in cancer tissues.

13.3 Inorganic Nanoparticles

The inorganic nanoparticles have the physicochemical properties similar to the inorganic components like metals.

13.3.1 Gold Nanoparticles

The nanoparticles that are mostly used for imaging are inorganic nanoparticles, which are widely used as versatile tools with unique optical properties. For instance, gold (Au) nanoparticles have potential biocompatibility, unique physical and chemical properties like higher X-ray absorption coefficient and less toxicity. They are synthesized in organic solvents and to make it stable, the stabilizing agents are added to the Au nanoparticles which chemically and physically are absorbed on to their surface. The Au nanoparticles are produced under benign conditions for safety and less toxicity in the application of cancer therapy. For instance, silica capped Au nanoclusters (AuNCs) are conjugated with folic acid-(FA) in order to target folate receptor (FR^+) in MGC-803 cell lines and were tested in vivo in nude mice models with gastric tumors; the results of this investigation showed highly efficient florescent imaging in CT scans (Zhou et al. 2013). Similarly, FA conjugated with another nanoprobe, AuNCs, was designed by Zhou et al. (2013). The designed nanoprobe, AuNCs@SiO_2-FA, was effectively targeted against GC cells that showed FR expression. Based on the in vitro studies, they suggested that FA nanoprobe is strongly attracted to FR^+ than FR^- cells. The designed nanoprobe was created with enhanced emitting fluorescence properties. Thus, this could lead to its potential application in targeting cancer cells and imaging in vivo in the future. Au nanorods coated with indocyanine green-loaded mesoporous silica (ICG-loaded Au@SiO_2) were reported by Luo et al. (2011) for efficient imaging in CT and fluorescence imaging. The X-ray CT scanning showed enhanced contrast and sustained the nanomaterial generated NI fluorescence for almost 12 h after intratumoral injections in GC tissue. Thus, this could be a novel dual mode contrasting agent for imaging in GC. Later, Tsujimoto et al. (2014) explored the diagnosis and therapeutic efficacy by using photodynamic therapy (PDT), specifically, using ICG lactosomes in xenograft nude mice which had peritoneal metastasis of GC which enhanced the permeation and retention effect. The report from their PDT images revealed existence of the peritoneal tumor over the abdominal wall in xenograft mice. Additionally, they also observed reduction in the size and weight of the tumor with improved survival rate. Thus, this could be a potential novel device used for the diagnosis and therapy of peritoneal dissemination in GC. Ferumoxtran-10 was used earlier to enhance MRI and to diagnose GC lymph node metastasis; however, its use was fatal with adverse side effects including hypersensitivity (Tatsumi et al. 2006). Recently, Chang et al. (2019) developed monodisperse nanoparticles such that they have grown the silver (Ag) shell on the surface of the Au nanobipyramid core, that is of high stability and uniformity. Additionally, it also possesses surface enhanced Raman scattering (SERS) activity. It was later altered with 4-mercaptobenzoic acid (MBA), a Raman reporter molecule, and conjugated with FA and reduced bovine serum albumin (rBSA) to finally develop nanoprobe, AU-NBP@Ag-NR-rBSA-FA, in order to spot the live MGC-803 and A-549 cancer cells in Petri dishes. The results suggested that with good stability under optimal condition, SERS activity could detect live cancer cells through SERS mapping. Thus, this novel core shell with high stability, bearing ultrasensitive SERS and with Raman imaging is both advantageous and efficient for targeting cancer cells and can be used in clinical and theragnostic applications. The Au nanoparticles are further made more specific to the target site by conjugating them with various targeted ligands including peptides, antibodies and proteins for a more efficient diagnosis, drug delivery and therapy. For instance, Liang et al. (2015) conjugated CD44v6 mAb of CD44 + gastric cancer stem cells (CD44 + GCSCs) as a targeting ligand with Au nanostars-PEGylated nanoprobe. Later, they evaluated the cytotoxicity, endocytosis and cellular affinity and the results suggested that they have better biocompatibility, excellent targeting efficiency and photothermal ablation ability. Additionally, this procedure is also an ideal novel strategy to overcome the resistance in PDT and moreover, the results show better prognosis and also reduced recurrence and metastasis.

13.3.2 Quantum Dots

Quantum dots (QD) are fluorescent semiconductor nanoparticles and are alternatives to organic fluorescent dyes. They have unique electrical and optical properties with broad absorption features and continued emission of fluorescence due to their quantum size affects that increases their longevity. These are highly advantageous in the research field for biological imaging. Peng et al. (2018) found a multiplexed QD probe based spectrum for molecular imaging to evaluate the stromal activation, expression of lysyl oxidase (LOX) and their spatial relationship histologically, which was proposed as stromal marker in GC. LOX is an amine oxidase enzyme that modifies elastin and collagens in ECM and mediates cross linking of fibers. However, its aberrant expression is invasive in various cancers like GC and was found to promote recurrence. Similarly, the expression of caveolin-1 (Cav-1) and light chain 3B (LC3B) are used as biomarkers for the detection of GC in the Epstein—Barr virus associated with GC (EBVaGC) and non EBVaGC using QDs. The loss of Cav-1 expression is a marker for metastasis and early recurrence of disease; whereas, LC3B is a biomarker of autophagy that localizes to the membrane of autophagosome. Later, He et al. (2012) performed QD based immunofluorescence histochemistry to determine the expression of markers LC3B and Cav-1, that are labelled with QDs in about 118 GC patients, and detected EBV encoded small RNA in situ hybridization in EBVaGC. The study concluded that the lower expression of Cav-1 could be a novel predictor for detection of GC progression and LC3B is correlated with lower invasion. Furthermore, to make the use of QD nanoprobes more specific, Ruan et al. (2012) prepared a nanoprobe coupling with HER2-mAb and RNase A (HER2-RDQs) against in-situ and subcutaneous GC mouse models by using MGC803 cell lines. The prepared QDs nanoprobes enhances the imaging and selectively kills GC cells immediately after 3 h and 6 h of post injection in subcutaneous and in-situ GC cells, respectively. They determined that RNase A in the nanoprobe separates and attacks the RNAs in cytoplasm and averts protein synthesis to ultimately induce apoptosis. Thus, this improves the survival time and inhibits growth of GC tissue.

Recently, researchers have been using luciferase-luciferin system to develop bioluminescence imaging of cancer cells; however, the visible bioluminescence released naturally from fireflies is harmful as it is absorbed strongly into haemoglobin and melanin. Thus, longer wavelengths of luminescence is only desirable for bio-molecular imaging. Alternatively, Tsuboi and Jin (2018) have employed bioluminescence resonance energy transfer (BRET), that is coupled with emission and developed from luciferin, to near infrared fluorophores for bioluminescence imaging, using QDs and fluorescent proteins as energy acceptors. The BRET coupled NIR-QDs is further conjugated with the recombinant Renilla luciferase (RLuc) protein on the QDs surface, that is already coated with glutathione through hexa histidine tag (His6) of recombinant protein, which has immunoglobulin binding domain of protein G (GB1). This is further conjugated with anti-EGFR mAb to detect EGFR. EGFR is a glycoprotein and its aberrant activation leads to various cancers including breast and GC. The designed nanoprobe is efficient in producing sensitive NIR images of EGFR in live GC cells. In recent decades immunotherapy that includes antigen presenting cells like dendritic cells would be a promising tool in developing immunity against cancer. Li et al. (2015) for the first time developed a fused vaccine by fusing both allogenic dendritic cells and inactive GC cells (MGC803) combined with cytokine induced killing cells. The fused cells were then labelled with $CuInS_2/ZnS$ NIR-QDs. They further determined prophylactic and immunotherapeutic efficacy by injecting the vaccine into three different mice which included tumor free, tumor suffering and post-surgery models (Li et al. 2015). The reports from the animal imaging studies suggested the migration of the vaccine through NIR-QDs. Additionally, the fused cells and fused cells along with cytokine induced killing cells targeted the cytotoxic T lymphocytes against the tumor cells that resulted in reduction in tumor size and promoted the prophylactic effect in vivo (Li et al. 2015). Thus, this could be helpful for future research studies to prevent metastasis and strategy for the early stage tumor therapy.

Like the QDs, fluorescent carbon dots in recent decades have been more encouraged for imaging however, they are more advanced than the QDs due to the later having limitations, such as photobleaching and hazardous properties due to the use of heavy metal elements. The carbon dots are with high biocompatibility, low photobleaching and better photoluminescence and are a promising nanoparticle for biomolecular imaging and drug delivery. The carbon dots based polyethylenimine (PEI) complex was developed by Wang et al. (2014a), as a siRNA carrier for targeting antiapoptotic protein survivin. The designed nanoprobe exhibited high quantum yield and biocompatibility. The reported images from confocal fluorescence clearly illustrated the uptake of siRNA-Cdots@PEI complex by the MGC-803 cells. Moreover, the presence of siRNA in the tumor cells that induced apoptosis and promoted cell cycle arrest at G1 phase were also observed in Westernblot analysis and qRT-PCR, suggesting that they have better gene silencing efficiency. Thus, the use of fluorescent carbon dots would be a successor for the QDs to serve as potential tools in the area of drug delivery and imaging.

13.3.3 Superparamagnetic Iron Oxide Nanoparticles

The magnetic nanoparticles have similar chemical and structural properties to QDs and Au nanoparticles; they are both highly biocompatible with magnetic saturation and have large functional surfaces. The superparamagnetic iron oxide nanoparticles (SPIONs) are widely used as contrasting agents in MRI and have large magnetic moments. Liu et al. (2012a), for the first time, used SPIONs coated with SiO_2, labelled them with near infrared fluorescence dye and monoclonal antibody/mAb (anti-CD146), and administered them into a xenograft GC model. The imaging results obtained after 30 min of administration were highly efficient. Later, Bao et al. (2011) used BRCAA1 mAb conjugated with fluorescent magnetic nanoparticles in vivo to diagnose the gastric tumor using magnetic resonance and fluorescent imaging. Furthermore, FDA have approved humanized mAb and trastuzumab or herceptin against the extracellular domain of HER-2 conjugated with superparamagnetic nanoparticles, which is used for therapy as well for imaging in GC (Kulhari et al. 2015; Kataoka et al. 2016; Fornaro et al. 2011). Later, HAI-178 mAb was developed by conjugating fluorescent magnetic, QD nanoparticles wrapped by silicon by Wang et al. (2014b) with the help of Dr. Jian et al. in both in vitro and in vivo GC cells. α-subunit of ATP synthase is over expressed in GC cells and tissues. HAI-178 antibody was prepared against the α-subunit of ATP synthase by Dr. Pan et al. (2011). They also suggested that this could be taken as a dual model both for imaging and therapy, even in the early stages of gastric cancer, which showed better biocompatibility with inhibition of GC cell growth within 2 h of post injection. Wu et al. (2012) developed a novel dual modal combining both optical and magnetic resonance bioimaging with carbonate nanoprobe fixed with rare earth ions, including silica coated magnetic nanoparticle $Gd_2(CO_3)_3$:Tb, of diameter 16 nm dispersed in water via the micro-emulsion method. The modal is advantageous with the delivery of nanoprobe into the gastric cancer cell lines, SGC7901 in a very short period with less toxicity and no adverse effects in the kidney—even if over doses were taken. The magnetic resonance imaging of the nanoprobe displays a clear T (1)-weighted effect that serves as a positive contrast agent with high potentiality. As another step in the cancer research, gene therapy is a challenging but promising therapeutic strategy for cancer therapy; however, this have been limited due to the improper delivery mechanisms. Thus, the molecular imaging that possesses tracking efficiency would be supportive for gene therapy. Chen et al. (2012) constructed a non-viral antibody directed vector called scFv (CD44v6)-PEG-g-PEI-SPION for the diagnosis and gene therapy both in vitro and in vivo. This was designed using a superparamagnetic iron oxide nanoparticle for siRNA delivery; the study also analysed the targeting ability for the cellular uptake and distribution of siRNA at the target site, which was evident in vitro and in vivo MRI imaging. They further determined that siRNA is effectively transferred to the target site and also mediates the silencing of genes in vitro (Chen et al. 2013). Thus, this could be a promising gene vector for cancer gene therapy. Similarly, Luo et al. (2017) also designed the SPIONs for the gene delivery system to deliver a programmed death ligand 1(PD-L1) siRNA to target PD-L1 cells into SGC-7901 cells, encapsulated with FA-PEG-SS-PEI polymer, where T2-weighted contrast was used to enhance the MRI imaging. PD-L1 is a trans-membrane protein where clusters of differentiation 274 (CD274) binds to PD-1 that exists on the surface of T-cells. PD-L1 suppresses the adaptive immune system and inhibits apoptosis in T-regulatory cells. These are highly expressed in tumor tissues resulting in poor prognosis. Moreover, the results suggested are as follows, with efficient MRI tracing imaging and after internalization, blocking of PD-L1 in vitro was exhibited. Additionally, they also determined that polyplex FA-PEG-SS-PEI-SPION/siRNA after internalization showed down regulation of PD-L1 at both the mRNA and protein level and also showed decrease in cytokine secretion levels when co-cultured with T-cells. Thus, this highlights the nanoparticle as an efficient gene nanocarrier and also restores the immune response. Additionally, Wang et al. (2015) designed another SPION, coated with dense silica ($dSiO_2$) and labelled with 800ZW. They conjugated the entire designed nanoparticle 800zw-SPION@$dSiO_2$ with YY146 to target CD146 in a MKN45 xenograft model, as a novel biomarker for detecting GC metastasis. CD146 is a cell surface protein, correlated with the progression of tumor cells that act aberrantly. Moreover, it also acts as unique biomarker for inducing EMT in GC cell progression and metastasis. YY146 is a mAb synthesised as anti-CD146. Furthermore, the NIRF imaging and histology of the GC tumor was evaluated for pharmacokinetics. The non-invasive in vivo monitoring results suggested binding of CD146 marker to be successful. Thus, this could be a novel strategy for image-guided therapeutic intervention in diagnosis and therapy of GC. Multi drug resistance is always a tedious hurdle for the therapy in cancer. miRNA-16 in GC plays a crucial role in altering the resistance developed against drugs like Adriamycin (ADR). Sun et al. (2014) developed a magnetic nanoparticle, coated with iron oxide nitrate, using PEG conjugated with miRNA-16, to enhance the sensitivity in GC cells to ADR. The developed nanoprobe delivery carrier showed better biocompatibility when administered into SGC7901/ADR nude mice. The reports suggested effective delivery of miRNA-16 as evident from imaging modalities including bioluminescence imaging. Additionally, reduction in the tumor size with the enhancement of apoptosis was probably due to the GC cells developing sensitivity against ADR, after the injection of nanoprobe was also observed. Thus, this could be a novel strategy in developing therapeutic application for multi drug resistance GC therapy.

13.4 Organic Nanoparticles

The organic nanoparticles are equally efficient for diagnosis and drug delivery; however, they have gained less attention when compared with inorganic nanoparticles. There are very few organic nanoparticles used in image based therapy; for instance, DPA-TSB/2,5,2′,5′-tetra (4-N, N-diphenylamine styryl) biphenyl is a water dispersible organic nanoparticle, has two photon absorption (TPA) properties that are efficient in producing 3D fluorescent imaging, which could be a promising tool in biological imaging, since it shows no surfactants and easy cytoplasm internalization. Chen et al. (2016) used DPA-TSB nanoparticle in SGC-7901 gastric cancer cell lines to evaluate certain characteristics including intracellular Ca^{2+}, MMP alterations, apoptosis and accumulation of ROS. They determined that the nanoparticle taken in the concentration of >4 μg/ml showed significant results affecting the discussed factors listed above and can be used as an imaging marker at an ideal concentration and for drug delivery.

The organic nanoparticles are solid particles and are capable of carrying molecules including drugs that are encapsulated within. The organic nanoparticles that are widely employed for drug delivery as drug vehicles or carriers include dendrimers, liposomes and micelles.

13.4.1 Dendrimers

The dendrimers are regularly branched synthetic polymers. The architecture is designed in layers that includes internal region that encloses a single compound, central core and terminal groups, that develop into a globular conformation. Polyamidoamine (PAMAM) was the first oral drug delivery hyper branched polymer-based dendrimer to be synthesized and commercialized. They can easily cross the intestinal epithelial barriers and pass through the paracellular pathway. In order to determine the efficiency of PAMAM, which is orally administered for its stability and transport through the gastrointestinal tract, Goldberg et al. (2011) conjugated 7-ethyl-10 hydroxy-camptothecin, SN-38, which is a topoisomerase-1 poison against CRC, GC and hepatic metastasis cell line to G3.5 PAMAM dendrimer. This is further ester-linked with glycine and βalanine. However, the G3.5-Gly-SN38 was found to be more effective with less stability in vitro; specifically, it showed the release of the drug within 48, 24 and 6 h into the liver, intestine and stomach, respectively, with an IC50 of 0.60. Additionally, the reports suggested that drug SN38 conjugated with PAMAM had a better and enhanced transepithelial transport than free SN38. Geraldo et al. (2011) developed a functionalized PAMAM dendrimer by combining it with FA coating it with 3-mercaptopropionic acid (MPA) and CdSe QDs cap. The Fourier transform infrared spectroscopy (FTIR) and high-resolution electron microscopy analysis showed the supramolecular complex had efficient transfer of biomarker to the selective GC cells. Later Nie et al. (2016) combined PAMAM dendrimer with QDs to act as a carrier for doxorubicin drug that attached with PEG and assembled with micelles to form vesicles. The entire vesicle was made with small conjugates and made to be pH sensitive under redox conditions. The reports suggested rapid drug release with enhanced cytotoxicity in MGC803 cells in vitro and in vivo; the vesicles showed better antitumor activity and systemic toxicity and no toxicity effects to healthy cells. In another work by Barzegar Behrooz et al. (2017) for cancer therapy, the study designed the PMAM dendrimer-PEG complex conjugated with AS1411 aptamer to deliver 5-FU and was observed with FTIR and atomic force microscopy (AFM). Their results suggested that the designed nanoparticle had more stability and accumulation of 5-FU in MKN45 GC cells. METase is L-methionine diamino-γ-mercaptomethane lyase enzyme destroys methionine (MET), where MET is an amino acid, essential for the self-growth of cancer cells. Recently, rMETase has been widely encouraged as a chemotherapeutic agent in GC therapy. Li et al. (2018) developed a synthetic nanoparticle G5-PAMAM-Au composed with hyaluronic acid, that acts as a CD44 receptor, a marker of CSC for carrying rMETase into xenograft nude mice. Increase in METase expression was determined with the inhibition of proliferation rate, increase in c-caspase 3 expression, Cyc C, Ros levels and inhibition of CD44 GC progression after administration into the mice, moreover, the nanoparticle designed was with better aqueous solubility and biocompatibility. The dendrosomes are another type of spherical and biodegradable nanomaterial used in anticancer therapy. Farnesiferol C is a natural anticancer drug; however, the use of this phytochemical is limited due to the lack of bioavailability. Aas et al. (2015), to improve the bioavailability of this phytochemical, combined farnesiferol C with dendrosome (DFC) against AGS GC cell lines. Their RT-PCR and MTT assay results suggested enhanced anti-proliferative and apoptotic activity with the increase in Bax/Bcl-2 ratio. The nanodrug developed was very significant in inhibiting the proliferation of GC cell in terms of dosage and in a time-dependent manner, when compared with the farnesiferol alone. The gene therapy was made much more successful with the use of gene carriers. PAMAM was also used in gene therapy as a carrier of miR-34a along with phenylboronic acid (PBA) and PEG-α-maleimide-ω-N-hydroxy succinimide. The resulted PPP conjugated with miR-34a, a tumor suppressor protein, was injected into the xenograft GC model to protect the gene from degradation by nucleases. The FACs analysis and confocal laser scanning microscopy revealed uptake of miR-34a by the GC cells through endocytosis. It was also suggested that the miR-34a showed an anti-proliferative effect via promoting apoptosis. On the other hand, metastasis was also regulated by downregulating Notch signalling pathway. Thus, these nanoparticles can be utilized as a novel strategy in the clinical field of cancer therapy through the delivery of genes, drugs or antibodies (Song et al. 2019). Thus, dendrimers are considered as an important tool for the therapy of cancer, since they are more functional when conjugated with ligands that have the ability to cross the barriers and reach the target site causing no effect to the healthy cells.

13.4.2 Liposomes

Liposomes are natural or synthetic phospholipids with an internal aqueous core and external lipid bilayer, surrounding it just like a biological membrane. They are amphiphilic in nature which enables them to carry both a hydrophilic drug, that is enclosed within the aqueous phase, and a hydrophobic drug that are absorbed into their outer membrane. Moreover, due to the cationic nature, they can easily attract and bind electrostatically with the anionic nucleic acids; additionally, they have vigorous diffusion, circulation and penetration properties. Previously, the liposome was used to target the gastric bacteria, *Helicobacter pylori*, which causes peptic ulcers, chronic gastritis and GC. Thamphiwatana et al. (2013) designed a robust chitosan modified with Au nanoparticle to stabilize liposome and use it as a carrier of antimicrobial drug doxycycline that rapidly triggers the bacteria at gastric pH. The gold stabilizer linked to the lysosome detaches from it at a neutral pH, and this then freely attaches to the bacteria and releases the drug. The results suggested that effective delivery of the drug, that remains inactive in the lumen of stomach later, binds to the bacteria at the mucus layer. This could be an efficient therapeutic strategy for the delivery of antimicrobial drugs for the treatment of gastric infections, which are one of the causes of GC. This is similar to the docetaxel drug which is used for chemohyperthermia therapy in GC. Yoshida et al. (2012) have developed a docetaxel drug encapsulated into magnetoliposome and injected it in the hind limb of balb mice, which had human MKN45 GC cells maintained at 43° C. The tumor progression rate gradually decreased after injection, when compared with the injection of doxetaxel drug alone, and showed an increase in the concentration of the tumor necrosis factor α. In the clinical field of gene therapy, liposome use was advantageous in the delivery of the gene to the target site in GC. For instance, Xin et al. (2015) developed a poly(D, L-lactide-co-glycolide acid) PLGA with PEGylated liposome that increases the capacity of drug loading and control drug release properties. This so developed nanoparticle was conjugated with ScFV, a single chain variable fragment that enhances the drug targeting efficiency in effective delivering. Furthermore, the stealth PLGA liposome CAGE scFv was encapsulated with rMETase and injected into a murine model to deliver it at the site of GC cells. They showed that the therapeutic efficacy increased in in vitro and in vivo with delivery of drug into SGC-7901 cells and also observed low toxicity in healthy cells. The cationic liposome nucleic acid (CL-NA) was used by Wonder et al. (2018) to determine the transfection efficiency of the exogenous nucleic acid-DNA or RNA in vivo in order to validate their early in vitro results. For this, they have stabilized the nanoparticle CL-NA by PEGylating it and injecting it into the human peritoneal metastatic GC mouse model. Biodistribution of DNA was assessed through confocal and whole organ imaging and showed positive results. Additionally, the accumulation of the nanoparticle by the healthy tissue was observed in low levels. Thus, this strategy can be employed for the delivery of peptides, including siRNA and proteins through peptide target nanoparticle therapy. Liposomes were also used to overcome the limitations of PDT in GC treatment. The PDT involves the wide use of Al (III) phthalocyanine chloride tetra sulfonic acid (AlPcS4) due to its low photobleaching, high quantum yields and easy penetration. However, in GC tissues, its penetration is limited due to its attraction towards serum albumin. In this regard, Xin et al. (2018) tested the use of nanoparticles including Au, liposome and nano micelles and its penetrating efficacy of AlPcS4. This report showed that the Au nanorods used for the delivery of AlPcS4 enhanced the anti-proliferative nature of the drug, along with the PDT effect and singlet O_2 generation. Furthermore, the liposome used induced the inhibition of GC cell and enhanced the efficacy by blocking the serum albumin whereas, the micellar nanoparticle developed a PDT effect with limited cytotoxicity. Currently, the drug delivery is now targeting the delivery of antibodies for the therapy of cancer. Yang et al. (2018a) for the first made use of gastric cancer initiating cells (CICs) to initiate GC and used CD44+ cells as a marker. Later siRNA loaded cationic liposomal nanoparticle was used for the delivery of siRNA to a special AT rich sequence and binding of protein 1 (SATB1), that promotes cancer metastasis, invasion and progression in various cancer like GC. They finally constructed SATB1 siRNA entrapped inside a cationic liposome and conjugated it with CD44 antibodies and triggered it against gastric CICs. The siRNA delivery to the CICs site was reported with the suppression of SATB1. Thus, this could be taken as a novel strategy for the down-regulating the expression or even knocking down tumor promoting genes.

13.4.3 Micelles

Like liposomes, micelles are amphiphilic polymers with a central hydrophilic shell and external hydrophobic core. They are extensively used as drug vehicles in the therapy of various cancers including GC. Paclitaxel is a chemodrug used for GC therapy; however, due to the failure in the first line therapy, when used for advanced stage of cancer, new formulation of paclitaxel, NK105, was taken with a recommended dose of 150 mg, per phase I. This is further encapsulated in the polymeric micelles nanoparticle and injected into GC patients and evaluated in Phase II trials. The overall response rate (25%), progression free survival (3.0 months), overall survival (14.4) were determined, that suggest for better efficacy of NK105 which warrants for future evaluation of phase III trial (Kato et al. 2012). Harada et al. (2012) developed a polymeric micelle encapsulated with docetaxel. They assessed NC-6301, PEG-poly aspartate encapsulated with docetaxel, linked via ester bonds with the native form of docetaxel, and injected this into the xenograft mice with human breast cancer and the orthotopic mice model with human scirrhous GC. They observed that the tolerated dose in mice of breast cancer was 50 mg/kg and was 10 mg/kg with the native drug. Moreover, mice with GC showed reduction in the size of metastatic lymph node. In another work from Wang et al. (2014c), they designed a pluronic-poly[α-(4-aminobutyl)-1-glycolic acid] (PAGA) micelle to load 5-FU through a self-assembly method against human GC cells. The MTT assay results suggested inhibition in the growth of tumor cells. The pluronic PAGA used for the delivery of 5-FU was highly effective with tunable drug release and could potentially be used as a drug delivery system in the anticancer therapy. Similarly, Vijayaraj Kumar and Venkata Subrahmanya Lokesh (2014) also designed amphiphilic PEGylated micelles to encapsulate and deliver rapamycin, against CRL 1739 in human GC and CRL 1658 NIH-3 T3 in mouse embryonic fibroblast cells. The results suggest enhanced antiproliferative activity in the GC cell. Later, NC-6300 that includes incorporation of epirubicin in micelle was designed by Yamamoto et al. (2015) was used to target against tissue factor (TF) cells. The tissue factor (TF) is the initiator of coagulation, which plays a key role in cancer progression and metastasis, thus, a marker for GC. They have developed mAb against TF and conjugated it with NC-6300. The reports indicated enhanced antitumor effect of the drug, when administered as anti-TF-NC-6300 than in the vital NC-6300 form. Micelles also serves as a nano-vehicle for the delivery of combinatorial drugs to overcome the obstacle of multidrug resistance. β-casein micelles are administered orally by Bar-Zeev et al. (2016) to deliver a combination of chemodrugs including paclitaxel and tariquidar, a P-glycoprotein inhibitor. The results assessed from the spectrophotometry revealed a high loading capacity of nanoparticle used; zeta potential analysis showed the solubility of the drug and the enhanced affinity of the drug via spectrofluorometry. MDR in GC was also controlled by the downregulation of P-gp expression. They also developed another β-casein nano-vehicle for the delivery of SN-38 and elacridar in order to treat multidrug resistance GC cells, that overexpresses breast cancer resistance protein (BCRP). The multi drug resistance reversal was observed with cytotoxicity developed in BCRP in GC cells. Thus, this can be used as a nano-vehicle for the delivery of combinational drugs to overcome the obstacle of resistance (Bar-Zeev et al. 2018). Thus, micelles are found to be more effective nano-vehicles; moreover, their conjugation with ligands could be a novel strategy for the delivery of drug.

13.5 Nanoparticles for Drug Delivery

The drug delivery system includes the delivery of chemodrugs, including 5-FU, docetaxel and cisplatin, to the target site. However, release of higher doses of drug at the site of healthy cells is a major obstacle, as it develops adverse effect, including high toxicity developed in systemic organs. Thus, improvements in the drug delivery system, that specifically limits the delivery only at the tumor site necessitates for a vehicle or a carrier. As discussed earlier, nanoparticles are used as nano-vehicles to carry the chemodrug more efficiently by conjugating them with the markers and mAb to potentiate the transfer of drug. Table 13.1 includes some of the nano-vehicles and chemodrugs worked on so far for the inhibition of GC cells.

13.6 The Preclinical and Clinical Trials of Nanoparticles

The biocompatible nano-vehicles that have been employed widely in clinical trials are polymer, lipid and protein based. The advances developed in the field of nanotechnology clinical trials that occur annually occur to evaluate their therapeutic efficiency, and their pharmacokinetics and pharmacodynamics, for the safety of the drug use. Below are the few nanodrugs used for the GC therapy as illustrated in Table 13.2. Furthermore, research scientists should collaborate with industries in order to upscale the trials and bring out the promising drug products to the clinic.

13.7 Conclusion and Future Perspective

The tremendous potentiality of the nanotechnology, that includes use of organic and inorganic nanoparticles, open doors to new applications and development for the therapy of dreadful diseases like cancers. GC is currently a serious, malignant disease with a high mortality rate. Therefore, development of nanoparticles for the diagnosis and therapy of GC is critical. The use of inorganic nanoparticles for the tumour imaging is a potential approach. However, the use of organic particles for imaging is still challenging due to the difficulties in tracking them. A well elucidation of the molecular mechanism involved in cancer progression supports the identification of diagnostic markers and therapeutic targets that are essential for

Table 13.1 Nanoparticles and chemotherapy

Nanovehicle	Nanoparticle	Chemodrugs	Cell lines	Result	Reference
mPEG-C(60)-5FU/ fullerene	PEG	5-FU	MCF-7, BGC-823	Improved antitumor efficacy and stable for 24 h	Dou et al. (2012)
mPEG-PLGA-PLL-cRGD-5FU	PEG	5-FU	SGC-7901	Release of drug is prolonged, enhanced the effectiveness of the drug	Liu et al. (2012b)
Intelligent gelatinases-DAC-5FU	Gelatinases	5-FU & deoxycytidine (DAC)	BGC823, SGC7901 and MKN45	Sensitization of AP-2 epsilon transcription factor by DAC, induced apoptosis, inhibit cancer cell proliferation	Wu et al. (2015)
PLGA-5FU-paclitaxel-sLeA	PLGA	5-FU, paclitaxel, sialyl-LewisA (mAb)	NCI-N-87, AGS	Successful delivery of 5-FU	Fernandes et al. (2019)
PEG-PCL-5FU-trastuzumab-antimRNA-21	PEG, PCL	5-FU, trsatuzumab, AMO-21	NUGC4, SGC7901, BGC823	PTEN upregulated, sensitizes HER2, miRNA expression downregulated	Hu et al. (2017)
HA-CH-IRN/5-FU-PLGA	Chitosan	5-FU, irinotecan	MGC803/cancer bearing mice	Effective antitumor activity	Gao et al. (2017)
Anti-TF-NC-6300	Micelles	Epirubicin	BxPC3, 44As3	Accumulation of drug is seen in the TF GC tumor cells	Yamamoto et al. (2015)
Exo-anti-214-DDP	Exosome	Cisplatin (DDP)	BALB (female nude mice) SGC7901	Reverse the resistance to cisplatin, promote apoptosis, inhibits migration	Wang et al. (2018)
HA-PEG-CDDP-SRF	PEG/lipid	Cisplatin (CDDP), Sorafenib (SRF)	MKN28, SGC7901/GC bearing mice	Combinational therapy/enhanced anti-tumor effect and reduced systemic toxicity	Yang et al. (2018b)
Oxaliplatin-Au-Fe3O4-Herceptin	Au, Fe3O4	Oxaliplatin, Herceptin (mAb)	SGC 7901	Her2 targeted therapy	Liu et al. (2018)
PLGA (TXT + LY294002)	PLGA	Docetaxel (TXT)/ LY294002	Orthotopic and xenograftGC mouse model	Decreased proliferation capacity, increases apoptosis, no toxicity in the organs	Cai et al. (2019)
DOC-PEG-PCL—antiPD-L1	PEG	Docetaxel, anti PD-L1 (mAb)	MGC803, MKN45, HGC27	Enhanced G2/M cell cycle arrest, inhibition of microtubule synthesis	Xu et al. (2019)

Table 13.2 Present status of nanoparticle drug in the clinical trials

Title of the study	Disease condition	Intervention used	Phase	Identifier	Status of the trial
Clinical trial of ABI-007 plus S-1 as second line chemotherapy in advanced GC patients	Gastric adenocarcinoma	Albumin bound paclitaxel	Phase I & phase II	NCT01336062	Completed
TKM 080301 for primary or secondary liver cancer	GC with hepatitis metastasis	TKM-080301	Phase I	NCT01437007	Completed
Genetic analysis guided dosing of FOLFIRABRAX to the patients of advanced gastro-intestinal cancer	Stage IIIA, stage IIIB and stage IV GC	Paclitaxel albumin, Leucovorin calcium, irinotecan HCL, 5-FU	Phase I and phase II	NCT02333188	Completed
Lapatinib and paclitaxel in therapy of advanced solid tumors	GC	Lapatinib, paclitaxel	Phase I	NCT00313599	Completed

Source: www.clinicaltrials.gov.in

the cancer therapy. Thus, nanotechnology could be a potential novel research technology in identifying these biological markers and targets for cancer therapy.

The development of nanoparticles with multifunctional system that includes imaging and therapy requires safe delivery of the synthesized nanoparticles. Thus, researches are now evaluating a better administration route for the safe delivery of these drugs for the most effective results. Moreover, the current researcher envisions, that within a decade or so, with the advancement in novel formulations, applications and evaluation of clinical trials could use nanotechnology routinely for clinical therapies to diagnose the cancer in its early stages. However, the use of nanoparticles made from carbon, copper, supermagnetic and platinum, which are highly toxic and easily accumulate in liver and spleen, may not be possible. Therefore, polymer-based drug carriers that are biodegradable and have little to no side effects need to be considered. Additionally, the need for research centres, like Nanotechnology characterization laboratory, are required for testing the toxicity of nanodrugs. Advances in the nanoformulation is achieved only by the collaborative work of oncologists, pharmacists, medical researchers and gastroenterologists. Their collaboration is necessary for the that would be the synthesis of novel nanoparticles to further transfer the formulation for the clinical therapy of the patient.

References

Aas Z, Babaei E, Feizi MAH, Dehghan G (2015) Anti-proliferative and apoptotic effects of dendrosomal farnesiferol C on gastric cancer cells. Asian Pac J Cancer Prev 16:5325–5329

Ahmed N (2005) 23 years of the discovery of helicobacter pylori: is the debate over? BioMed Central

Bao B, Ali S, Kong D, Sarkar SH, Wang Z, Banerjee S, Aboukameel A, Padhye S, Philip PA, Sarkar FH (2011) Anti-tumor activity of a novel compound-CDF is mediated by regulating miR-21, miR-200, and PTEN in pancreatic cancer. PLoS One 6:e17850

Bar-Zeev M, Assaraf YG, Livney YD (2016) β-Casein nanovehicles for oral delivery of chemotherapeutic drug combinations overcoming P-glycoprotein-mediated multidrug resistance in human gastric cancer cells. Oncotarget 7:23322

Bar-Zeev M, Kelmansky D, Assaraf YG, Livney YD (2018) β-Casein micelles for oral delivery of SN-38 and elacridar to overcome BCRP-mediated multidrug resistance in gastric cancer. Eur J Pharm Biopharm 133:240–249

Barzegar Behrooz A, Nabavizadeh F, Adiban J, Shafiee Ardestani M, Vahabpour R, Aghasadeghi MR, Sohanaki H (2017) Smart bomb AS 1411 aptamer-functionalized/PAMAM dendrimer nanocarriers for targeted drug delivery in the treatment of gastric cancer. Clin Exp Pharmacol Physiol 44:41–51

Cai J, Qian K, Zuo X, Yue W, Bian Y, Yang J, Wei J, Zhao W, Qian H, Liu B (2019) PLGA nanoparticle-based docetaxel/LY294002 drug delivery system enhances antitumor activities against gastric cancer. J Biomater Appl 33:1394–1406

Chang J, Zhang A, Huang Z, Chen Y, Zhang Q, Cui D (2019) Monodisperse Au@ Ag core-shell nanoprobes with ultrasensitive SERS-activity for rapid identification and Raman imaging of living cancer cells. Talanta 198:45–54

Chen Y, Wang W, Lian G, Qian C, Wang L, Zeng L, Liao C, Liang B, Huang B, Huang K (2012) Development of an MRI-visible nonviral vector for siRNA delivery targeting gastric cancer. Int J Nanomedicine 7:359

Chen Y, Lian G, Liao C, Wang W, Zeng L, Qian C, Huang K, Shuai X (2013) Characterization of polyethylene glycol-grafted polyethylenimine and superparamagnetic iron oxide nanoparticles (PEG-g-PEI-SPION) as an MRI-visible vector for siRNA delivery in gastric cancer in vitro and in vivo. J Gastroenterol 48:809–821

Chen F, Ehlerding EB, Cai W (2014) Theranostic nanoparticles. J Nucl Med 55:1919–1922

Chen G, Li S Y, Malik HT, Ma Y-G, Xu H, Sun L-K (2016) Organic two-photon nanoparticles modulate reactive oxygen species, intracellular calcium concentration, and mitochondrial membrane potential during apoptosis of human gastric carcinoma SGC-7901 cells. Biotechnol Lett 38:1269–1276

Dou Z, Xu Y, Sun H, Liu Y (2012) Synthesis of PEGylated fullerene–5-fluorouracil conjugates to enhance the antitumor effect of 5-fluorouracil. Nanoscale 4:4624–4630

Fernandes E, Ferreira D, Peixoto A, Freitas R, Relvas-Santos M, Palmeira C, Martins G, Barros A, Santos LL, Sarmento B (2019) Glycoengineered nanoparticles enhance the delivery of 5-fluoroucil and paclitaxel to gastric cancer cells of high metastatic potential. Int J Pharm 570:118646

Fornaro L, Lucchesi M, Caparello C, Vasile E, Caponi S, Ginocchi L, Masi G, Falcone A (2011) Anti-HER agents in gastric cancer: from bench to bedside. Nat Rev Gastroenterol Hepatol 8:369

Fortunato L, Rushton L (2015) Stomach cancer and occupational exposure to asbestos: a meta-analysis of occupational cohort studies. Br J Cancer 112:1805

Gao Z, Li Z, Yan J, Wang P (2017) Irinotecan and 5-fluorouracil-co-loaded, hyaluronic acid-modified layer-by-layer nanoparticles for targeted gastric carcinoma therapy. Drug Des Devel Ther 11:2595

Geraldo DA, Duran-Lara EF, Aguayo D, Cachau RE, Tapia J, Esparza R, Yacaman MJ, Gonzalez-Nilo FD, Santos LS (2011) Supramolecular complexes of quantum dots and a polyamidoamine

(PAMAM)-folate derivative for molecular imaging of cancer cells. Anal Bioanal Chem 400:483–492

Goldberg DS, Vijayalakshmi N, Swaan PW, Ghandehari H (2011) G3. 5 PAMAM dendrimers enhance transepithelial transport of SN38 while minimizing gastrointestinal toxicity. J Control Release 150:318–325

Harada M, Iwata C, Saito H, Ishii K, Hayashi T, Yashiro M, Hirakawa K, Miyazono K, Kato Y, Kano MR (2012) NC-6301, a polymeric micelle rationally optimized for effective release of docetaxel, is potent but is less toxic than native docetaxel in vivo. Int J Nanomedicine 7:2713

He Y, Zhao X, Gao J, Fan L, Yang G, Cho W, Chen H (2012) Quantum dots-based immunofluorescent imaging of stromal fibroblasts caveolin-1 and light chain 3B expression and identification of their clinical significance in human gastric cancer. Int J Mol Sci 13:13764–13780

Hu N, Yin JF, Ji Z, Hong Y, Wu P, Bian B, Song Z, Li R, Liu Q, Wu F (2017) Strengthening gastric cancer therapy by trastuzumab-conjugated nanoparticles with simultaneous encapsulation of anti-MiR-21 and 5-fluorouridine. Cell Physiol Biochem 44:2158–2173

Imano M, Yasuda A, Itoh T, Satou T, Peng Y-F, Kato H, Shinkai M, Tsubaki M, Chiba Y, Yasuda T (2012) Phase II study of single intraperitoneal chemotherapy followed by systemic chemotherapy for gastric cancer with peritoneal metastasis. J Gastrointest Surg 16:2190–2196

Ishigami H, Kitayama J, Kaisaki S, Hidemura A, Kato M, Otani K, Kamei T, Soma D, Miyato H, Yamashita H (2009) Phase II study of weekly intravenous and intraperitoneal paclitaxel combined with S-1 for advanced gastric cancer with peritoneal metastasis. Ann Oncol 21:67–70

Ishigami H, Kitayama J, Kaisaki S, Yamaguchi H, Yamashita H, Emoto S, Nagawa H (2010) Phase I study of biweekly intravenous paclitaxel plus intraperitoneal cisplatin and paclitaxel for gastric cancer with peritoneal metastasis. Oncology 79:269–272

Kataoka H, Mori Y, Shimura T, Nishie H, Natsume M, Mochizuki H, Hirata Y, Sobue S, Mizushima T, Sano H (2016) A phase II prospective study of the trastuzumab combined with 5-weekly S-1 and CDDP therapy for HER2-positive advanced gastric cancer. Cancer Chemother Pharmacol 77:957–962

Kato K, Chin K, Yoshikawa T, Yamaguchi K, Tsuji Y, Esaki T, Sakai K, Kimura M, Hamaguchi T, Shimada Y, Matsumura Y, Ikeda R (2012) Phase II study of NK105, a paclitaxel-incorporating micellar nanoparticle, for previously treated advanced or recurrent gastric cancer. Investig New Drugs 30:1621–1627

Key J, Leary JF (2014) Nanoparticles for multimodal in vivo imaging in nanomedicine. Int J Nanomedicine 9:711

Kulhari H, Pooja D, Rompicharla SV, Sistla R, Adams DJ (2015) Biomedical applications of trastuzumab: as a therapeutic agent and a targeting ligand. Med Res Rev 35:849–876

Li R, Xie L, Zhu Z, Liu Q, Hu Y, Jiang X, Yu L, Qian X, Guo W, Ding Y (2011) Reversion of pH-induced physiological drug resistance: a novel function of copolymeric nanoparticles. PLoS One 6:e24172

Li R, Wu W, Liu Q, Wu P, Xie L, Zhu Z, Yang M, Qian X, Ding Y, Yu L (2013) Intelligently targeted drug delivery and enhanced antitumor effect by gelatinase-responsive nanoparticles. PLoS One 8:e69643

Li C, Liang S, Zhang C, Liu Y, Yang M, Zhang J, Zhi X, Pan F, Cui DJB (2015) Allogenic dendritic cell and tumor cell fused vaccine for targeted imaging and enhanced immunotherapeutic efficacy of gastric cancer. Biomaterials 54:177–187

Li Y-F, Zhang H-T, Xin L (2018) Hyaluronic acid-modified polyamidoamine dendrimer g5-entrapped gold nanoparticles delivering metase gene inhibits gastric tumor growth via targeting cd44+ gastric cancer cells. J Cancer Res Clin Oncol 144:1463–1473

Liang S, Li C, Zhang C, Chen Y, Xu L, Bao C, Wang X (2015) CD44v6 monoclonal antibody-conjugated gold nanostars for targeted photoacoustic imaging and plasmonic photothermal therapy of gastric cancer stem-like cells. Theranostics 5:970

Liu W-F, Ji S-R, Sun J-J, Zhang Y, Liu Z-Y, Liang A-B, Zeng H-Z (2012a) CD146 expression correlates with epithelial-mesenchymal transition markers and a poor prognosis in gastric cancer. Int J Mol Sci 13:6399–6406

Liu P, Wang H, Wang Q, Sun Y, Shen M, Zhu M, Wan Z, Duan Y (2012b) cRGD conjugated mPEG-PLGA-PLL nanoparticles for SGC-7901 gastric cancer cells-targeted delivery of fluorouracil. J Nanosci Nanotechnol 12:4467–4471

Liu D, Li X, Chen C, Li C, Zhou C, Zhang W, Zhao J, Fan J, Cheng K, Chen L (2018) Target-specific delivery of oxaliplatin to HER2-positive gastric cancer cells in vivo using oxaliplatin-au-fe3o4-herceptin nanoparticles. Oncol Lett 15:8079–8087

Luo T, Huang P, Gao G, Shen G, Fu S, Cui D, Zhou C, Ren Q (2011) Mesoporous silica-coated gold nanorods with embedded indocyanine green for dual mode X-ray CT and NIR fluorescence imaging. Opt Express 19:17030–17039

Luo X, Peng X, Hou J, Wu S, Shen J, Wang L (2017) Folic acid-functionalized polyethylenimine superparamagnetic iron oxide nanoparticles as theranostic agents for magnetic resonance imaging and PD-L1 siRNA delivery for gastric cancer. Int J Nanomedicine 12:5331

Mirkin CA, Meade TJ, Petrosko SH, Stegh AH (2015) Nanotechnology-based precision tools for the detection and treatment of cancer. Springer, Cham

Muthu MS, Leong DT, Mei L, Feng S-S (2014) Nanotheranostics-application and further development of nanomedicine strategies for advanced theranostics. Theranostics 4:660

Nie J, Wang Y, Wang W (2016) In vitro and in vivo evaluation of stimuli-responsive vesicle from PEGylated hyperbranched PAMAM-doxorubicin conjugate for gastric cancer therapy. Int J Pharm 509:168–177

Pan J, Sun L-C, Tao Y-F, Zhou Z, Du X-L, Peng L, Feng X, Wang J, Li Y-P, Liu L (2011) ATP synthase ecto-α-subunit: a novel therapeutic target for breast cancer. J Transl Med 9:211

Peng C, Liu J, Yang G, Li Y (2018) Lysyl oxidase activates cancer stromal cells and promotes gastric cancer progression: quantum dot-based identification of biomarkers in cancer stromal cells. Int J Nanomedicine 13:161

Rawla P, Barsouk A (2019) Epidemiology of gastric cancer: global trends, risk factors and prevention. Przeglad Gastroenterologiczny 14:26

Ruan J, Song H, Qian Q, Li C, Wang K, Bao C, Cui D (2012) HER2 monoclonal antibody conjugated RNase-A-associated CdTe quantum dots for targeted imaging and therapy of gastric cancer. Biomaterials 33:7093–7102

Song Z, Liang X, Wang Y, Han H, Yang J, Fang X, Li Q (2019) Phenylboronic acid-functionalized polyamidoamine-mediated miR-34a delivery for the treatment of gastric cancer. Biomater Sci 7:1632–1642

Sun Z, Song X, Li X, Su T, Qi S, Qiao R, Wang F, Huan Y, Yang W, Wang J (2014) In vivo multimodality imaging of miRNA-16 iron nanoparticle reversing drug resistance to chemotherapy in a mouse gastric cancer model. Nanoscale 6:14343–14353

Tatsumi Y, Tanigawa N, Nishimura H, Nomura E, Mabuchi H, Matsuki M, Narabayashi I (2006) Preoperative diagnosis of lymph node metastases in gastric cancer by magnetic resonance imaging with ferumoxtran-10. Gastric Cancer 9:120–128

Thamphiwatana S, Fu V, Zhu J, Lu D, Gao W, Zhang L (2013) Nanoparticle-stabilized liposomes for pH-responsive gastric drug delivery. Langmuir 29:12228–12233

Tsuboi S, Jin T (2018) Recombinant protein (luciferase-IgG binding domain) conjugated quantum dots for BRET-coupled near-infrared imaging of epidermal growth factor receptors. Bioconjug Chem 29:1466–1474

Tsujimoto H, Morimoto Y, Takahata R, Nomura S, Yoshida K, Horiguchi H, Hiraki S, Ono S, Miyazaki H, Saito D (2014) Photodynamic therapy using nanoparticle loaded with indocyanine green for experimental peritoneal dissemination of gastric cancer. Cancer Sci 105:1626–1630

Vijayaraj Kumar P, Venkata Subrahmanya Lokesh B (2014) Designing and in-vitro characterization of micelle forming amphiphilic PEGylated rapamycin nanocarriers for the treatment of gastric cancer. Curr Drug Deliv 11:613–620

Wang L, Wang Y, Li Z (2013) Nanoparticle-based tumor theranostics with molecular imaging. Curr Pharm Biotechnol 14:683–692

Wang Q, Zhang C, Shen G, Liu H, Fu H, Cui D (2014a) Fluorescent carbon dots as an efficient siRNA nanocarrier for its interference therapy in gastric cancer cells. J Nanobiotechnol 12:58

Wang C, Bao C, Liang S, Zhang L, Fu H, Wang Y, Wang K, Li C, Deng M, Liao Q (2014b) HAI-178 antibody-conjugated fluorescent magnetic nanoparticles for targeted imaging and simultaneous therapy of gastric cancer. Nanoscale Res Lett 9:274

Wang Q, Liu P, Sun Y, Wu H, Li X, Duan Y, Zhang Z (2014c) Pluronic-poly [α-(4-aminobutyl)-l-glycolic acid] polymeric micelle-like nanoparticles as carrier for drug delivery. J Nanosci Nanotechnol 14:4843–4850

Wang P, Qu Y, Li C, Yin L, Shen C, Chen W, Yang S, Bian X, Fang D (2015) Bio-functionalized dense-silica nanoparticles for MR/NIRF imaging of CD146 in gastric cancer. Int J Nanomedicine 10:749

Wang X, Zhang H, Bai M, Ning T, Ge S, Deng T, Liu R, Zhang L, Ying G, Ba Y (2018) Exosomes serve as nanoparticles to deliver anti-miR-214 to reverse chemoresistance to cisplatin in gastric cancer. Mol Ther 26:774–783

Wonder E, Simón-Gracia L, Scodeller P, Majzoub RN, Kotamraju VR, Ewert KK, Teesalu T, Safinya CR (2018) Competition of charge-mediated and specific binding by peptide-tagged cationic liposome–DNA nanoparticles in vitro and in vivo. Biomaterials 166:52–63

Wu Y, Xu X, Tang Q, Li Y (2012) A new type of silica-coated Gd(2)(CO(3))(3):Tb nanoparticle as a bifunctional agent for magnetic resonance imaging and fluorescent imaging. Nanotechnology 23:205103

Wu F-l, Li R-T, Yang M, Yue G-F, Wang H-y, Liu Q, Cui F-b, Wu P-y, Ding H, Yu L-X (2015) Gelatinases-stimuli nanoparticles encapsulating 5-fluorouridine and 5-aza-2′-deoxycytidine enhance the sensitivity of gastric cancer cells to chemical therapeutics. Cancer Lett 363:7–16

Xin L, Cao J-Q, Liu C, Zeng F, Cheng H, Hu X-Y, Shao J-H (2015) Evaluation of rMETase-loaded stealth PLGA/liposomes modified with anti-CAGE scFV for treatment of gastric carcinoma. J Biomed Nanotechnol 11:1153–1161

Xin J, Wang S, Wang B, Wang J, Wang J, Zhang L, Xin B, Shen L, Zhang Z, Yao C (2018) AlPcS4-PDT for gastric cancer therapy using gold nanorod, cationic liposome, and Pluronic® F127 nanomicellar drug carriers. Int J Nanomedicine 13:2017

Xu S, Cui F, Huang D, Zhang D, Zhu A, Sun X, Cao Y, Ding S, Wang Y, Gao E (2019) PD-L1 monoclonal antibody-conjugated nanoparticles enhance drug delivery level and chemotherapy efficacy in gastric cancer cells. Int J Nanomedicine 14:17

Yamamoto Y, Hyodo I, Koga Y, Tsumura R, Sato R, Obonai T, Fuchigami H, Furuya F, Yasunaga M, Harada M (2015) Enhanced antitumor effect of anti-tissue factor antibody-conjugated epirubicin-incorporating micelles in xenograft models. Cancer Sci 106:627–634

Yang F, Zheng Z, Zheng L, Qin J, Li H, Xue X, Gao J, Fang G (2018a) siRNA-encapsulated immunoliposomes conjugated with CD44 antibodies target and eliminate gastric cancer-initiating cells. OncoTargets Ther 11:6811

Yang F, Li A, Liu H, Zhang H (2018b) Gastric cancer combination therapy: synthesis of a hyaluronic acid and cisplatin containing lipid prodrug coloaded with sorafenib in a nanoparticulate system to exhibit enhanced anticancer efficacy and reduced toxicity. Drug Des Devel Ther 12:3321

Ye W, Chow W-H, Lagergren J, Yin L, Nyrén O (2001) Risk of adenocarcinomas of the esophagus and gastric cardia in patients with gastroesophageal reflux diseases and after antireflux surgery. Gastroenterology 121:1286–1293

Yoshida M, Sato M, Yamamoto Y, Maehara T, Naohara T, Aono H, Sugishita H, Sato K, Watanabe Y (2012) Tumor local chemohyperthermia using docetaxel-embedded magnetoliposomes: interaction of chemotherapy and hyperthermia. J Gastroenterol Hepatol 27:406–411

Zhou Z, Zhang C, Qian Q, Ma J, Huang P, Pan L, Gao G, Fu H, Fu S, Song H (2013) Folic acid-conjugated silica capped gold nanoclusters for targeted fluorescence/X-ray computed tomography imaging. J Nanobiotechnol 11:17

The manufacturer's authorised representative in the EU is Springer Nature Customer Service Centre GmbH, Europaplatz 3, 69115 Heidelberg, Germany. If you have any concerns regarding our products, please contact ProductSafety@springernature.com

Printed and bound by CPI Group (UK) Ltd, Croydon, CR0 4YY
27/07/2026
02177419-0002